MURMURATIONS

MURMURATIONS

JAMES CROMBIE

Foreword by **SEÁN RONAYNE**
Words by **JOHN FALLON**

THE LILLIPUT PRESS
DUBLIN

First published 2024 by

THE LILLIPUT PRESS

62–63 Sitric Road,
Arbour Hill,
Dublin 7,
Ireland
www.lilliputpress.ie

Images © James Crombie, 2024
Foreword © Seán Ronayne, 2024
Main text © James Crombie & John Fallon, 2024

10 9 8 7 6 5 4 3

A CIP record for this title is available from The British Library.

Hardback ISBN 978 1 84351 911 9
eBook ISBN 978 1 84351 9294

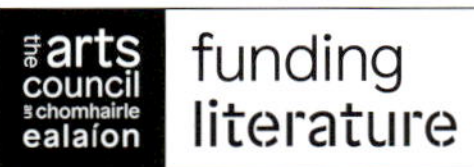

Lilliput gratefully acknowledges the financial support of the Arts Council /
An Chomhairle Ealaíon.

Set in Field Gothic and Adobe Garamond Pro by Niall McCormack
Printed and bound in Czechia by Finidr

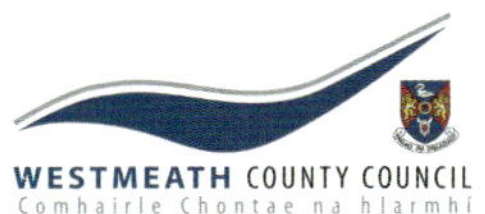
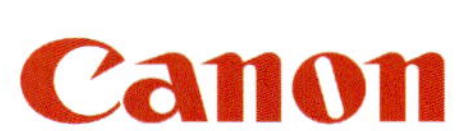
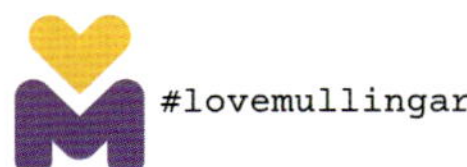

To my wife Ann

FOREWORD
by Seán Ronayne

COMMON STARLINGS (*Sturnus vulgaris*) are a widespread and often overlooked species throughout the nation of Ireland, and indeed Europe. Starlings are a medium-sized (approx' 20cm long) passerine bird. At first glance they look a plain black, but look again! On closer inspection their black plumage shimmers in waves of green and purple iridescence. A further sprinkling of white speckles stand out like sparkling sequins.

If their subtly wonderful plumage isn't enough to woo, then their incredible song and their uncanny vocal mimicry surely makes them one of our most amazing bird species. Starlings are lifelong vocal learners. Many species only learn in the first few months of their lives, after which their song is set in stone. Starlings, however, can continue to learn and add to their song throughout their lives, which, paired with a sharp ear for accurate mimicry, makes them one of our most impressive songsters.

I had the pleasure of hosting a family of Starlings in my roof in North Cork some years back. At the time I was passively recording the sounds of nocturnally migrating birds overhead. I strategically angled my microphone so that it not only pointed at the night sky, but also the nest hole in the roof. This meant that my recorder would capture all of the birds that passed overhead while I slept, but then also captured the song of the Starling as it woke each morning. What I learned over the following months was nothing short of jaw-dropping.

As different wildlife events unfolded at night, the Starling began to reflect this in its mimetic repertoire. In late January the local foxes began to echo their banshee-like screams along the fields behind the house, and a week or two later this sound was added to the Starlings' song. Later on, in late March, when Snipe and Golden Plover began to migrate overhead at night in numbers, these too were added to the song. But best of all was hearing 'our' Starling begin to speak in a Cork accent. Each day I would let our dog Toby out in the garden to do his business and have a roll in the grass. After some time, I'd open the door and call him back in, with a Cork lilt: 'c'mere'. I nearly fell off my chair one morning when I heard the Starling echo my command from the roof above me, accent and all.

Despite their many charms, Irish Starlings are in trouble. Once common throughout the country, their numbers have suffered a serious decline, and they are now featured in the Amber-list of Birds of Conservation Concern Ireland. This often comes as a surprise to people when they learn it for the first time, because our Starlings have always been such a familiar sight in a plethora of both rural and urban habitats throughout the land.

However, our agricultural intensification, coupled with a staggering increase in pesticide and herbicide use, as well a large decline in plant species diversity on our farms and in our townlands has meant that insect populations have plummeted. As insects form a major part of the diet of Starlings, this loss has in turn impacted their ability to raise young and thrive as they once did. A reduction in nesting sites has also had a big impact on their ability to successfully breed and raise young.

Starlings are most typically cavity nesters – like House Sparrows, Swifts and Barn Swallows, they have traditionally taken advantage of human-made dwellings, nesting in irregularities in the stone walls and straw or loose-fitting roof tiles of our older buildings. However, modern buildings are now much less irregular, with clean, straight, plastered walls, and solid-plastic fascia and soffit borders to our roofs. This has left many Starlings and other species without homes.

In many countries, Starlings also nest in old Woodpecker holes, where the latter are common. On a recent visit to the heart of the wild Danube Delta in Romania, where up to eight species of Woodpecker occur, I noticed something quite special. Not only were many trees peppered with old and new Woodpecker holes, many of these holes were also occupied by Starlings! And so, with the recent colonisation of Ireland by the Great Spotted Woodpecker, perhaps these wonderful ecosystem engineers will provide homes for our Starlings, too, as they continue to increase in numbers and spread throughout the country.

Although impressive in many ways as individuals, it's what Starlings do in large groups that truly captures people's imaginations. Every autumn, Starlings from around Europe gather in Ireland, to escape the oncoming harsher winter conditions in Northern and Eastern Europe. The much colder winter weather conditions there mean that birds cannot find the food and warmth they need to persist until spring, and thus they migrate and seek out countries with milder climates, until the winter passes. Starlings typically migrate in August and September, and in October, we can begin to truly see a rise in our numbers here in Ireland.

When we see a Starling here in winter, it may well be a local bird, or it could have travelled from as far as Russia to be here. During the day both resident and winter migrants will feed side-by-side, scattered throughout the landscape. However, as the light begins to fade, the magic begins to unfold. Starlings from a 30km radius or more will begin to fly from all directions towards their selected roosting locations – the place where they will sleep together for the night. Before they go to ground, they gather together and form a large aerial dancing flock, which is collectively known as a murmuration.

Many of these roosting locations have been used for as long as people can remember and are clearly very important sites for these birds. One such example, occurs along the shoreline of Lough Ennell, in Co. Westmeath. This site is home to Ireland's most spectacular murmuration and has been known to host up to 250,000 birds at its peak in early March. Its popularity is probably down to its central location, and large reed-fringed shorelines. These wet, soft-edged reedbeds offer protection from land-based predators and allow the birds to roost together safely. However, they still need to be vigilant of otters, mink, and even hunting owls. Recent advances in infrared technology have shown that Barn Owls sometimes target unsuspecting roosting Starlings with great success. In the roost, safe from most, but not all predators, they sleep closely and are collectively warmed by the significant body heat generated by the great numbers of neighbours, providing enough of a temperature rise to survive sleeping in the relatively mild Irish winters.

Before going down to roost however, these magnificent birds put on a show that rivals even the world-renowned aerial display of the great Aurora Borealis. Despite their staggering numbers, birds shimmer in the sky, twisting and turning as if they were one organism. But how do they coordinate themselves like this? Well, amazingly, a recent study has found that Starlings keep an eye on their six to seven nearest neighbours in order to move in unison. Even more impressive is the fact that Starlings' lateral vision is so extensive that it allows them to see all around their bodies, almost like a human driver looking at the mirrors of their car. This makes them perfectly adapted to cohesive flock flight, and further lends to their incredible dusk performances.

Although these wonderfully coordinated murmurations are mesmerisingly beautiful, they do serve a much more pressing function. This dance in the twilight sky is a matter of life or death – these large gatherings attract aerial predators in numbers, looking to take advantage of a concentrated abundance of prey. Birds of prey – mostly Sparrowhawks and Peregrine Falcons, from the same townlands the Starlings feed in by day – come to pick an evening snack from the skies before they too hunker down for the night. In moving as one unit the Starlings make it as confusing and as difficult as possible for their aerial hunters to pick one out.

I had the great pleasure of meeting James Crombie at Lough Ennell in the winter of March 2023, where I saw and sound-recorded the indescribable beauty of Ireland's most spectacular murmuration, thanks to James's expert guidance. James's knowledge and understanding of this murmuration is nothing short of remarkable. He has put untold hours into capturing the story of this magical spectacle of Irish nature, which is blatantly evident in the photos to follow. It was an absolute honour to be asked to introduce his stunning work, which pays tribute to this great natural wonder far better than any words could.

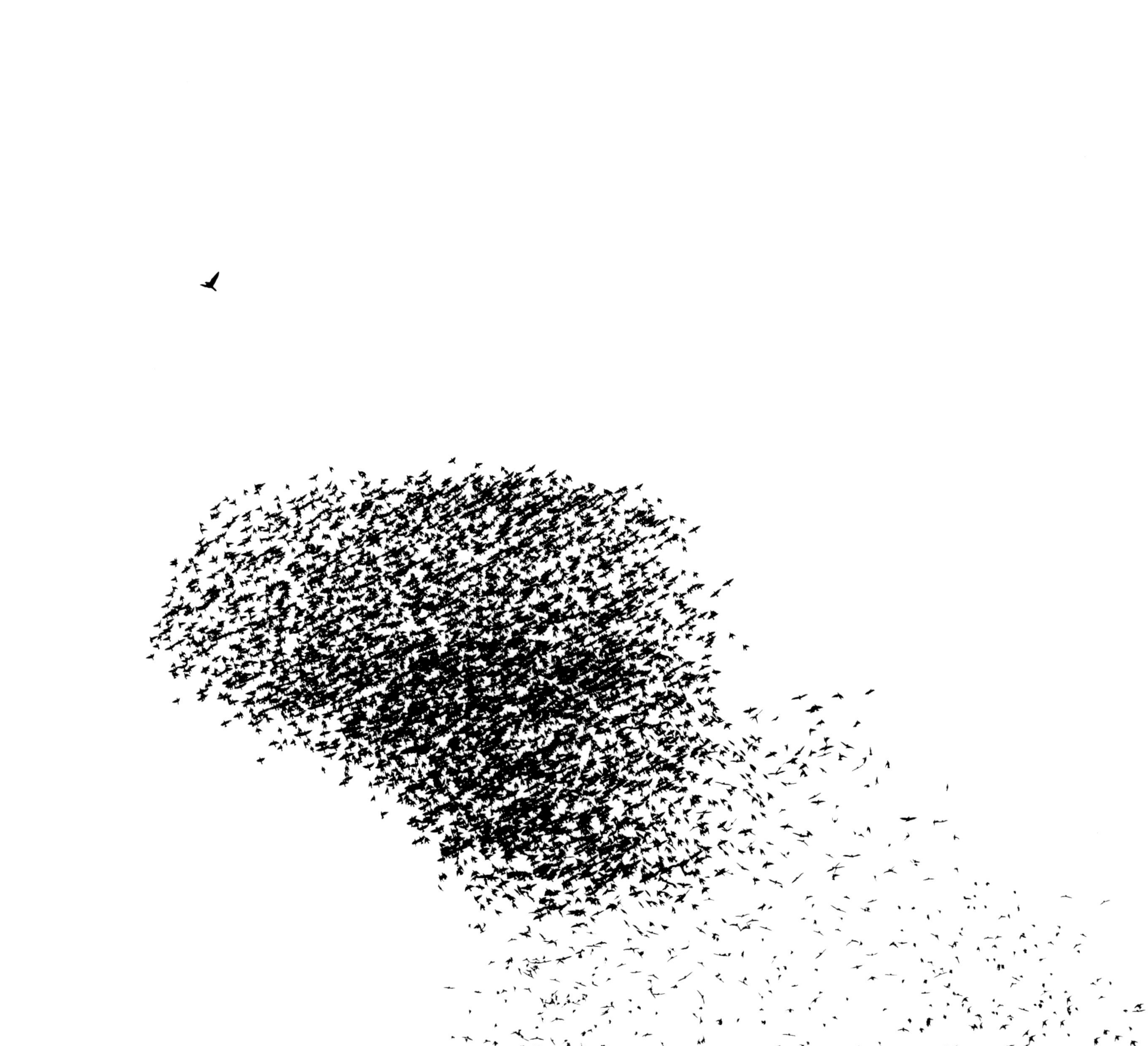

MURMURATIONS

James Crombie *with* John Fallon

WHAT DRIVES US in the pursuit of excellence, in the pursuit of beauty, of the next great moment, the next great shot? What drives one man to photograph the greatest sporting moments of our time? What drives that same man to head out into the dark and cold of three-hundred winter evenings, to kneel by the edge of a lake and wait, camera in hand, for something special to take place, away from the roar of crowds and the clash of bodies. What drives that man to wait, and wait, learning the patterns of nature and the slow shift of the seasons and finally to capture glimpses of one of Ireland's greatest natural phenomena.

James Crombie has never been sure what drives him back to the shores of Lough Ennell, back to the murmurations of starlings who make the lake a temporary home during the winter months, before many of them make their annual migration to Scandinavia. Yet James has returned night after night, for over four years, to watch the weft and flow of the birds, reading images into their movements and capturing like no photographer before him the natural life cycles of the lake, the birds, and the environment that sustains this amazing natural event.

Over those four years, alongside friends new and old, James learned the ways of the lakeshore – both its wildlife and its people. From the murmurations whose movements could darken the sky, to the subtle beauty of the lake's swans, its majestic birds of prey, ethereal horses and much more, James came to know the place and its seasons in ways he couldn't have imagined. He also came to know the people of the lake – the farmers, swimmers, bird-watchers and locals – and how their lives were entwined with the lake and the animals who lived on it.

Four years and half a million photographs later, this is the story of one photographer's relationship with a lake, and the wonders that he found there.

A WESTMEATH NATIVE, James, who now lives in Offaly with his wife Ann and their young family, is one of Ireland's top photographers, covering sporting events throughout Ireland and all over the world for the Dublin-based Inpho Photography agency. James has travelled the globe in pursuit of photographing sporting greatness, from the Olympics to the All-Ireland, and works with Inpho as an official photographer for a number of international rugby organisations including the IRFU, EPCR, Six Nations and the British and Irish Lions.

It can be a hectic working lifestyle for professional photographers on that international beat. Aside from the continual travel, matchdays are particularly demanding. A photographer covering a Six Nations rugby international will snap somewhere between 3,000 and 4,000 photographs per game. Time is of the essence, and a photographer must have the photo of a try being scored submitted by the time the conversion is taken. Crombie and his colleagues were well used to those demands, travelling from one major event to another and juggling endless deadlines and a pace that never let up.

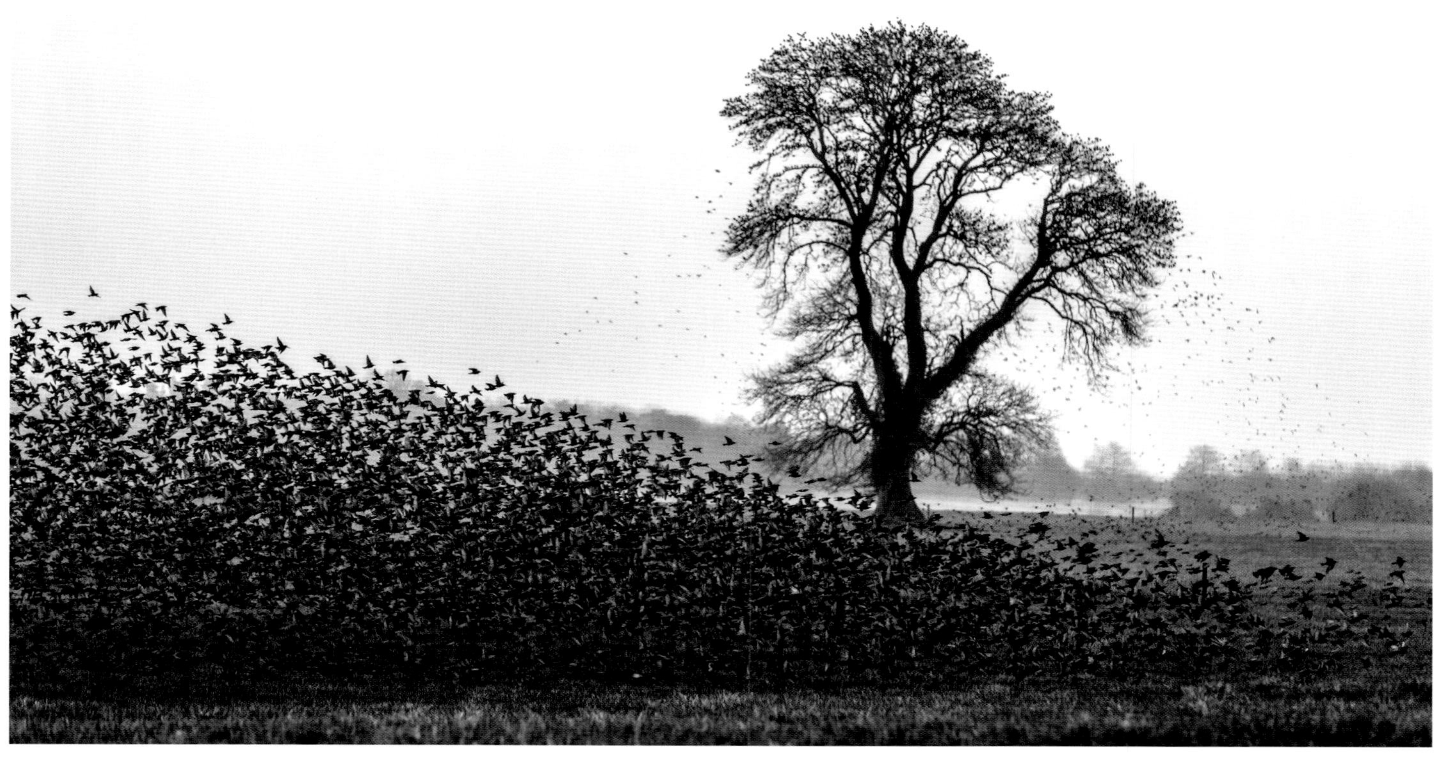

That is, until March of 2020, when the Covid pandemic struck and, initially at least, meant that all sports were called off. Used to travelling the world, working at a relentless pace, this was something James, or anyone else, could never have foreseen. The hectic work-life he was so accustomed to came to a shuddering halt in that moment. Far from the maddening crowds of Croke Park or the Stade de France, he found himself confined to the environs of his rural home in the Irish midlands.

'There was absolutely nothing on, which meant I had nothing to photograph. We had an initial schedule worked out that was just an endless list of matches, press conferences, commercial shoots, flights, car hire, accreditation, train tickets and so on. The usual week-to-week hectic schedule, and overnight it just ground to a halt.

'I suppose, like most people, you welcomed the fact that there was time to spend with family, even allowing for all the uncertainty.

'But as the days rolled into weeks and the weeks into months, there was still no sign of sport resuming and I started looking around for other things to photograph.

'The fact that I was based in the midlands gave me a chance to explore things there and I was looking to do stuff which would interest the front pages of newspapers as much as the sports pages. There was a chance to slow down, to think out projects rather than be always chasing flights and matches and all the other madness.'

This was the first time in as long as he could remember that James could turn his lens away from sport, in search of something else – something new – and something different to anything he'd ever done before.

One project which had been brewing in his mind for several years was photographing 'a hurler on the moon'. He needed a few things to fall into place for this to happen, not least a full moon on a clear night early in April 2020. He identified Croghan Hill in Offaly, not far from where he lived, as the ideal location to position the hurler. The 234-metre hill, the remains of an extinct volcano, is located between Rhode and Tyrrellspass and rises from the Bog of Allen.

'The photo is a silhouette. It works if you use a very long lens and put someone on a high hill with nothing in between you and them at a full moon. It's called forced perspective, but basically, it looks like they are bigger than they are and that they are the same size as the moon. But it needs a lot to fall into place to get it exactly right.

'I needed a hill with a flat surface so that the hurler could be seen running off the edge of it, so Croghan Hill was perfect for that. That was the location sorted. Obviously I needed the weather and the moon to be right, but I also needed to work out where I needed to be to get the perfect shot. And I knew just the person to help me,' said Crombie.

Colin Hogg and James had gone to school together at the Mercy Secondary School in Kilbeggan. They were good mates in the days before social media, but didn't keep in touch after completing their Leaving Certificate, with Hogg moving to Dublin to study geophysics. Colin and James hadn't been in touch with each other for close on 20 years apart from the odd response to a Tweet here and there.

In 2019 Colin married his fiancée Nora Hopkins from Lixnaw in Kerry at the Adare Manor in Limerick and asked his old schoolmate Crombie to photograph the wedding. They kept in touch after that and Colin was delighted to assist in setting up the hurler photo on Croghan Hill.

James recruited his twin brother David to be 'the hurler' in the photo and while he reckoned he needed to be about one kilometre away to get the perfect shot with the moon in the background, he knew his positioning had to be exact. 'James needed to be in a position to see the moon rise up on the clean surface piece of the hill,' said Colin. The image was a stunning success, exhibited not just across a huge tract of media in Ireland but much further afield, and provided a welcome relief from the hardship of the Covid pandemic.

When James went on to win the Press Photographer of the Year later that year, that photo was a key part of his portfolio, with adjudicators acknowledging the level of forensic detail which had gone into capturing it.

With his attention now free to roam beyond the confines of sport, to the natural environment of the place he was from, James was discovering a trove of photographic potential, and was about to chance upon the place and a project that would consume him for years to come.

AFTER THE SUCCESS of that first photograph, James was keen to find a new project to focus on. It was around this time however that tragedy struck for Colin and Nora. The couple were by now expecting their first child, but early on discovered that there were major complications with the pregnancy, and the outlook was not good. The couple had, in a sad twist of fate, discovered the sex of the baby on the day in April that Colin and James teamed up to shoot the hurler on the moon. 'That was the day I found out I was going to have a son and I was kind of on a high and then I really wanted to get this right for James because I knew we had one go at it and I was buzzing,' said Colin. Not long afterwards they received the prognosis for their unborn son.

'We found out he had diaphragmatic hernia,' added Colin. 'It's where the diaphragm doesn't form. We were told at the time it's 50-50. But I kind of knew it was not 50-50. We knew the lungs weren't forming and the heart wasn't forming. We knew that. We knew it would require extreme surgery to try to make him survive. But as the pregnancy went on, the percentage went down and down.

'So we knew that he was going to die when he was born. So we had some small amount of time with him. Daniel was born on September 25th in 2020 and died that day.' Before Daniel was born, James had taken photographs of Nora and Colin at Lough Ennell, though the couple have not been able to look at them since.

Happily, the couple have since become parents to a beautiful daughter, Hannah, who was born in February 2022. In September of 2020 however, the young couple were going through every parent's worst nightmare.

'I left them alone for a while,' said James, himself a father of four daughters. 'To be honest, I didn't know what to say to them. I just felt they needed space to grieve. After some weeks I sent a camera over to Colin. I knew he always had a big interest in photography and that it might help take his mind off things.'

Colin, who was originally from Castletown Geoghegan, not far from the southern shores of Lough Ennell, had mentioned to James in the weeks leading up to this that there was a spectacular phenomenon that took place over the lake each year, which may just be worth their time. The idea had been put to one side in the grief of the intervening months, until Colin once again brought it up with James.

'We kept in touch as the weeks passed and then Colin suggested that maybe we should do something with the murmurations around Lough Ennell. I had no idea what he was talking about so I went home and started Googling. It was obvious from the start there was potential there, thousands and thousands of birds swooning through the air in endless formations.

'That said, I really didn't know what sort of photograph we were looking for or, indeed, how to go about getting it. But we had time on our hands and I think Colin just needed something to take his mind off what he and Nora had gone through, so we set about planning what to do.'

'For me, this was therapy,' said Colin. 'The pubs and restaurants were closed and here was this on my doorstep. It was a good distraction.'

THE PAIR MADE for the lake for the first time in the evening hours of the 7th of December 2020. Initially they weren't sure where to begin – the lake is over 6km long and 2km wide, with only a few points of public access to the water's edge. They settled on Lilliput, a popular local access point which hugs the southern edge of the lake. Between the vast reed beds at either edge of Lilliput is a small pier, tucked into a public swimming area with a view clear across the lake. That first trip James had no idea of what to expect, but was immediately drawn to the calm tranquillity of the open water. That first night was quiet, with damp in the air and little light in the gloom of the evening. Then, they waited in the silence. After about half an hour, there were the first signs of what was to come – the first small groups of starlings flying overhead, mirrored in the water of the lake. James recalls how, within minutes, there were a couple of thousand birds swerving and gliding in unison over the lake.

'I knew in that moment that this was an amazing phenomenon, I was instantly hooked.'

In retrospect, he sees that first night as just that – the first in a series of experiences at the lake, the catalyst for the next four years. Looking back, the murmuration that evening wasn't even especially large, and would soon be overshadowed by what was to follow, and yet it stirred something in him and he knew this wasn't going to be a one-evening project.

IT DIDN'T TAKE long to work out that the movement of the starlings happens each evening as they return from where they were feeding to roost in the reeds along the edge of the water. To catch sight of them was a matter of being there for that short window of opportunity each evening.

They started on a damp, calm evening on Monday, 7th of December 2020 and returned each evening for the next three or four weeks. It was difficult to establish any pattern that the birds were following other than when the movement occurred around the same time before nightfall. It was also difficult to get access to the shoreline to get a clear shot of the birds as the lakeside is densely knotted with trees and shrubbery.

'There are only three or four access points to the lake, the whole way round,' added James. 'There is Ladestown, Lilliput, Bloomfield and Tudenham, each of them with a car park, and also Butler's Bridge. But other than that there is really no other public access to the lake.'

It quickly became apparent that to get the best vantage on the birds, they needed to know the lands around the lake, and to do that they needed the intimate knowledge – and permission – of a local, someone who knew those shores like the back of their own hand. This was where Enda Maguire came in.

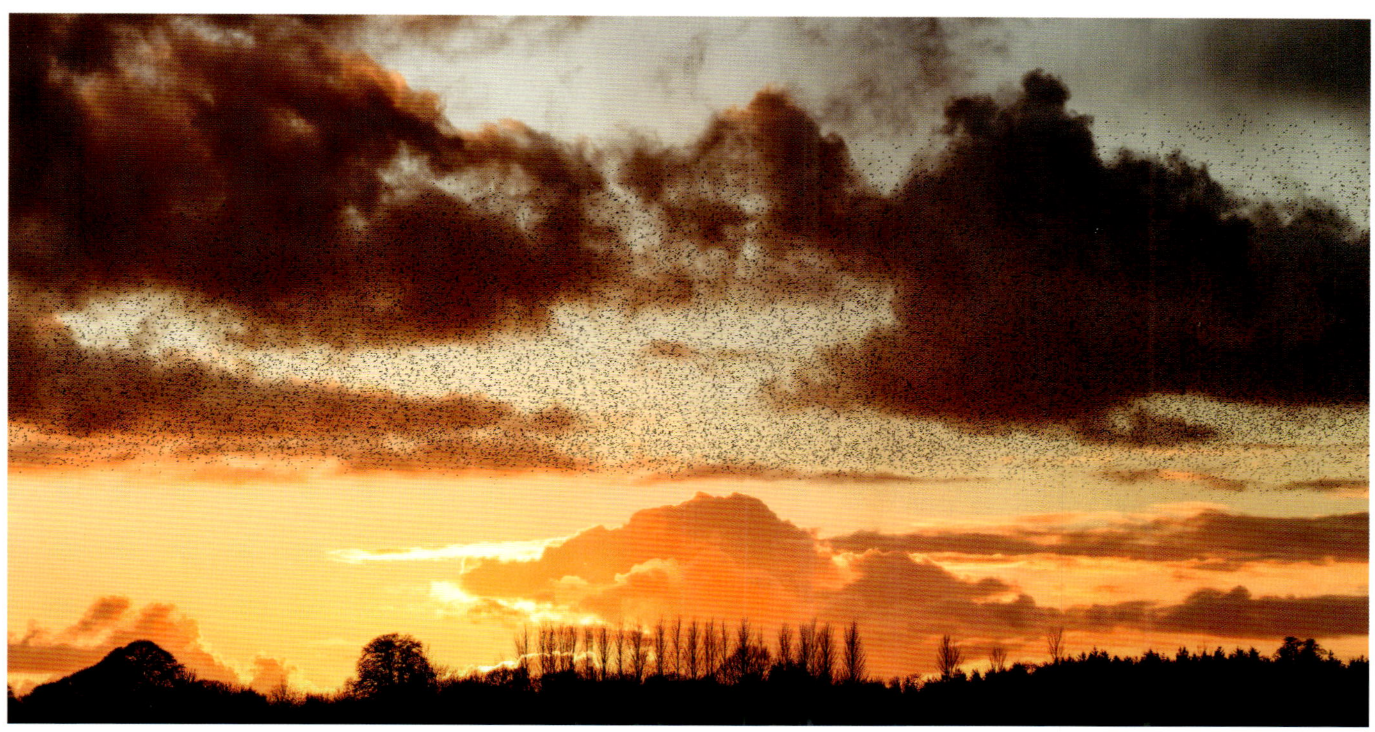

ENDA MAGUIRE presumed when he was growing up that every farm was graced annually by thousands of birds. He didn't pay particular heed each winter when the sky was darkened by untold numbers of starlings descending on their dairy farm on the banks of Lough Ennell in County Westmeath. Like all farmers, the Maguires ebbed and flowed with the seasons. The birds were just part of the environment, the farmers had as much control over them as they had over the weather.

'I thought they were that way everywhere. They were there in the same numbers as now. I never noticed any change but we didn't pay much heed to them. They were just there, they were part and parcel of life on this farm.

'They were probably more of a nuisance to us than anything else. They would be nesting here, there and everywhere, in hay, in roofs, or wherever they could. It was harmless though.

'They wouldn't do any harm. They wouldn't be around during the summer, it would only be during the winter and there was a limit to the amount of harm birds could do then. They were just part of life on the farm and they would have contributed by the insects they preyed on. But we just presumed it was the same on every farm in the country. It was a while before I realised it was so unique to here.'

Enda is the second generation of his family to farm this land. His father Michael was from Carrigallen in Leitrim and moved to Ladestown near Lough Ennell when the Land Commission divided up the old estates which had belonged to the landed gentry. The Maguire farm was part of two estates, Ladestown House and Keoltown House. There were large estates throughout the region. Fishing and hunting were very popular down through the generations. Most of the old estates were taken over by the Land Commission and divided into smaller units, but as many of the new farmers came from other areas of the country, a lot of local knowledge died.

'Most of the farmers around here would all be Land Commission farmers,' added Enda Maguire. 'So there's nobody around here that's been farming for seven or eight generations or anything like that. A lot of local knowledge in farming gets passed down through the generations but that's not the case here because our history only began, say, in the sixties.

'The starlings were here in huge numbers when my father arrived but it's hard to tell how long they had been here. I just remember them from the start. They were part and parcel of life here the same as the cows, the trees or whatever. You just didn't take any notice of them other than the fact that they travelled in such huge numbers.

'And there were two things you learnt very quickly as a child if they were flying overhead — don't scare them or startle them or they would pebble-dash you with droppings! And if that happened, don't look up! We weren't long learning that I can tell you.'

The starlings had attracted some film crews and photographers over the years. Most of them paid a visit or two to Lough Ennell and departed after getting some good footage or shots.

Then in 2020, as the world adapted to the Covid pandemic, along came James.

JAMES AND COLIN were in Ladestown, trying to get a better view of the murmuration swirling before them over the water. In the distance, on the shore, they spotted a farmer in the field, driving his tractor. 'We approached him and quickly figured out he was a keen wildlife enthusiast and had massive local knowledge of the murmurations as he'd been living there all his life.' What started off as a chance encounter soon grew into a lasting friendship, with Enda showing James and Colin all he knew of the lake. 'We now have a WhatsApp group called *Murmurations* where we constantly share information, jokes, ideas, stories.' The project of capturing the birds on film was growing into something greater now, and James was finding connections with the place he hadn't expected.

'In those first weeks we drove around and knew the birds were going towards Ladestown, but we knew we needed to get access to the land to get the best viewing positions. There are lots of trees around the car parks so we needed access to the higher ground where there were no trees. It turned out one of the days we were there Enda Maguire was out farming in the fields and we approached him and explained what we were trying to do.'

From the outset Enda and his wife Trudy and their daughter Anna were welcoming and helpful to the two murmuration seekers. James recalls how 'it's been the same with all the other farmers around here. They have all been so helpful. We wouldn't have been able to do it without them.' The access offered by Enda and some of the other farmers around the shore helped James and Colin map out the movement of the murmurations with more accuracy, although the exact behaviour of the starlings would remain unpredictable. After researching what they could, James discovered that the birds roosting in Lough Ennell returned there after feeding in a 30-kilometre radius. Some 'pioneer starlings' go up in the sky and start the movement which attracts in all the other starlings in the general area as they return home to roost in the reeds.

The reeds around Lough Ennell are ideal for roosting starlings, but the birds need to keep moving every so often as the reeds bend under the sheer weight of so many birds resting on them. 'Roosting on the reeds kept them safe from foxes and other land-based predators, but it is the threat from the sky which has led to the wonderful movements in the air,' added James. 'There is still a lot of mystery around these birds and their habits. They are very loud when they initially roost but then they go dead quiet. It's like it's "lights out" time.

'During December there wasn't a whole lot for us to shoot. We noticed the numbers kept increasing as the months passed and we knew that they would be gone by the end of March.

'It was a bit eerie being there. There was nobody else around. We are looking at these thousands and thousands of birds during these wonderful evasive movements and we were the only two people in the world looking at them.'

James would reel off a few hundred frames each evening and Colin would sometimes video the movements, but it was haphazard and unpredictable, 'The weather changed a lot from night to night and we were trying to discover how their behaviour changed in different weather,' added Colin.

'Fortunately sport equipment and wildlife equipment are very similar. Both require patience and anticipation. Unfortunately, murmurations happen at the worst time of day regarding light.' Accustomed to photographing sporting events under the crisp lighting

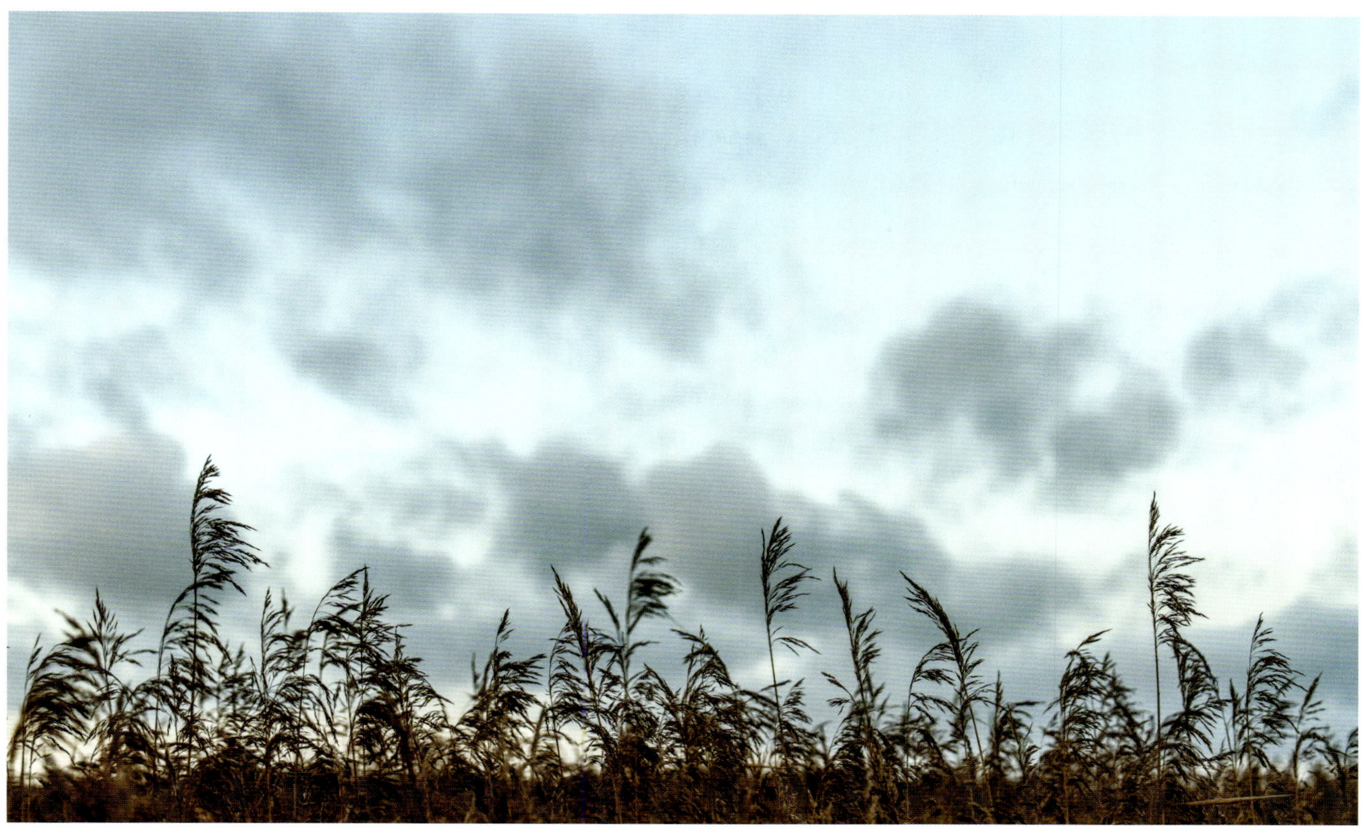

of summer afternoons or calibrated floodlights, the conditions by the lake in the twilight of winter evenings brought entirely new challenges. The incredible phenomenon was taking place right there, but capturing it on film remained the challenge. Even twenty or thirty years ago, James notes, photographing the birds at this hour would have been incredibly difficult. 'Thankfully with the innovations in technology since, the equipment is able to handle this a lot better than it would have done in the past.'

'We were trying to see if there was any trend so we could predict where to go. There was no trend on how long they would stay in an area but we began to identify the three or four different areas that they would come back to and that helped us get better positions. Sometimes they would do movements near the ground but unless you were in the right place you wouldn't see even though it could be happening a short distance away.'

James and Colin kept coming back every evening through Christmas and into January. The weather was grim a lot of the time but they persevered. It was very difficult at times, especially when they had to trudge across the wetlands, and it wasn't without mishaps either. One night James' car keys fell into the water. His twin brother David — the 'hurler on the moon' — stripped to his underwear and dived in to retrieve them. 'I have a photo of him in his jocks but I don't think he is too amused,' noted James.

THE MAIN LOCATIONS they chose in order to view the birds were Lilliput, Dysart Island, Ladestown, Butler's Bridge and Bloomfield House.

LILLIPUT

'A location we all grew up visiting and knew well, as it has been a popular local swimming and bathing area for generations. The murmurations generally start their annual routine there. It's a great location to view the whole lake – even when you don't know where the murmurations are, from there you can get a great view of the entire lake and it enables you to figure out which reedbed they're using at any given time.'

DYSART ISLAND

'An island that touches the land surrounded by reedbeds of Kilcooley Bay and Keoltown Reeds – a very popular spot for the murmuration roosts due to the vast area of reeds that surround it. They also use the land around One Tree Hill, named locally thought it may not be the actual name of the hill.'

LADESTOWN

'This is a popular parking and camping site for locals and tourists alike. Easily accessible, it doesn't offer great views of the murmurations but works as a great starting point to discover where they may take place. This was one of our early vantage points, before we began to venture further along the lake's edge, and away from some of the more heavily-populated areas.'

BUTLER'S BRIDGE AND BLOOMFIELD HOUSE

'Butler's Bridge is about a kilometre's walk along Lacey's Canal and leads to a viewing point of the lake – perfect for sunset views. This is the location of that now famous "bird murmuration" photograph I was lucky enough to capture in that first year. Both Butler's Bridge and Bloomfield House offer spectacular views of the murmurations during February and March and is one the most popular sites for bird watchers and photographers alike.'

JAMES AND COLIN persevered, returning to the lake day after day, and they reckoned they visited the lake in excess of 60 times before they got a shot, possibly *the shot*, they'd been hoping for. They continued throughout January and February but with the local people telling them the large murmurations were usually finished by St Patrick's Day, they decided they would keep going for another week into March.

"After a while we began to notice a bit of a pattern," added James. "The birds would start at Liliput at the south of the lake and they would move up along the west of the lake. They could spend every night for three weeks in one spot, maybe a week in another, then a few weeks in another. We didn't know where exactly but we realised that the best spot for us was at the north of the lake near Butler's Bridge as that gave us a clear shot down the lake.

'So that's where we positioned ourselves for a lot of what we thought would be our last week on this project. We had got a lot of good photos and video footage but there was nothing extraordinary.

'But we felt the whole project had been worthwhile, not least in helping Colin deal with the loss of his baby. We noticed the number of birds had increased as the months went by and by the end of February the sky was black with birds some evenings. It really was spectacular.'

It was during this initial swell of starlings that Crombie snapped a spectacular photograph of a murmuration which resembled the leaves on a tree on the lakeshore.

'I remember showing that one to my family and everyone saying, "well you have done it now, you have got your picture", but, I don't know, I just kept thinking there was something else here.'

THERE WAS NOTHING special about the evening of the 2nd of March 2021 as the pair arrived back at Lough Ennell around 5.45pm that evening. 'We didn't really use a video camera very much,' said James. 'But for some reason I said to Colin, I had three cameras, we'll put a video on one. Neither of us is an expert in shooting video but just decided to have a go that evening.

'We had been on a hill a couple of kilometres away the night before and I was thinking we should go to Butler's Bridge the next evening and give it another go. The number of birds that evening were massive, I had about 40 good pictures from that night before.'

The audio from the video which Colin was shooting picks up James immediately spotting the significance of what was unfolding in front of his eyes as the birds executed a series of spectacular movements in front of them.

'That's a bird! That looked like a bird,' exclaimed James.

The formation lasted just half a second, yet James' camera captured 14 frames a second, seven of the utterly unique bird formation. Of those, he knows one is the image he's been searching for. An image lasting half a second which will last a lifetime.

THE PHOTOGRAPH HIT the Irish media initially and then went global with CNN, Al Jazeera, National Geographic, New Scientist and Netflix among the international organisations to run with it. Something about this image had struck a chord with people around the world. As people in every nation were caught in the grip of the pandemic, the sight of this murmuration offered a moment of joy, of awe and wonder, in the midst of so much fear and uncertainty.

James was oblivious to the furore initially as he was on a shoot in Castlebar when the photo first appeared in the Irish media. When he turned his phone back on it didn't stop beeping for two days.

The family of a man who was deeply attached to Lough Ennell requested to have it emblazoned on his headstone in Kilbeggan cemetery. The image found its way on to memoriam cards, the annual reports of banks, on shutter doors on a building in Norway and on to a café mural in St Kilda's in Australia. Around the world there was a shared sense of identification with this image, and with this phenomenon.

'The response was unbelievable, it was just everywhere. It was phenomenal,' said James. 'On one hand you just wait for a moment like that, but in reality actually you don't. You just can't expect something like that to happen. I have often wondered how many frames did I shoot by the lake, no doubt tens of thousands, and then in the blink of an eye something like that happens in front of you and you are fortunate to have your finger on the button.

'I know we went back night after night hoping to get something special but never in your wildest dreams can you imagine something like that happening. Just that one in a million moment and Colin and I were fortunate enough to be there to record a spontaneous moment of nature.'

The night the photo broke in the Irish media, there was a stream of cars driving down past the entrance to Enda Maguire's farm in Ladestown, hoping to get a glimpse of the starlings doing something special. A short distance away in Butler's Bridge, around 50 people arrived to witness them and for many nights afterwards there was a steady stream of visitors.

'Once the photograph went out, there was fair traffic around here then,' said Enda Maguire. 'We got a fair few people driving up and down the road. They're all thinking they were going to see something weird. But it was never going to happen.

'It certainly created a new interest and it certainly did bring people to the area. I'm not sure what their expectation levels were. I remember meeting one family there by the lake one day and they said they had travelled a fair distance and there was nothing to see. It was as if they expected an evening show and were disappointed when that didn't happen. But it doesn't work like that.

'The numbers coming hoping to see the birds eased back again after a while, although to this day you'll still get a few who will swing by to see where it all happened.

'The starlings are still there, appearing on the scene around November every year and staying until March. Their numbers are as great as ever, they are still doing all their shapes and moves and carrying on as they always did.'

JAMES WAS CROWNED Ireland's Press Photographer of the Year on the back of the murmuration photo that year, for the second time in a row. Deep down he felt he would likely never get another photo like it, and yet he continued to return to the lake, night after night, season after season. Far from quitting after that initial photo, James would return every winter for over four years, while Colin maintained the almost daily presence that following winter.

'We didn't start this with a target in mind," said James. "The murmurations are fascinating and you just don't know what you will witness. It's been a four-year project now. Yes, *that* photograph came in the first year, but there have been unbelievable shots ever since as well.

'Nature photography is all about anticipation, speed and opportunity. You just have to be ready when that splinter of a moment arrives. You can't make it happen, that's outside your control. But being there in the first instance, and being ready with the right equipment in your hand, is within your control. Colin and I were just fortunate that all of those things aligned and a project which started out without a target in mind, ended up hitting the bullseye when we least expected it.'

AFTER THE INITIAL successes of the murmurations project, with the images finding their way around the world, James found himself returning to the lake, season after season – this time with no goal in mind. Times were changing and the conditions that had lead to that first fateful trip to the lake were changing with them. 'After Covid, Colin, his wife and their new baby relocated to Galway and Colin's workload increased. Myself and Enda continued the journey which probably became more about friendship than the pictures at times. Enda was able to keep an eye on the murmurations when I was away with work so we always had a sense of where it was happening. They move around the lake every 2-3 weeks, hence, you need to stay on top of where they will be.' James knew he had found something – a theme, a place – that he wasn't finished exploring yet.

From the shoreline, James began venturing out onto the water, and onto the tiny islands that dot the lake. 'Enda has a close friend named Packie Kelly who owns a boat and suggested that maybe we could get a different view of the murmuration on the water. With this new vantage we were able to open up our repertoire of locations which led to more interesting angles and really opened up the lake to a new perspective.' Shooting in waning hours of late winter and early spring, conditions were always challenging, with fog sometimes rising from the water and frost glittering the reeds, 'The boat trips were absolutely freezing … the key was to wait for very calm nights as the water added a new element to the compositions in terms of reflections, and also the ability to access locations that were inaccessible via land. At all times we were wary of getting too close to the murmuration roosts though, as we didn't want to interfere with the birds or disturb them'.

Months passed, then years, and the lake and birds continued to provide new wonders, new sights, and even on days when there were no murmurations – James estimates that from over 300 days at the lake, he has been able to photograph murmurations on only 100 – there was beauty and intrigue everywhere he looked. 'As the project continued I noticed how I was seeing more and more images and compositions in the same location. Even if I went back today I would capture a new image and still feel that there were thousands of new images to be taken. I enjoy that narrow focus of one location, the ability to create different images every time. I started to expand my focus to the whole lake capturing images of sunsets, sunrises, moonrises, auras, the abundance of wildlife, people using the lake, kayakers, canoeists, fishermen, swimmers, paddle boarders, etc. There was so much life in one contained place.

'A lot of people ask me why I keep going. I suppose it's like sports players not knowing when to retire … A little bit of me didn't want it to be a fluke that I captured that one special image so early. I also think that there are greater and better images to be taken there. I don't think it was always about that one shot and I was probably a little bit disappointed that it happened on day 50 as opposed to day 300 – that might have been a little easier. Although some of the other images didn't garner the same notoriety as the bird-shaped murmuration did, I think I'm prouder of some of them.' Matters of chance and matters of intuition, of photographic alignment of place and image, these have become core to James's view of what he does there. The shot of the birds passing, like fallen leaves, before the local tree that became such a focal point for him – that remains special to him. 'Waiting endlessly for the birds to pass that leafless tree in winter. It took 14 nights for that shot to come together … to me that's an image I'm nearly more proud of.

'I suppose the reason I continued there, after the "big" picture, was more about the addiction of witnessing the murmuration, and also the sense of peace it gave me. Once you experience a truly impressive murmuration, it's like a drug – you want to experience it again and again. It's hard to explain why I continued going, but that's as close to it as I can come.'

All of the images that follow were taken between 2020 and 2024 at Lough Ennell in Co. Westmeath in the Irish midlands. From the hundreds of thousands of images taken over those years, these have been selected to show the beauty of the cyclical seasons of the murmurations and the lake. For the most part these photographs were taken between the months of October to March when the starlings are most active.

MURMU

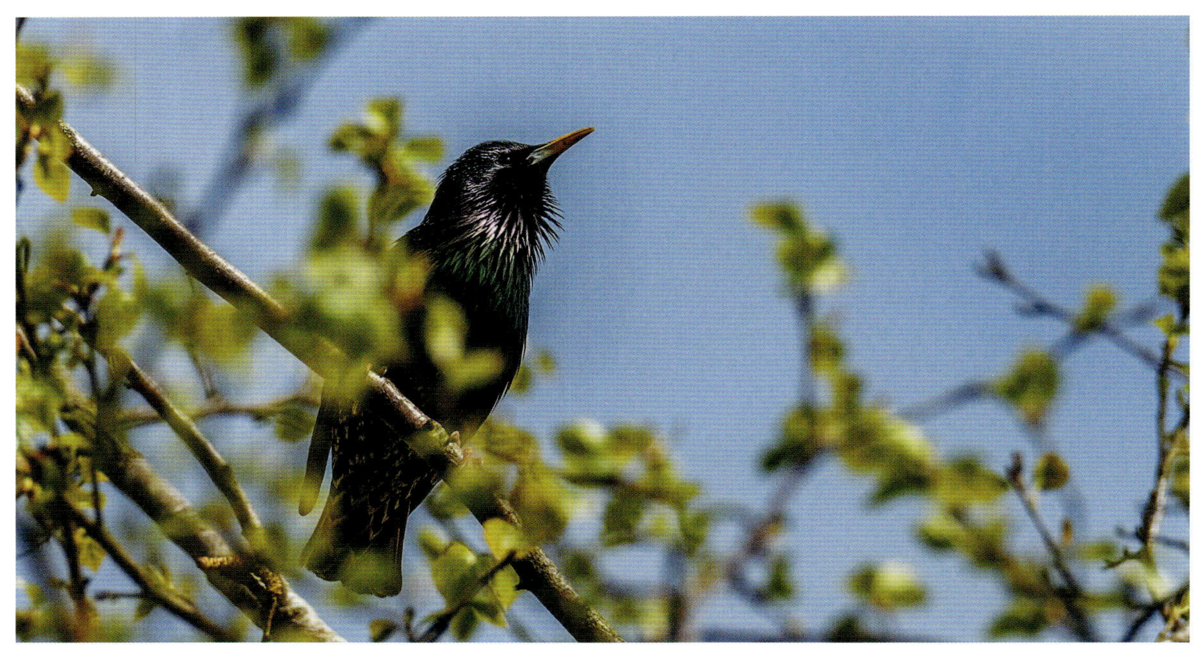

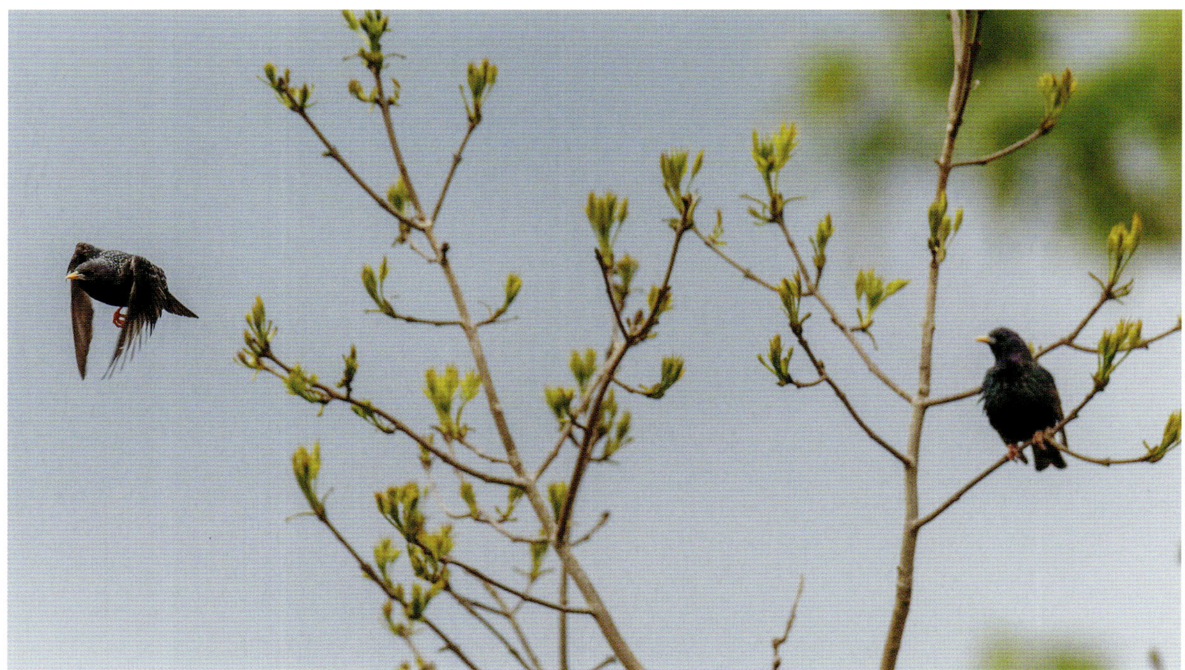

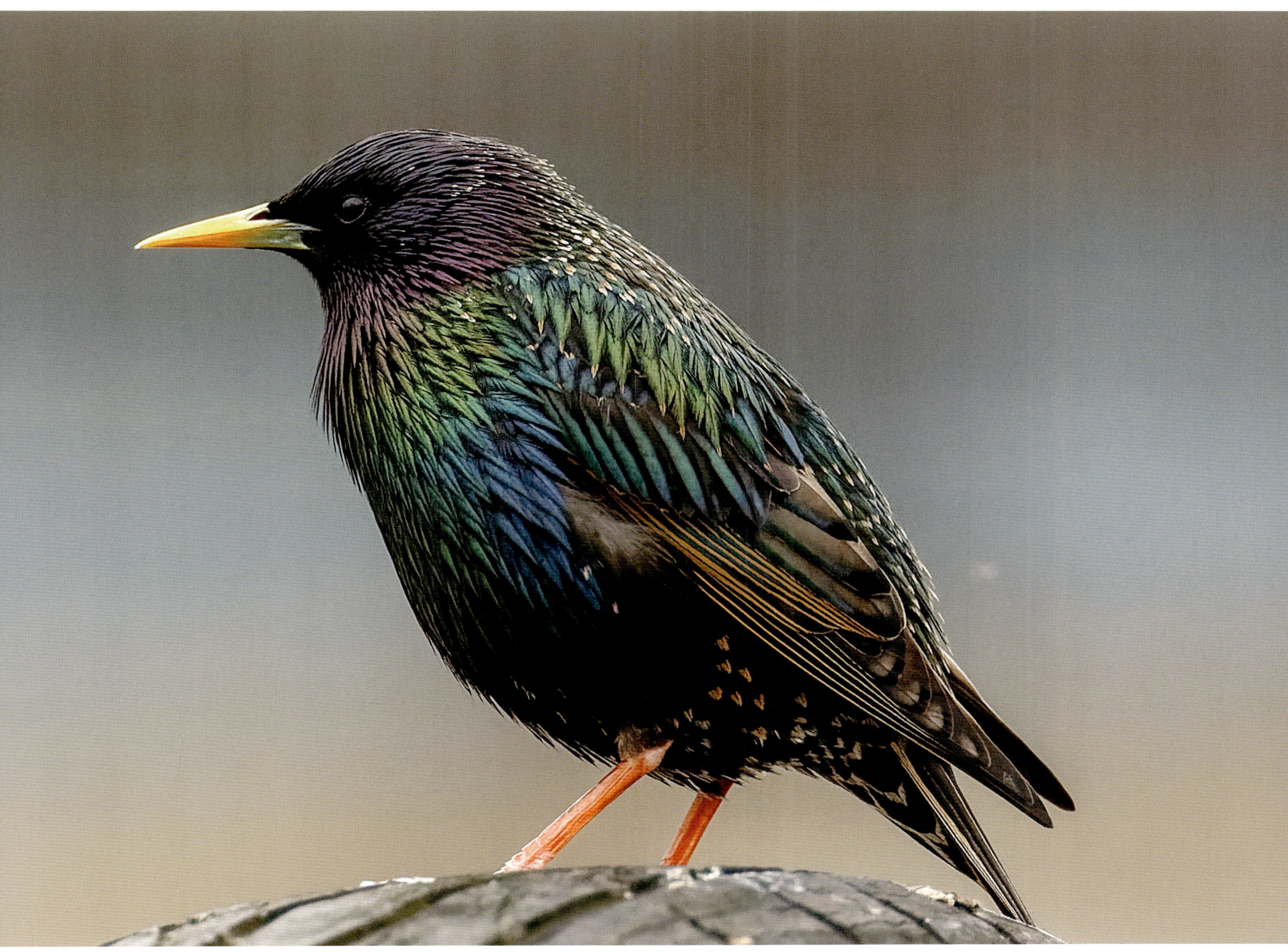

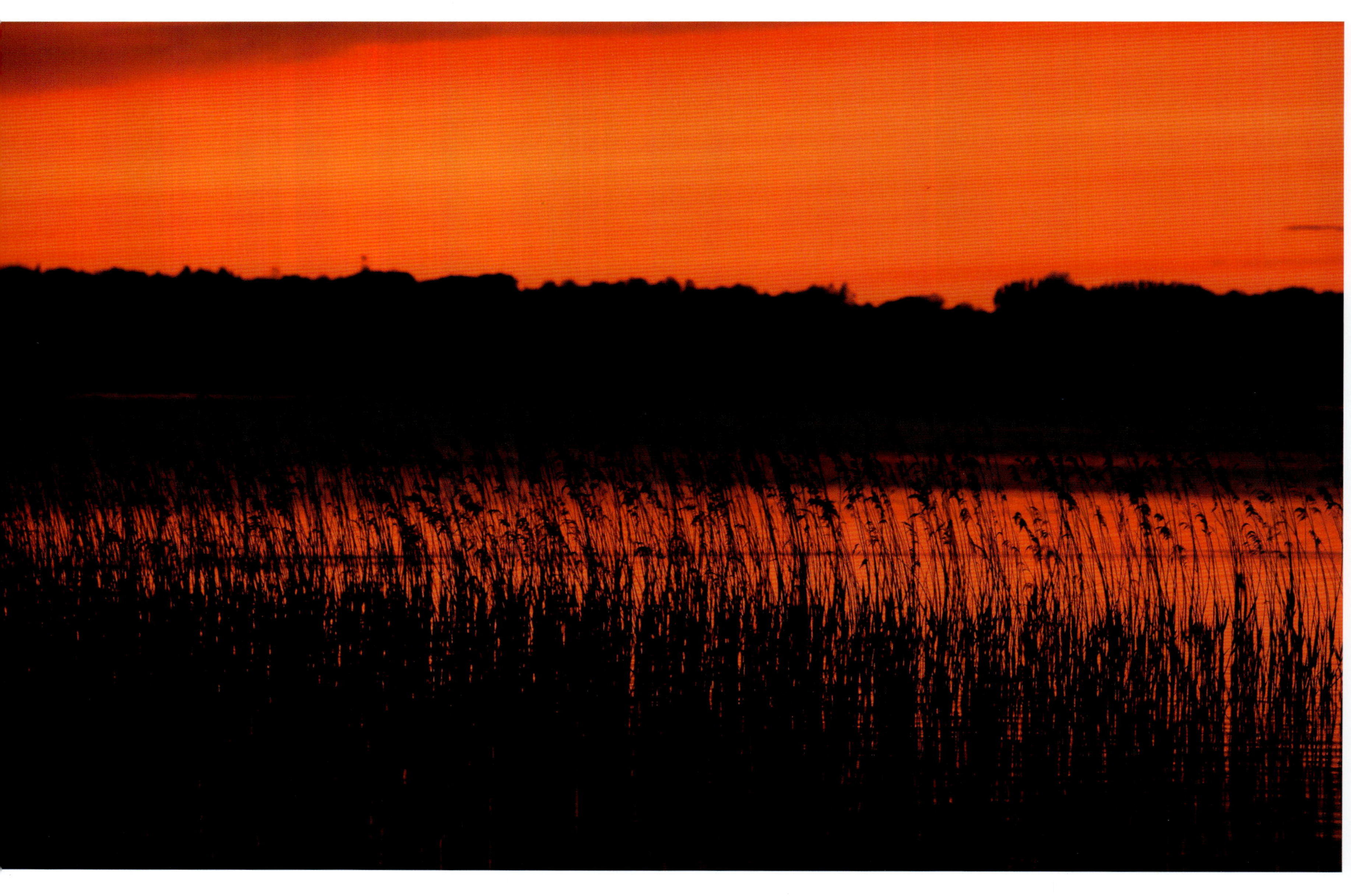

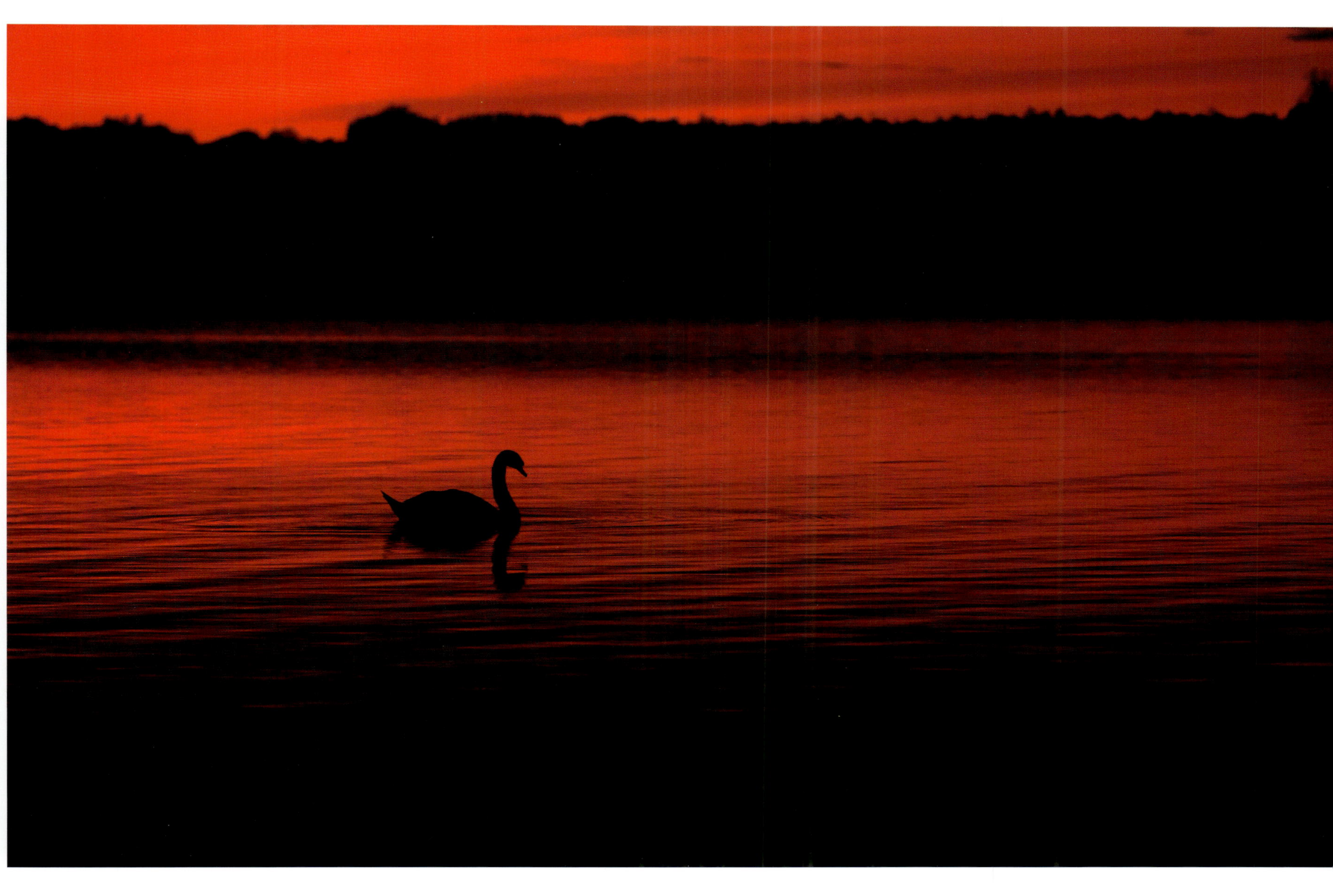

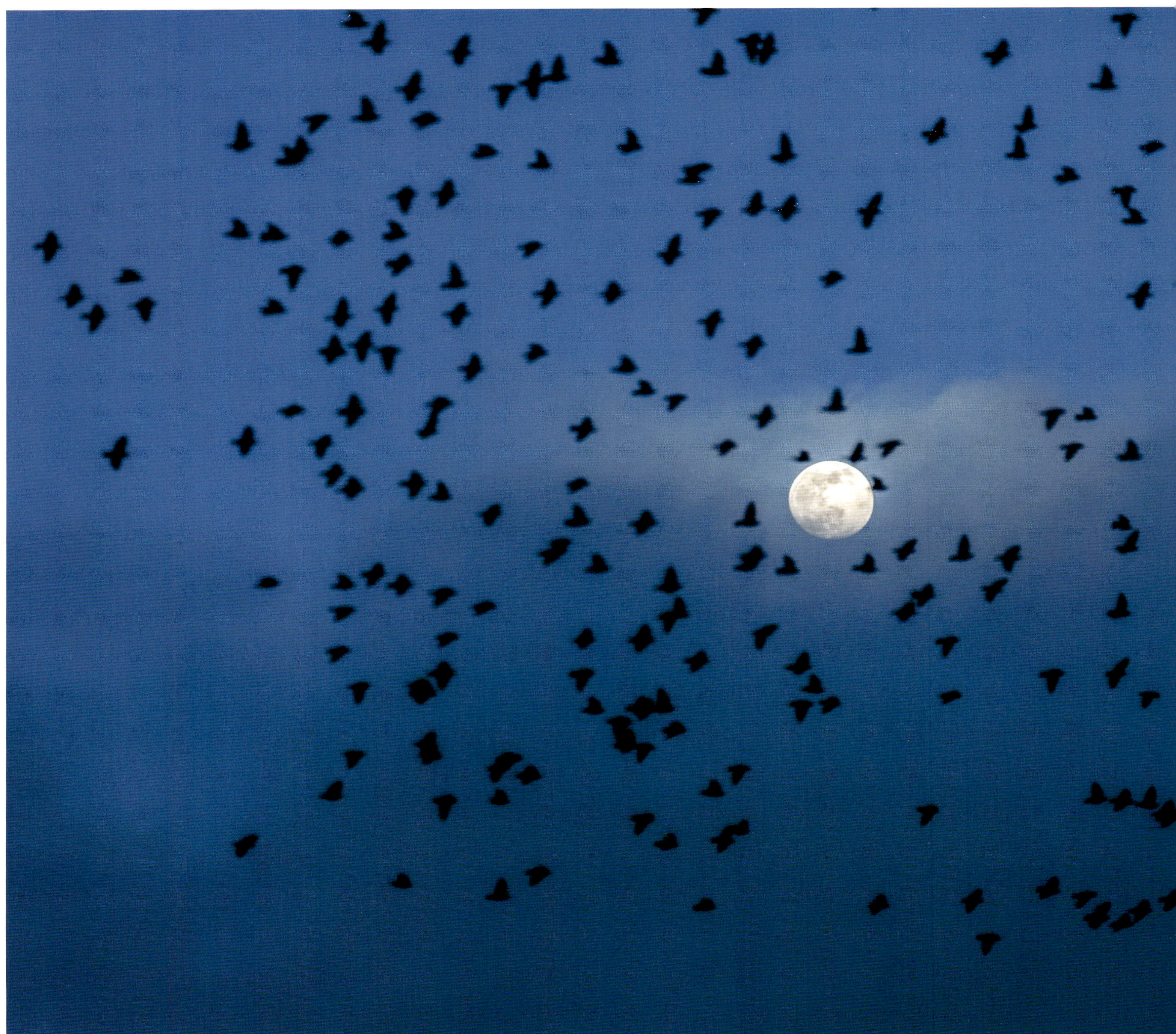

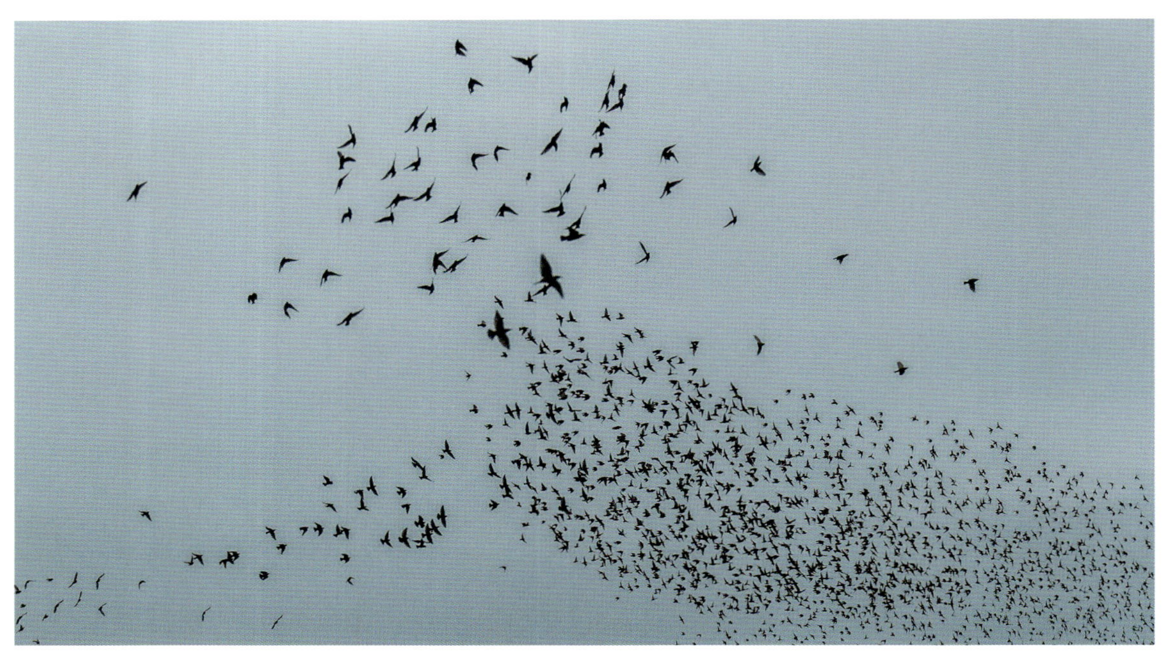

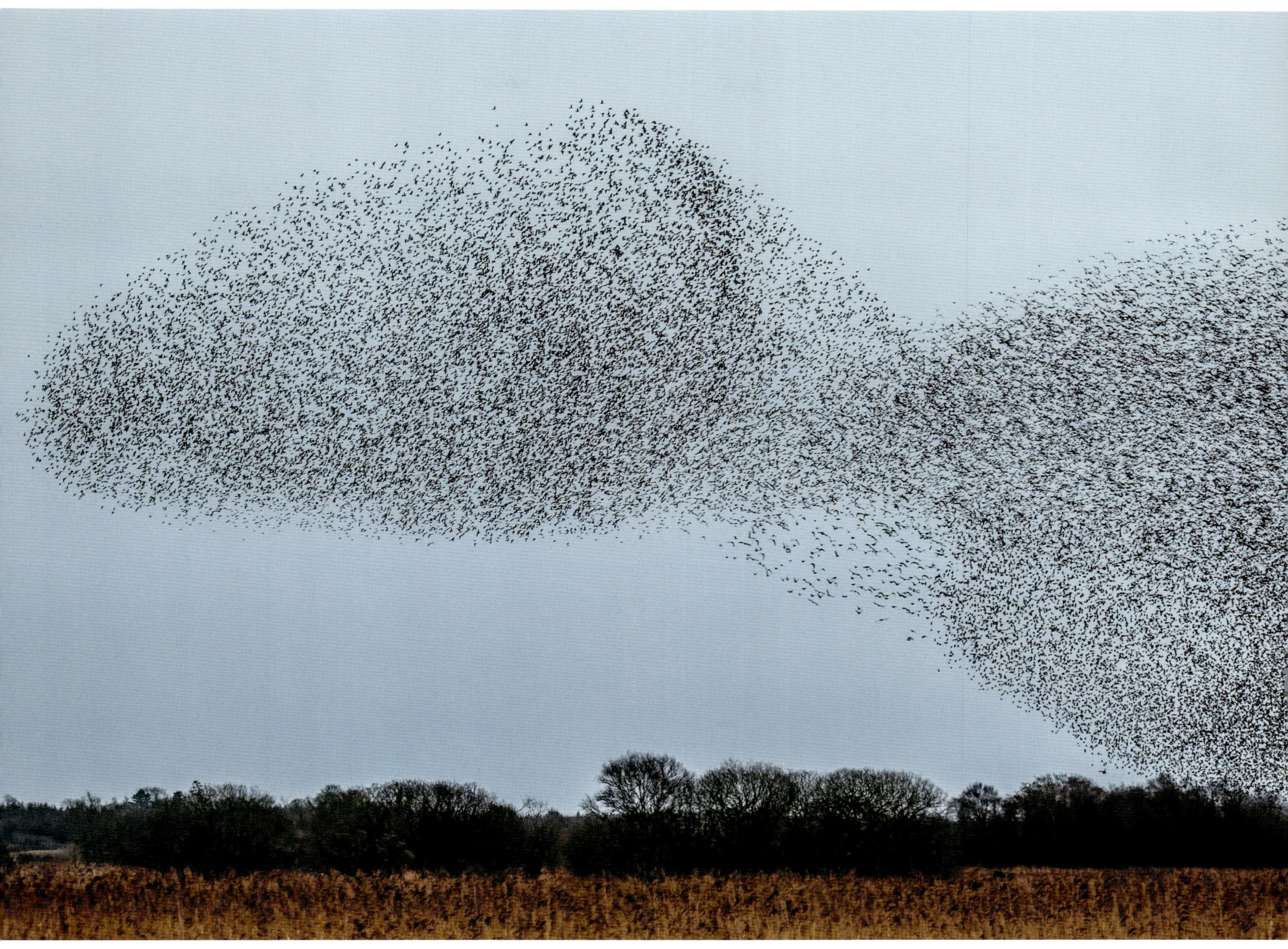

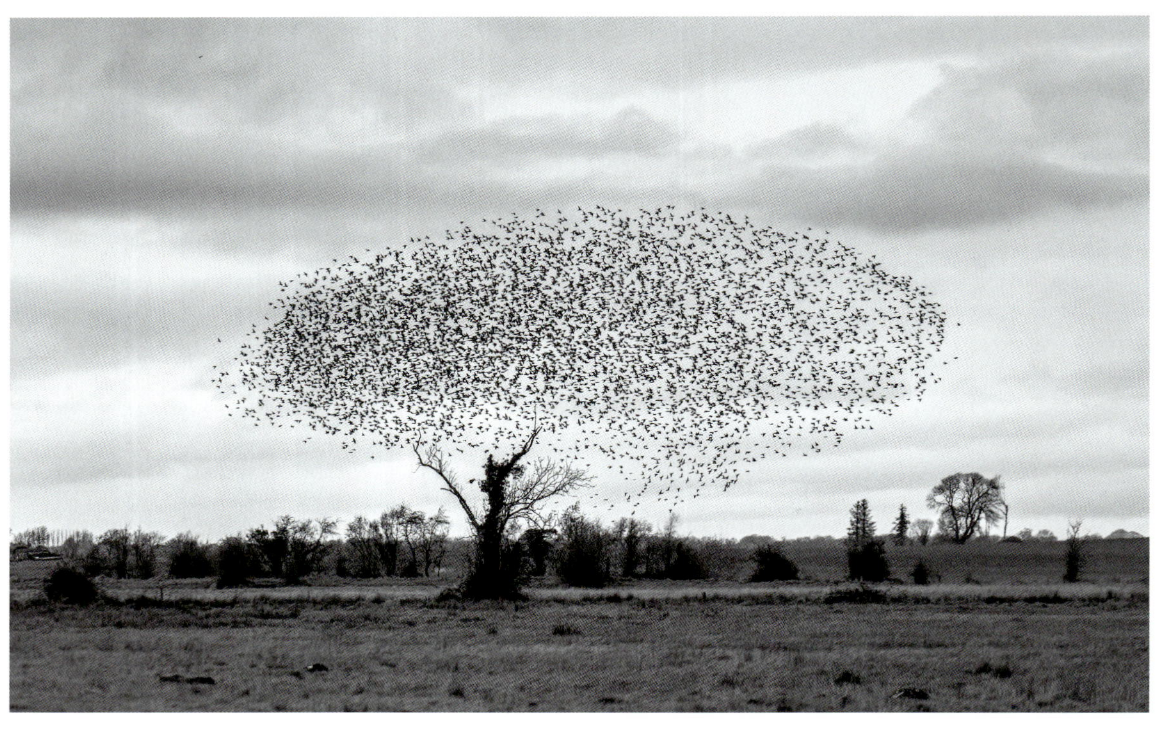
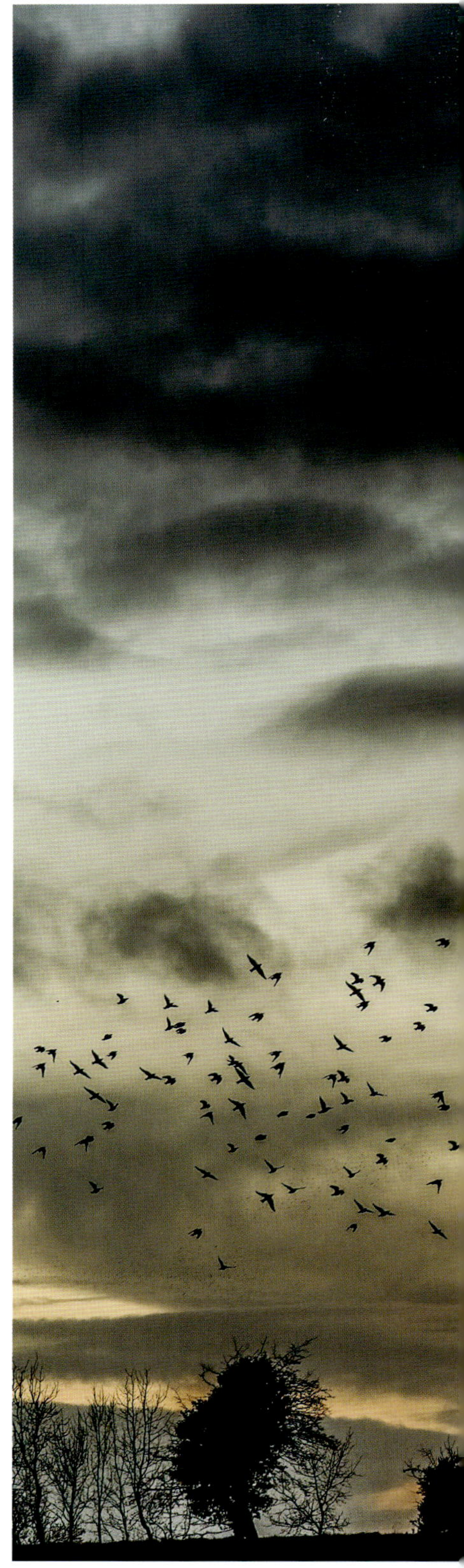

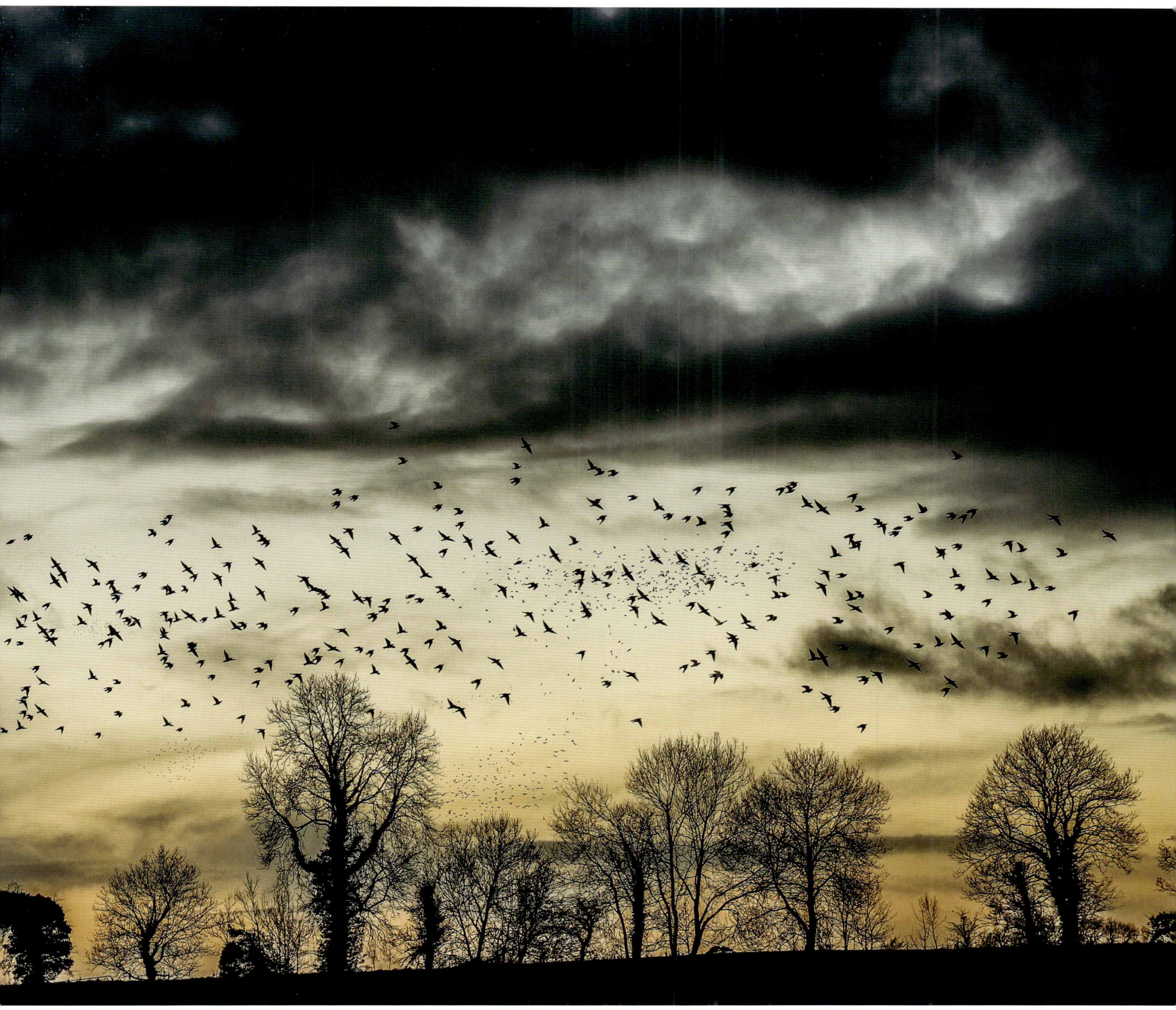

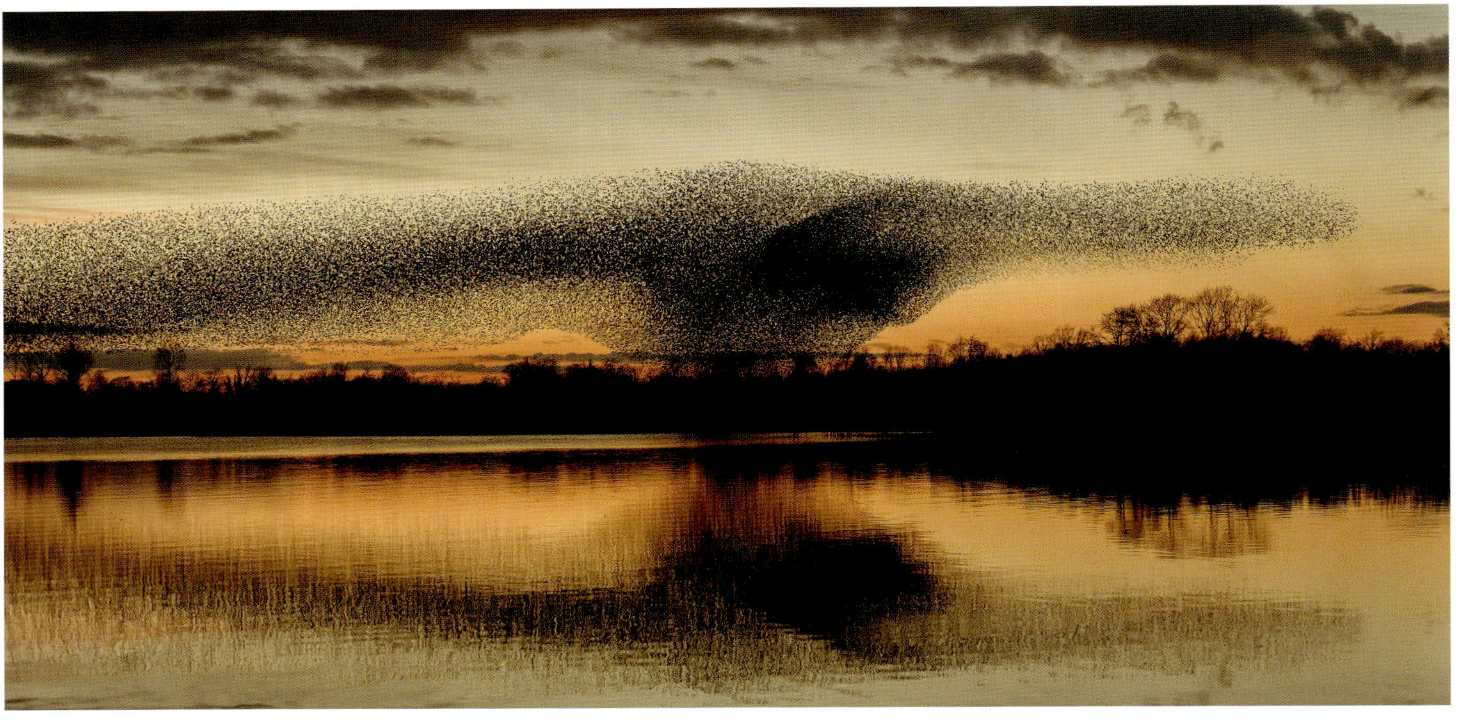

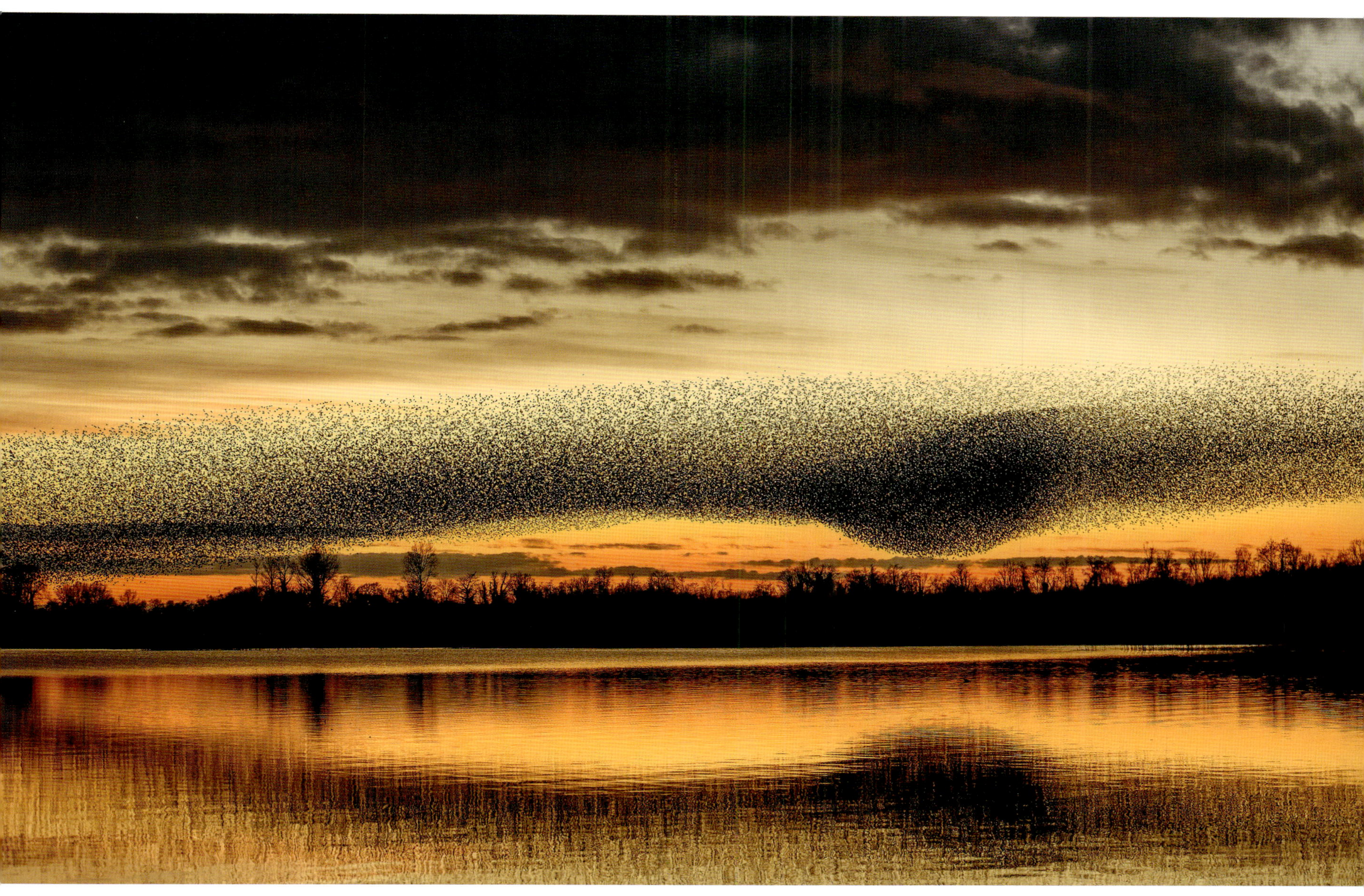

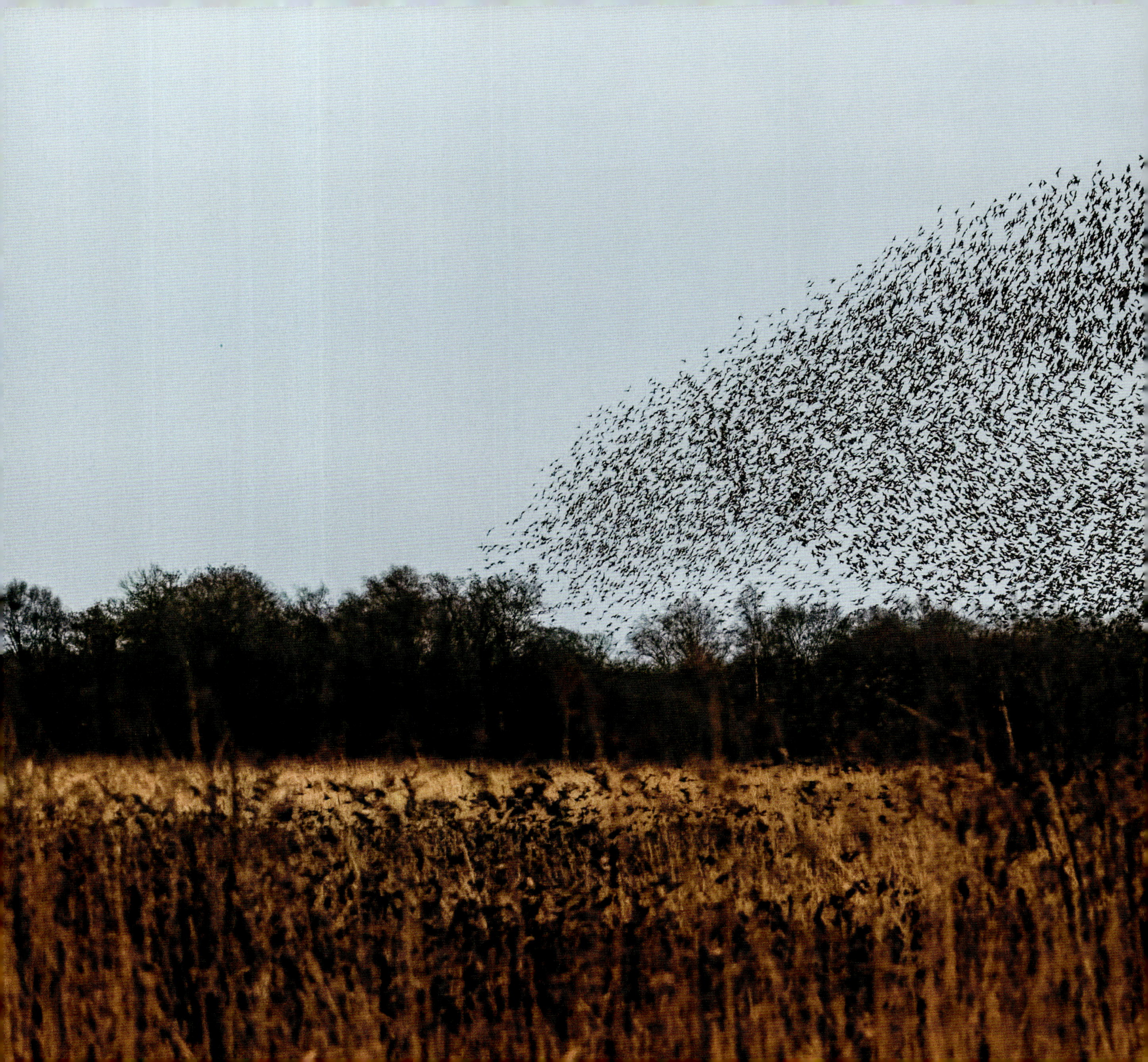

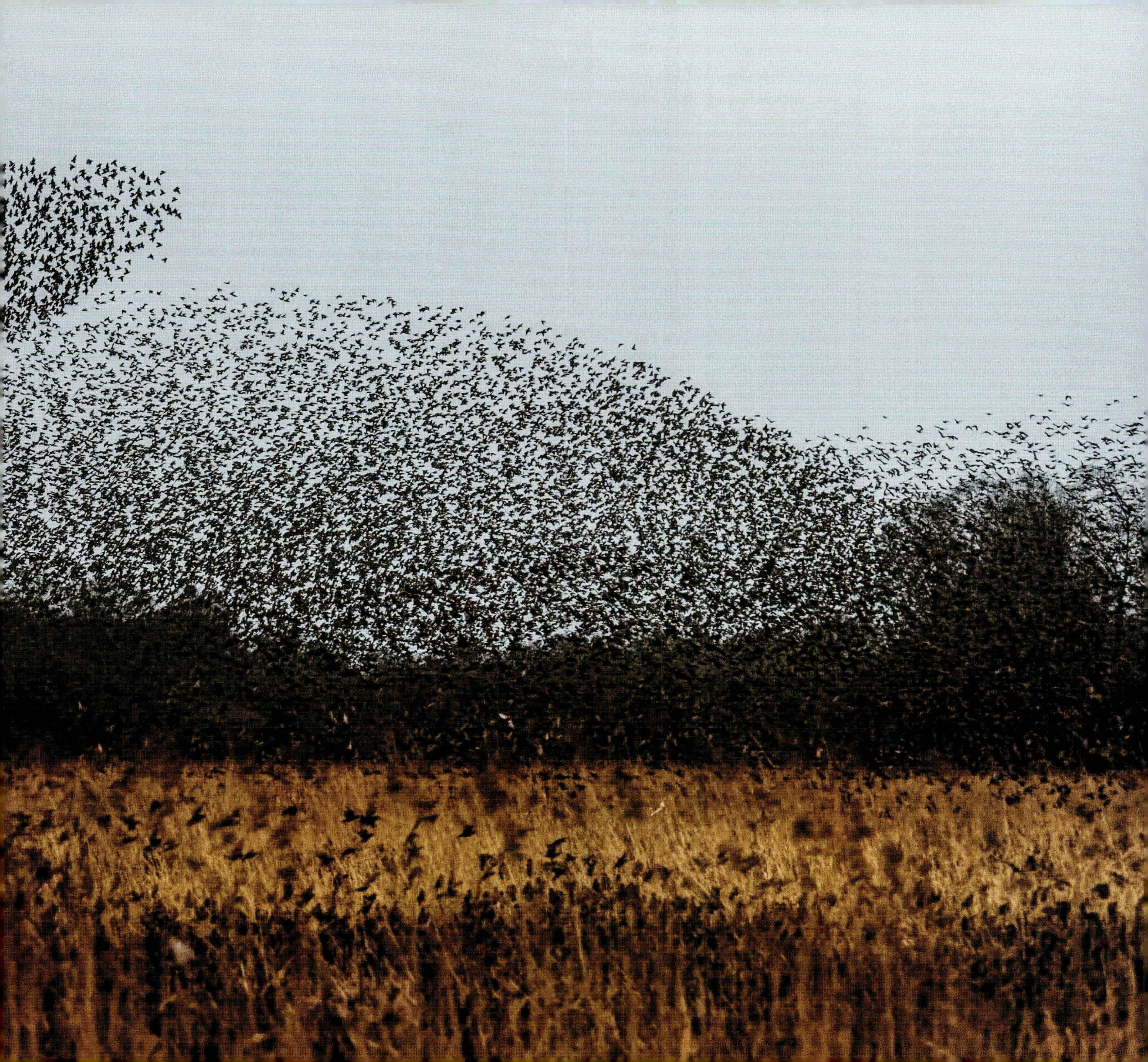

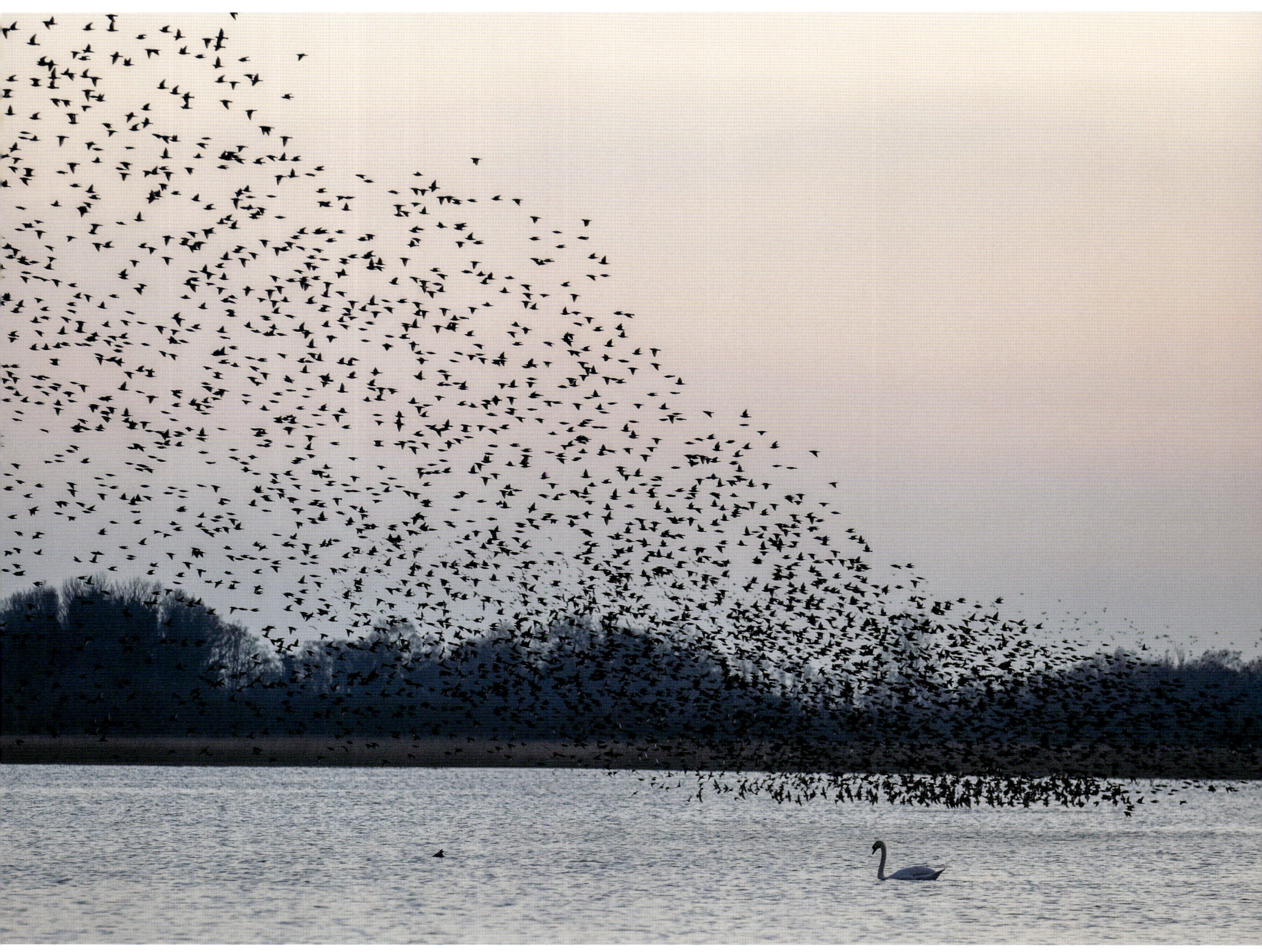

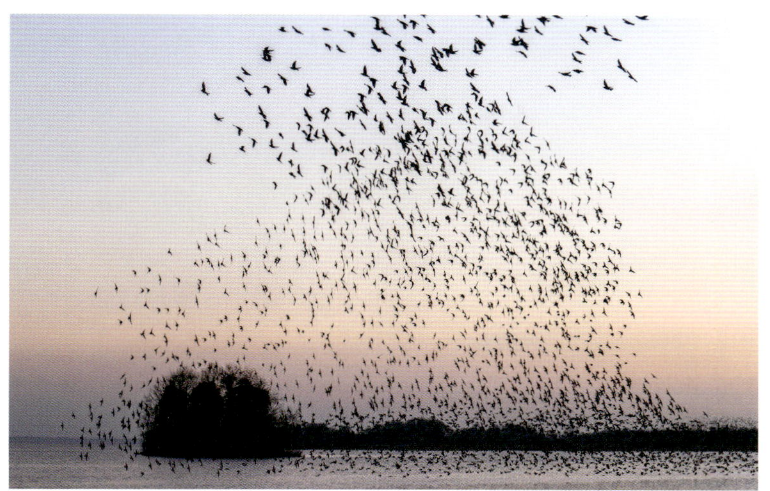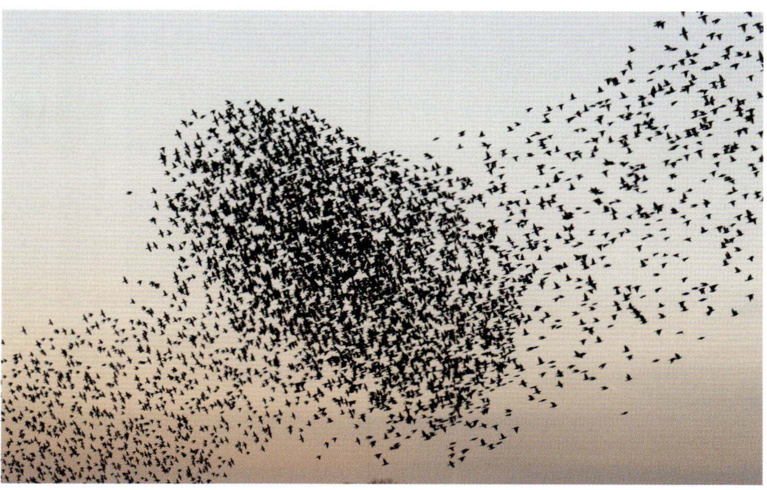

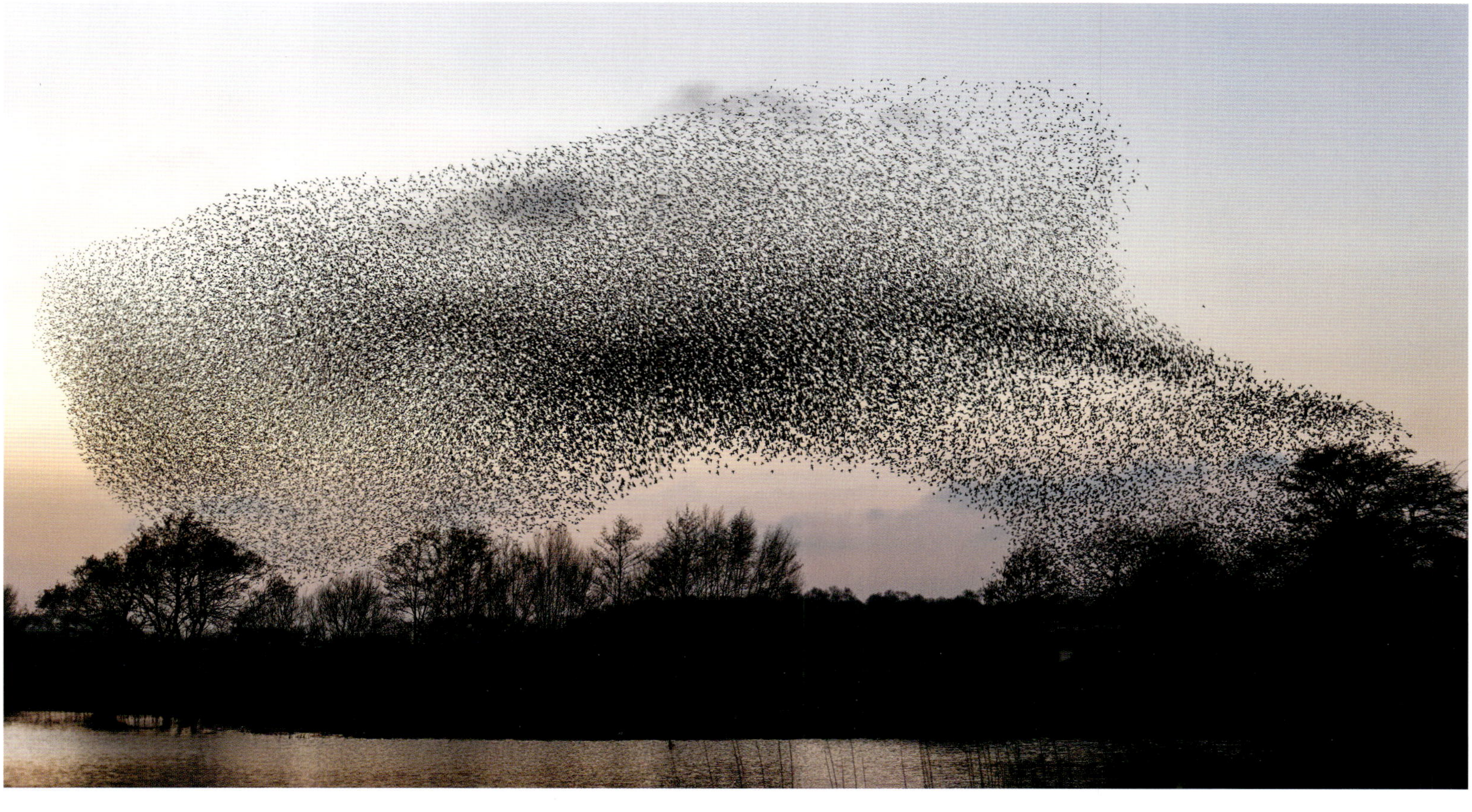

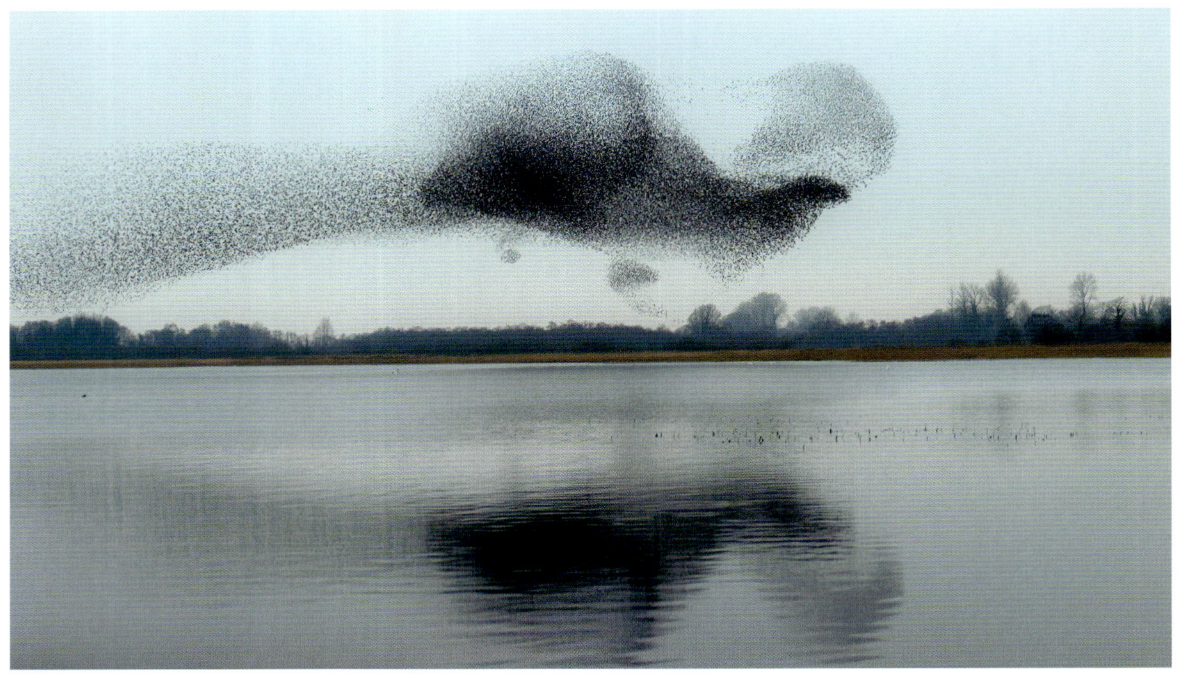

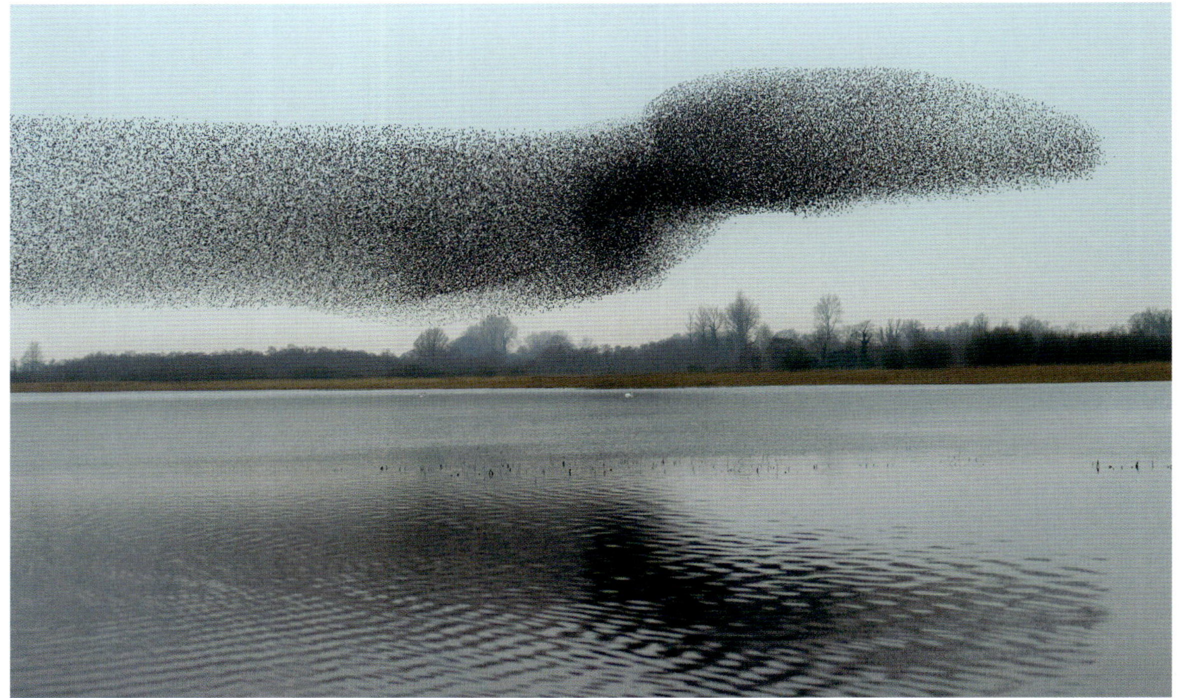

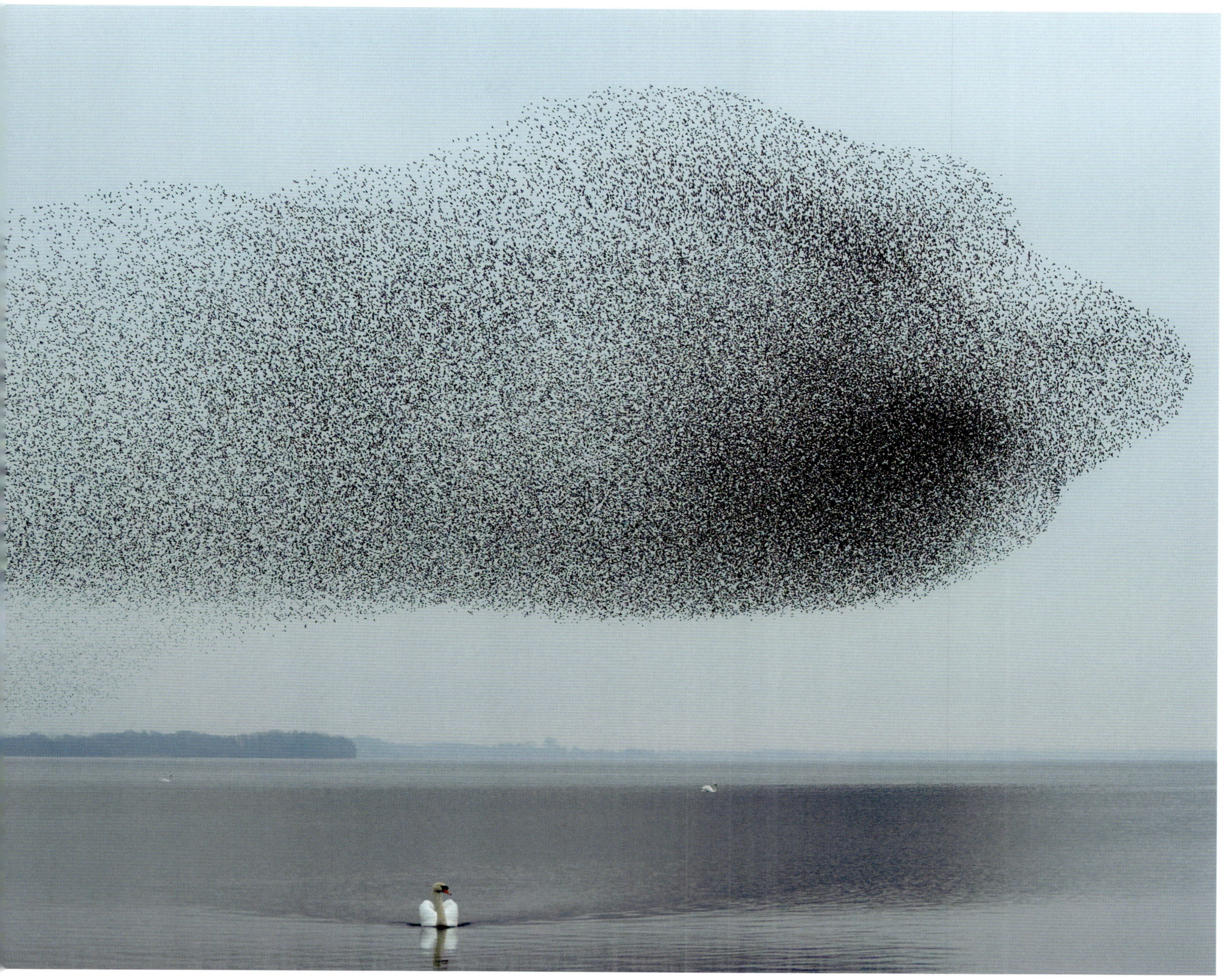

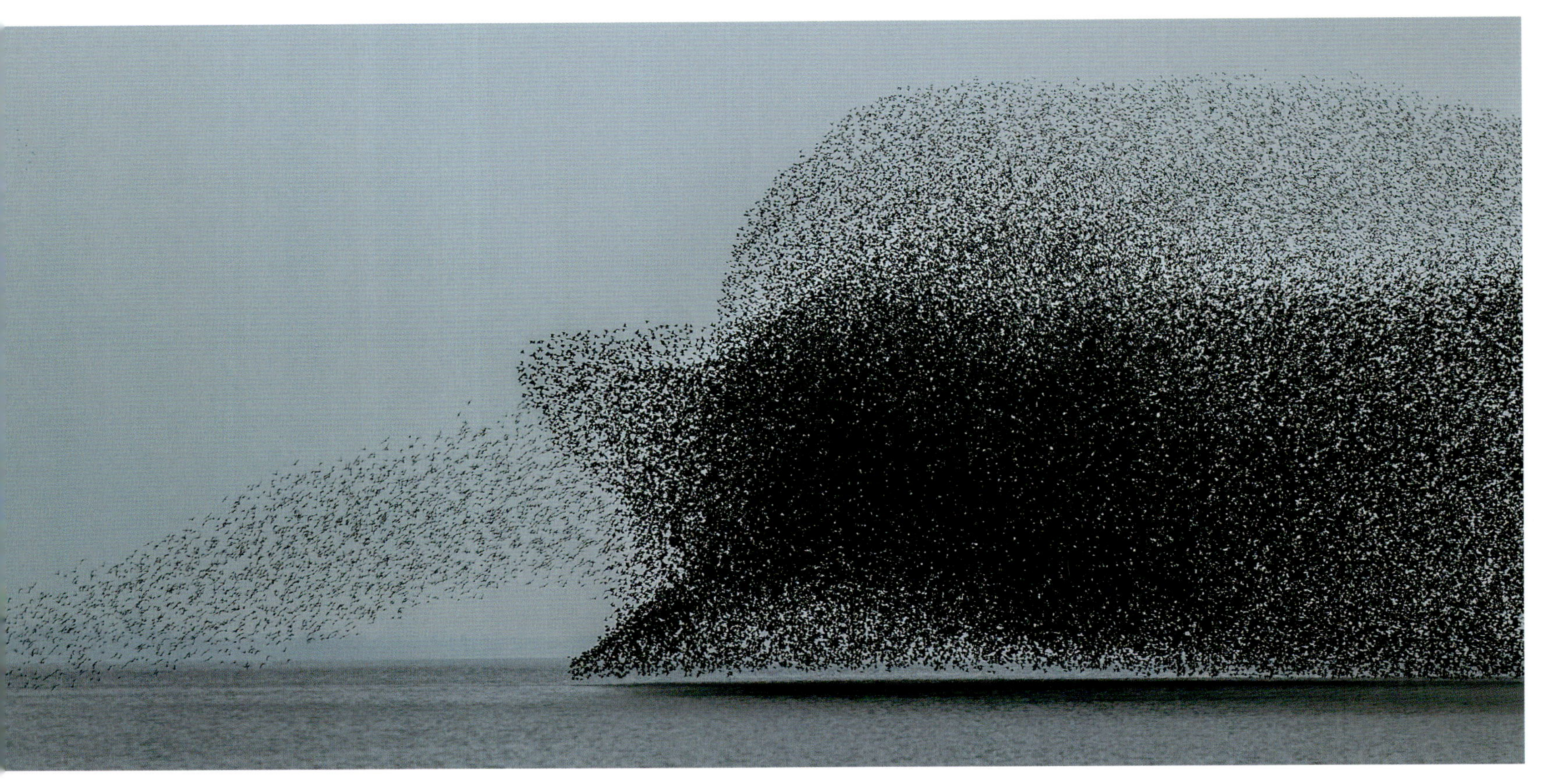

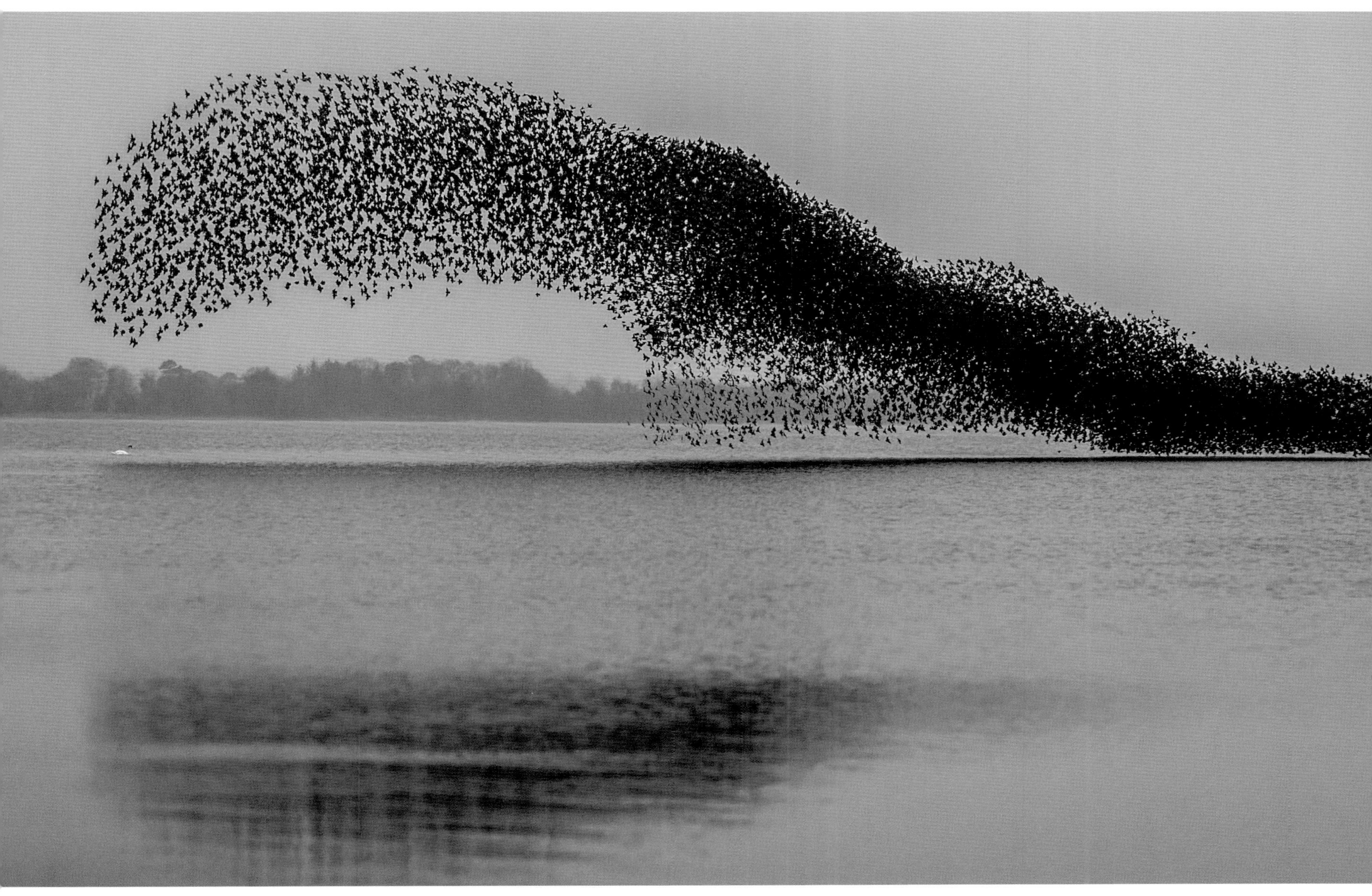

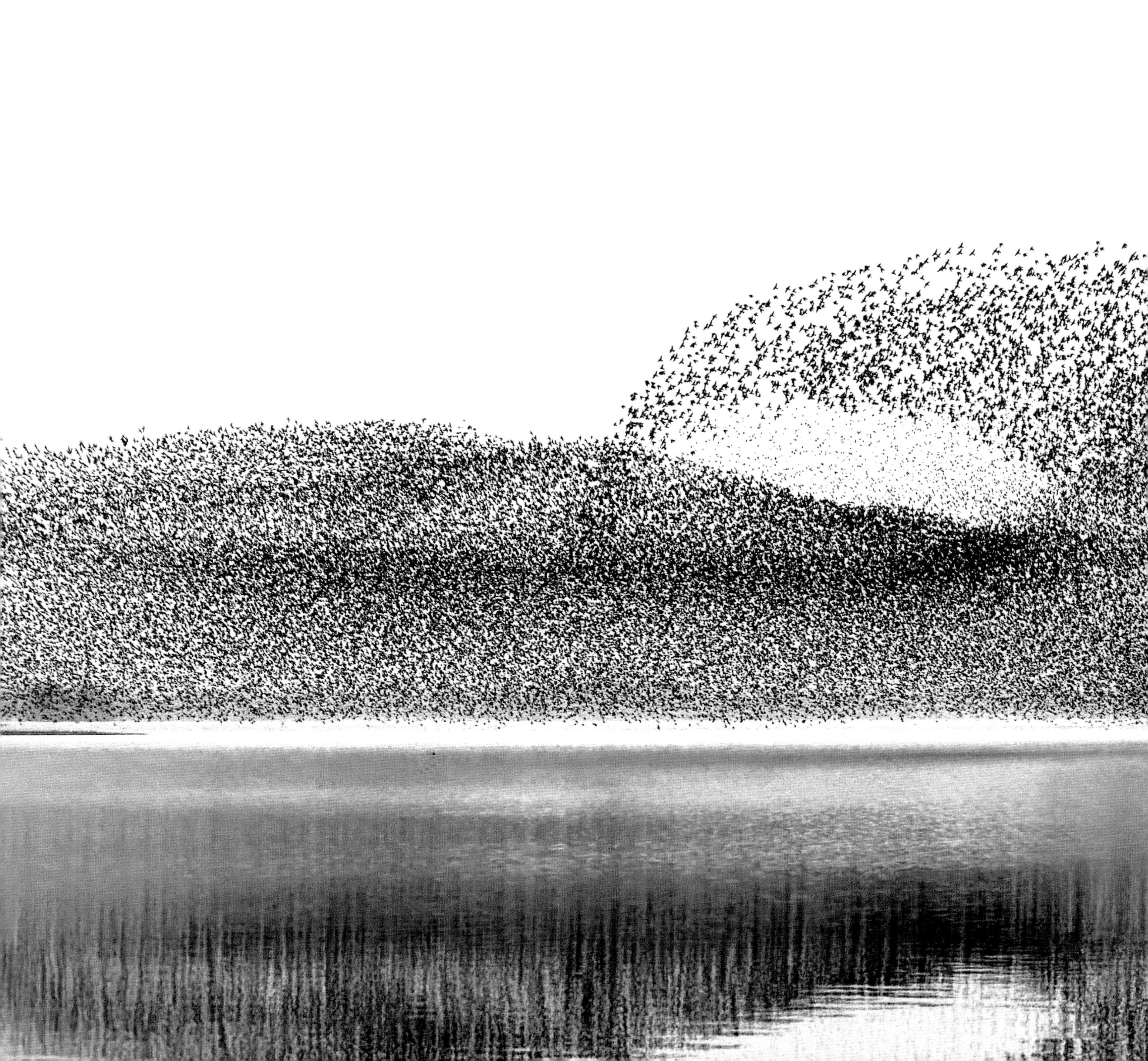

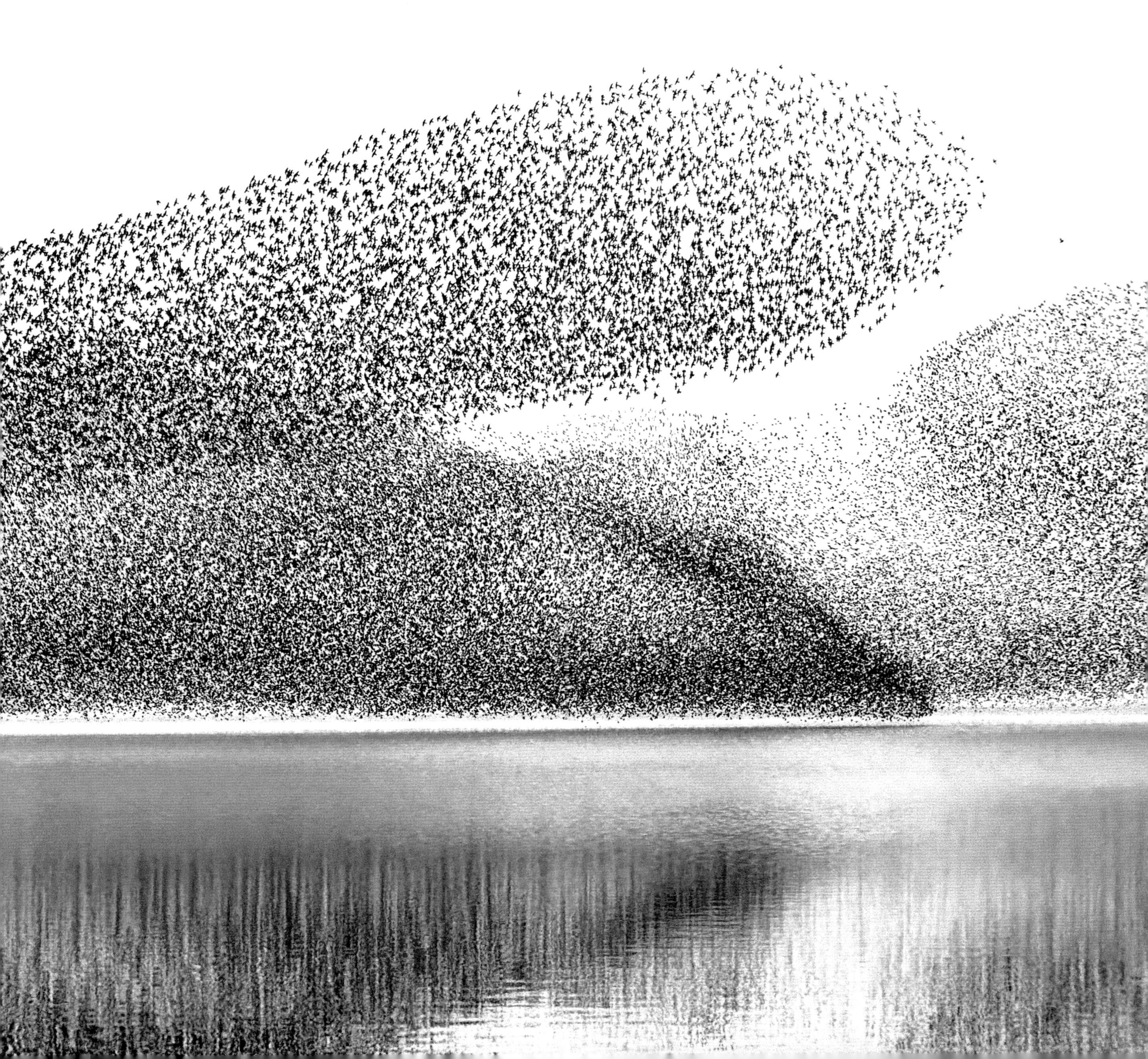

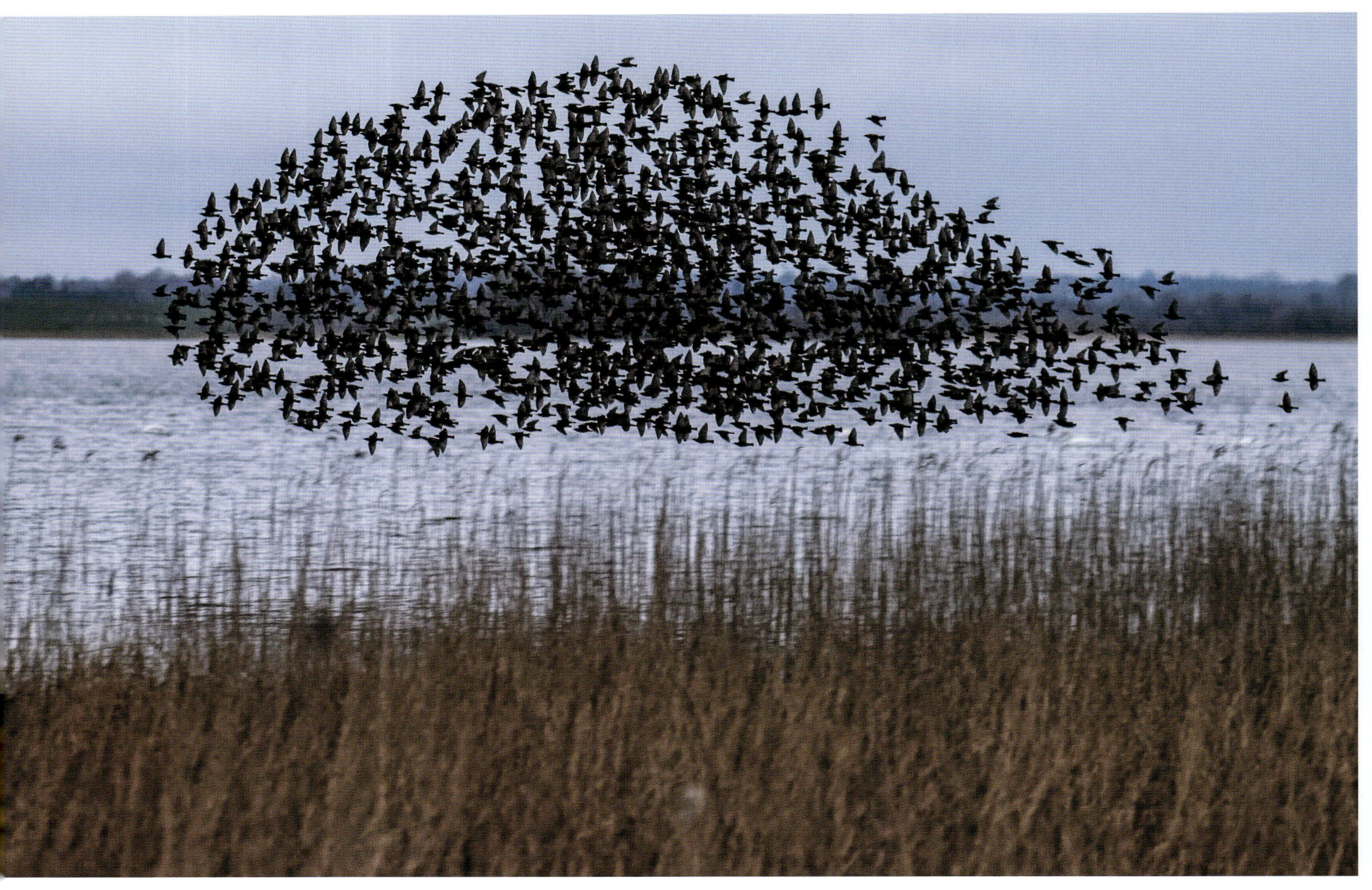

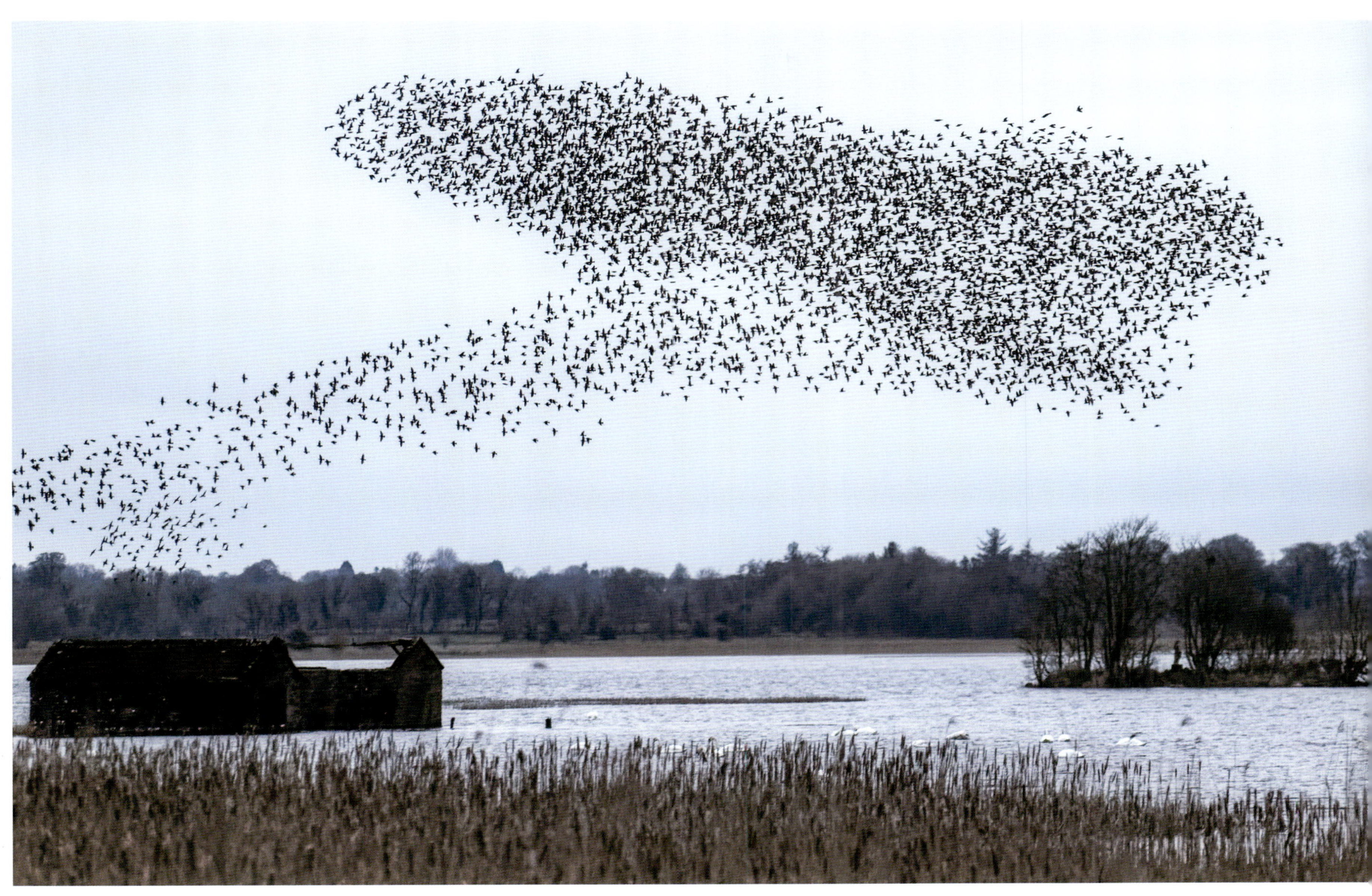

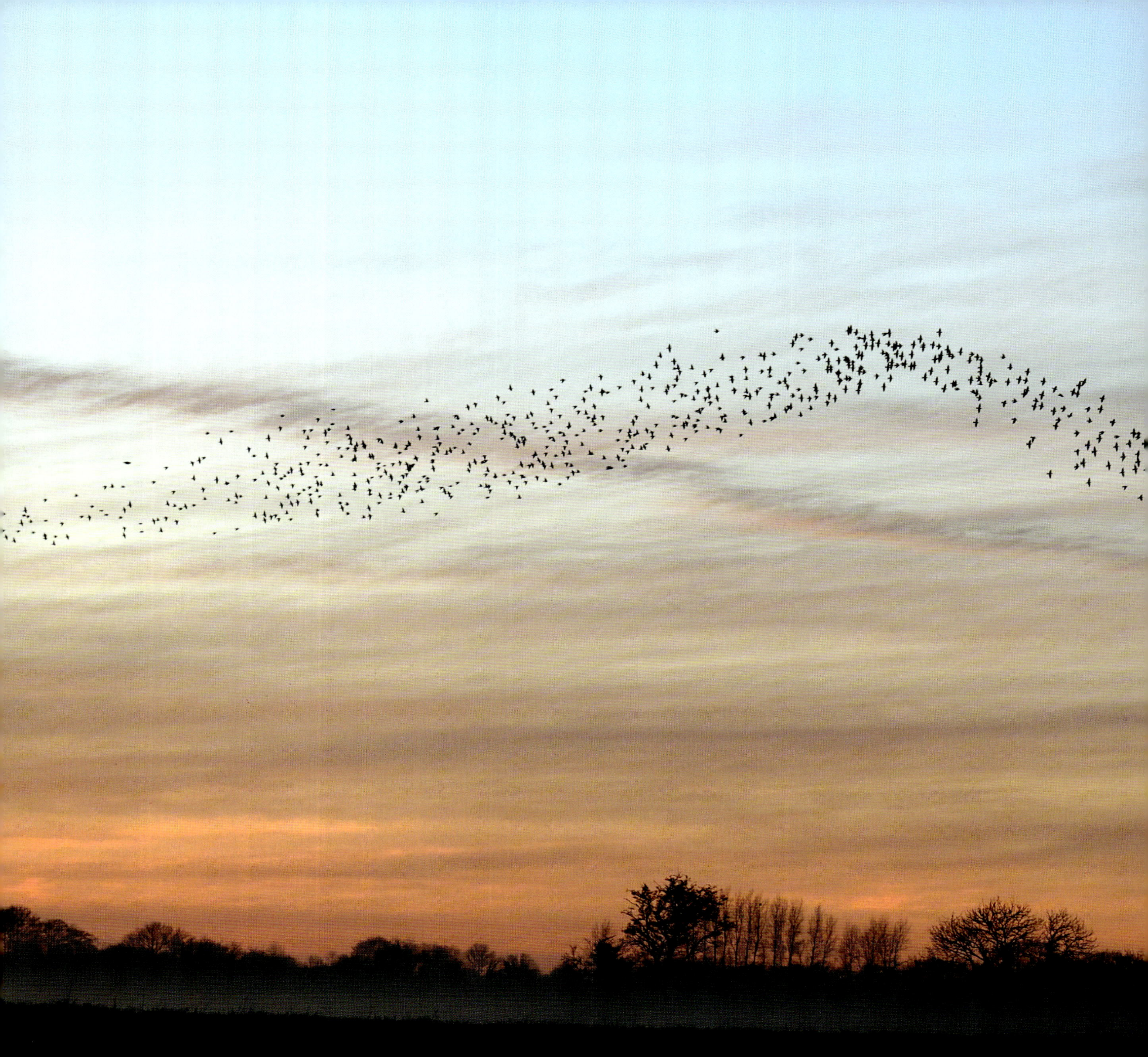

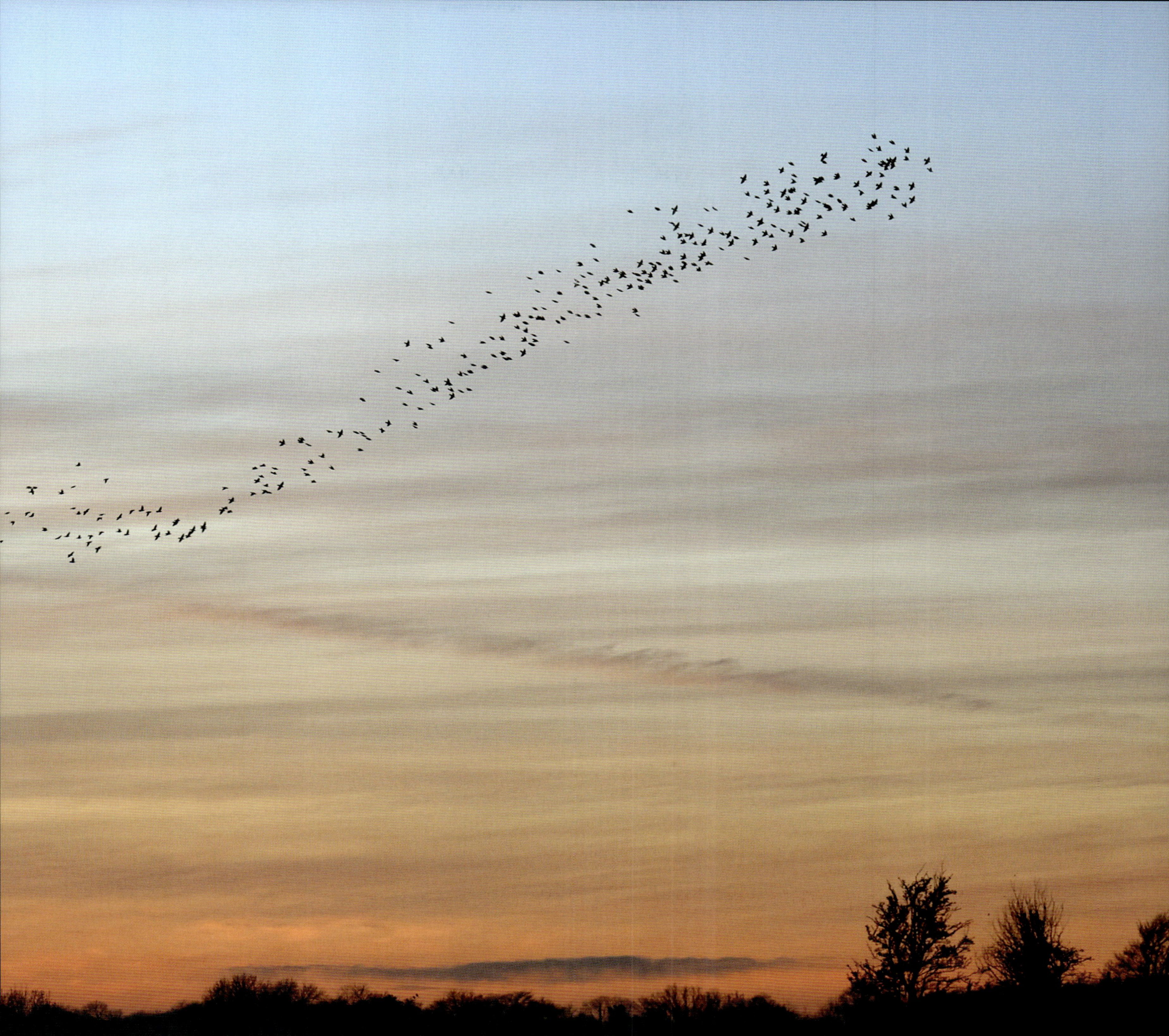

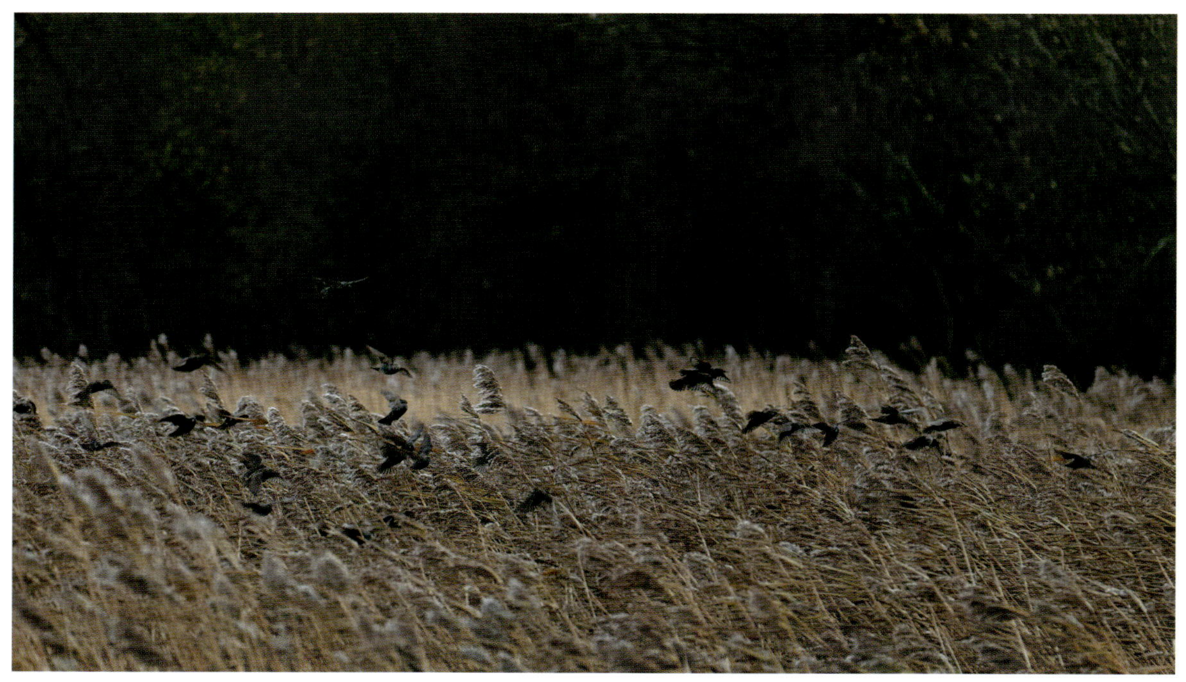

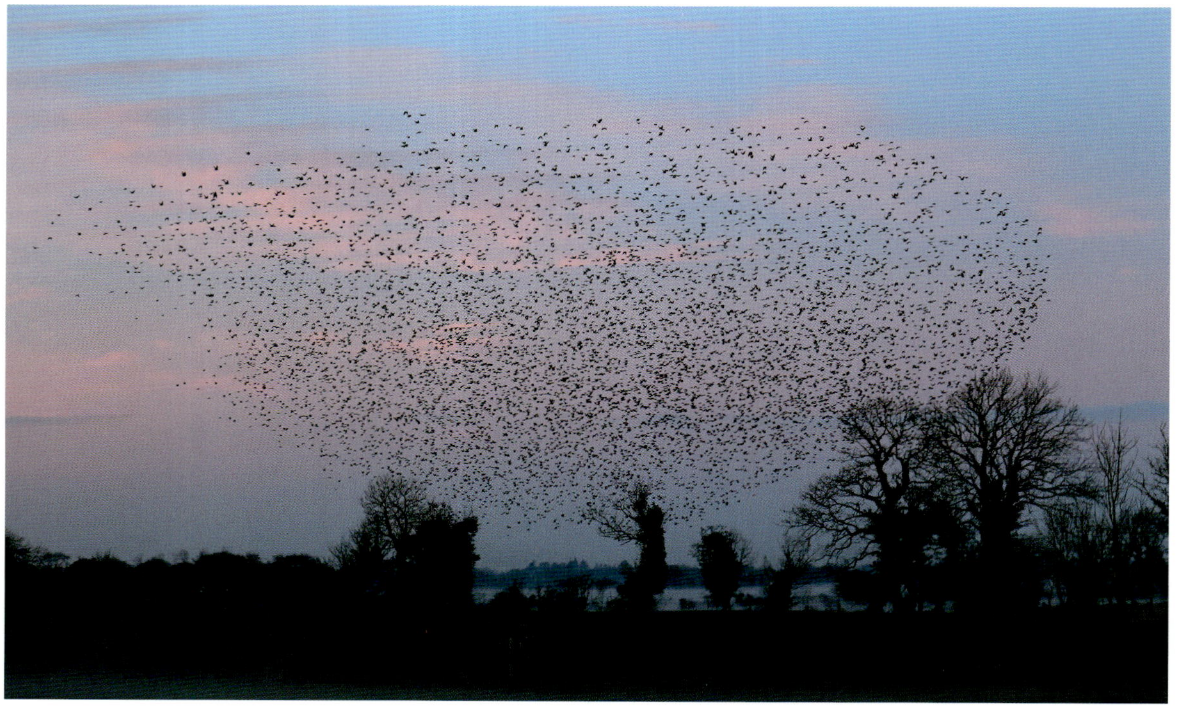

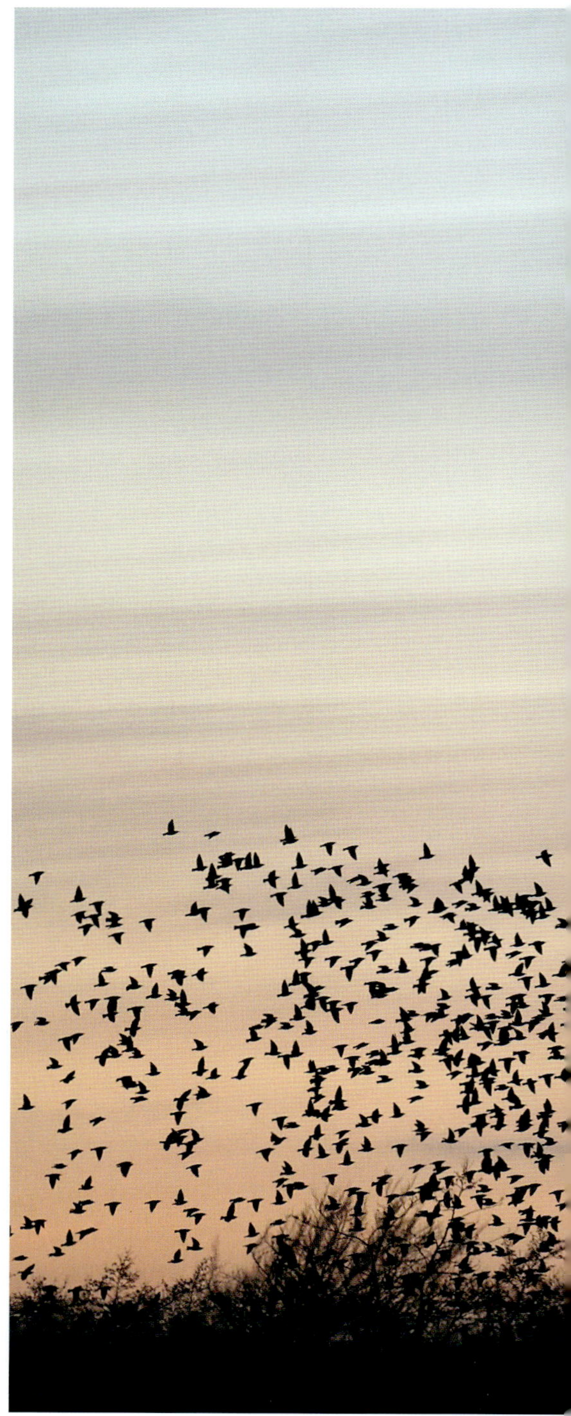

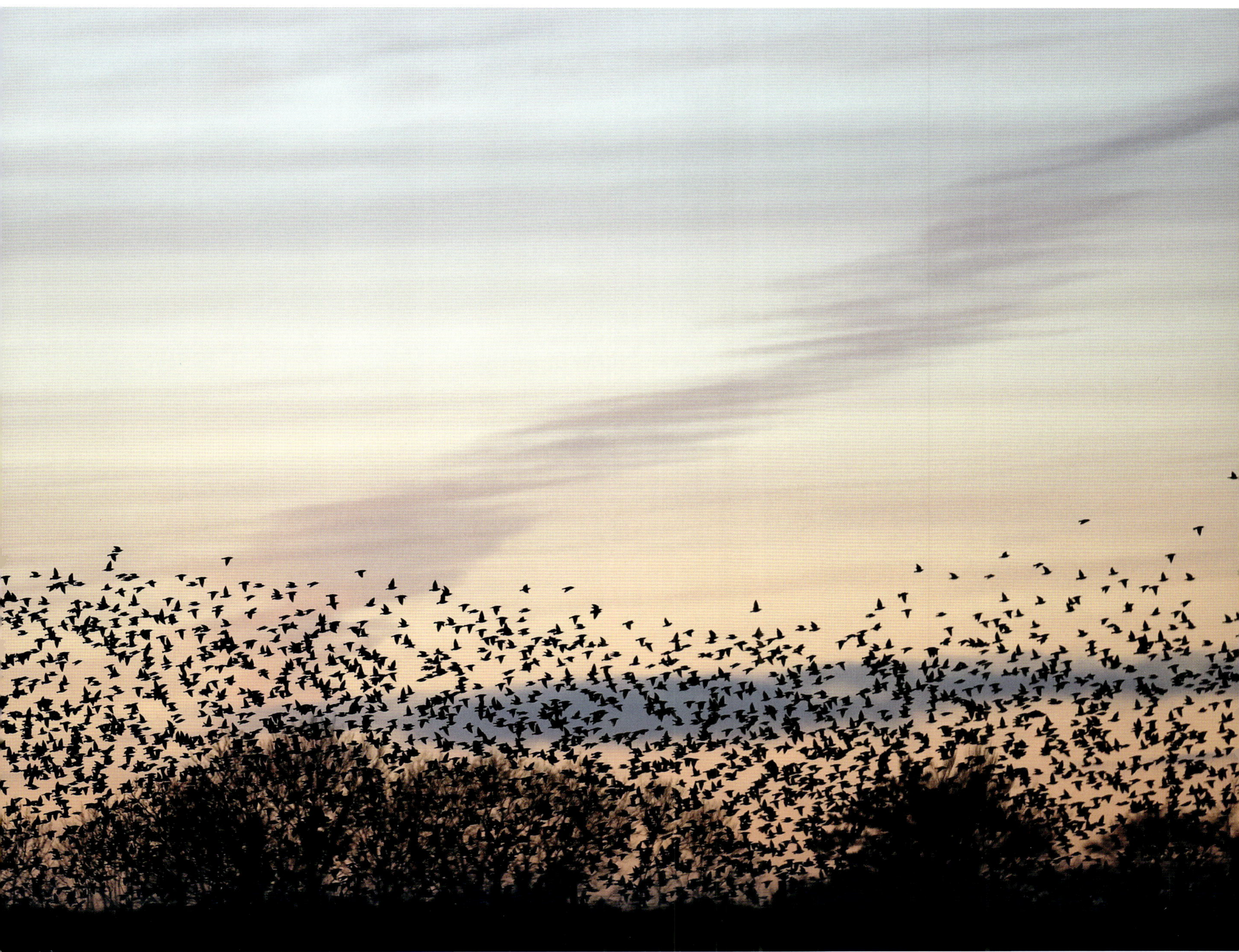

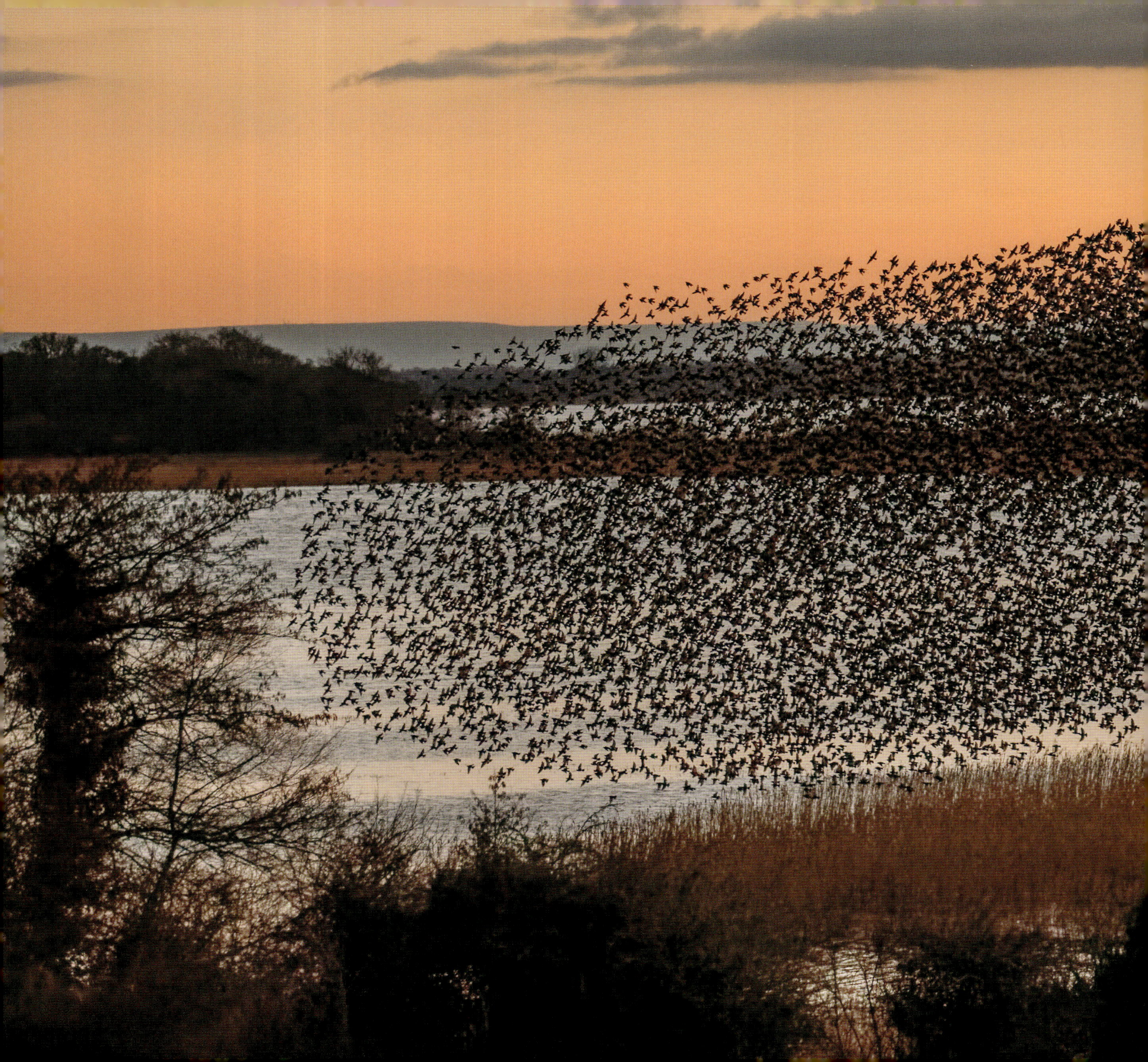

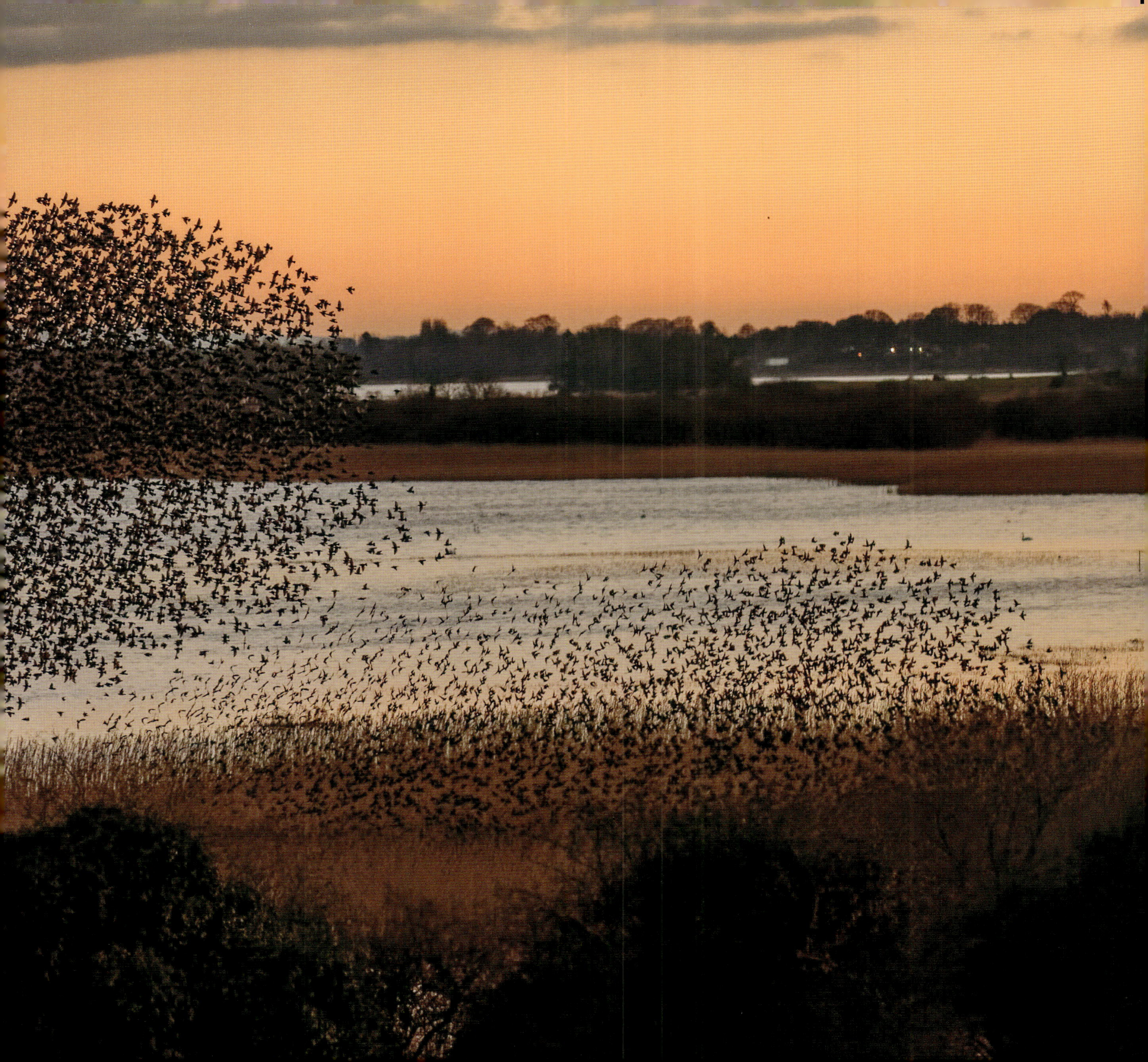

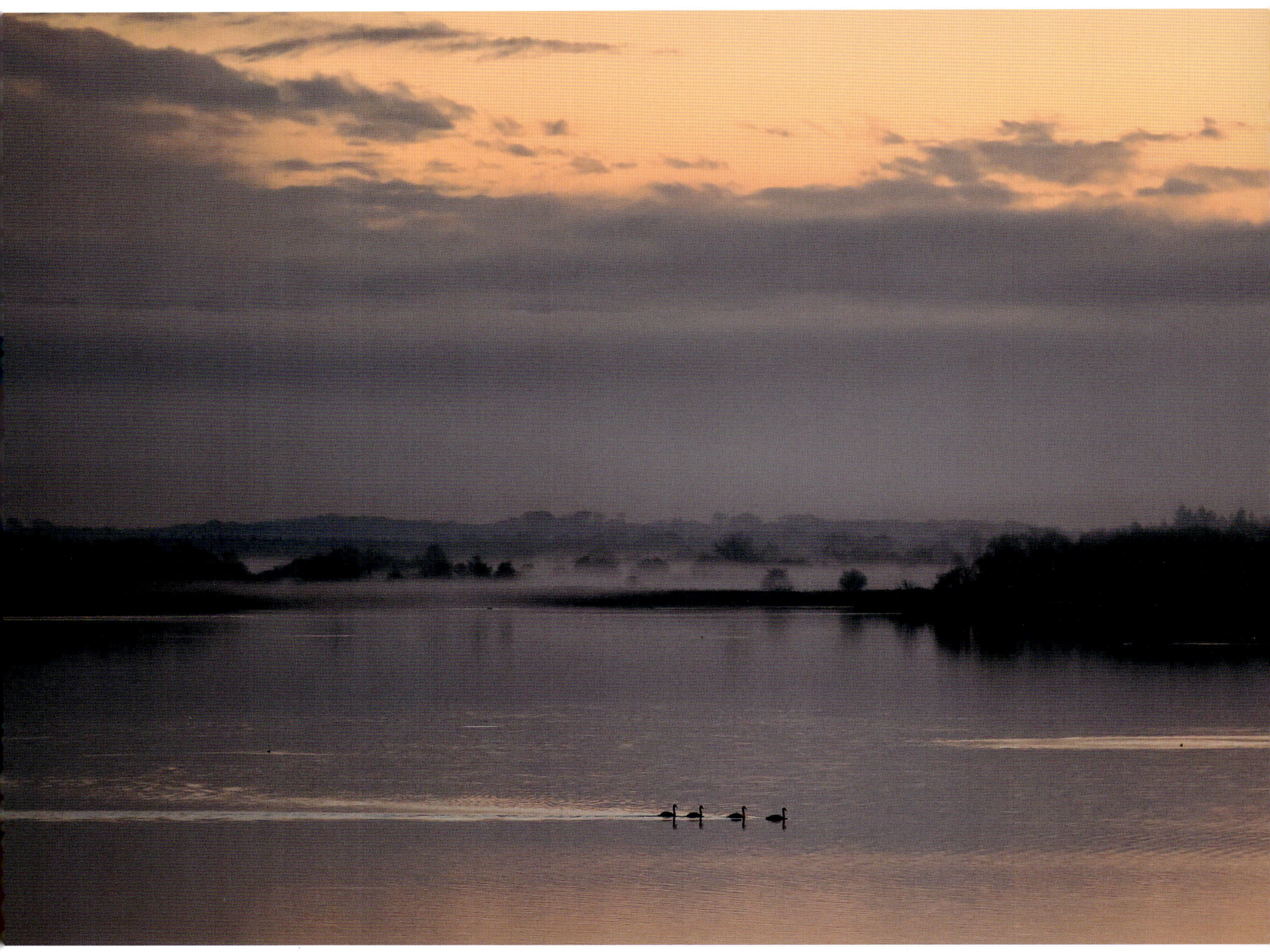

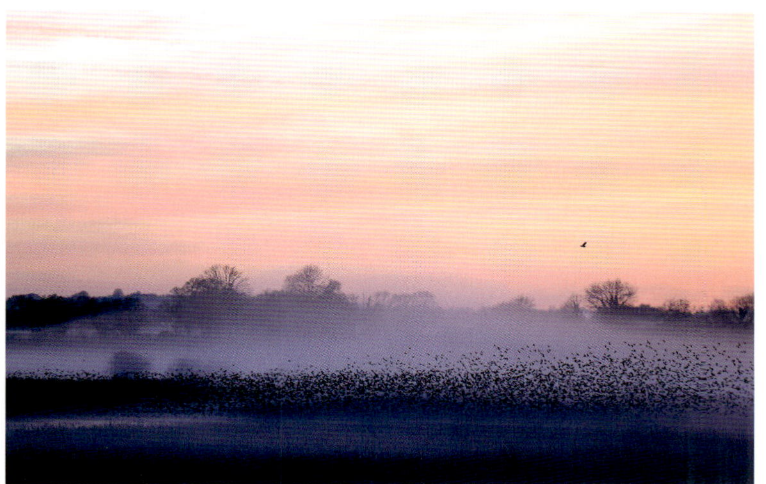
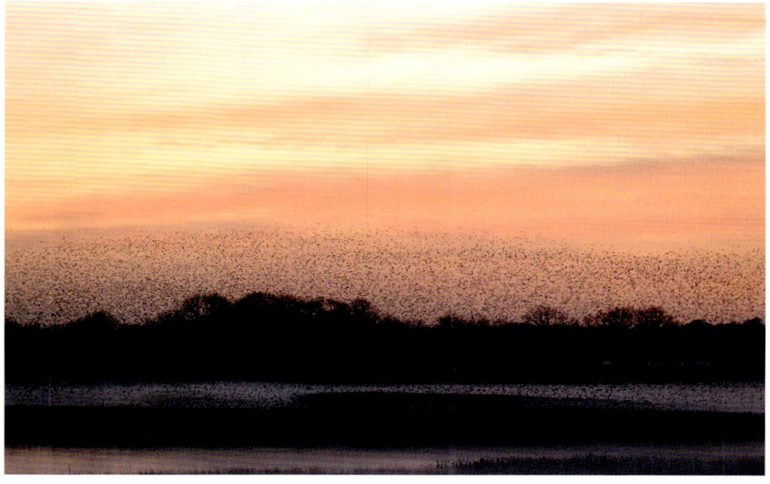

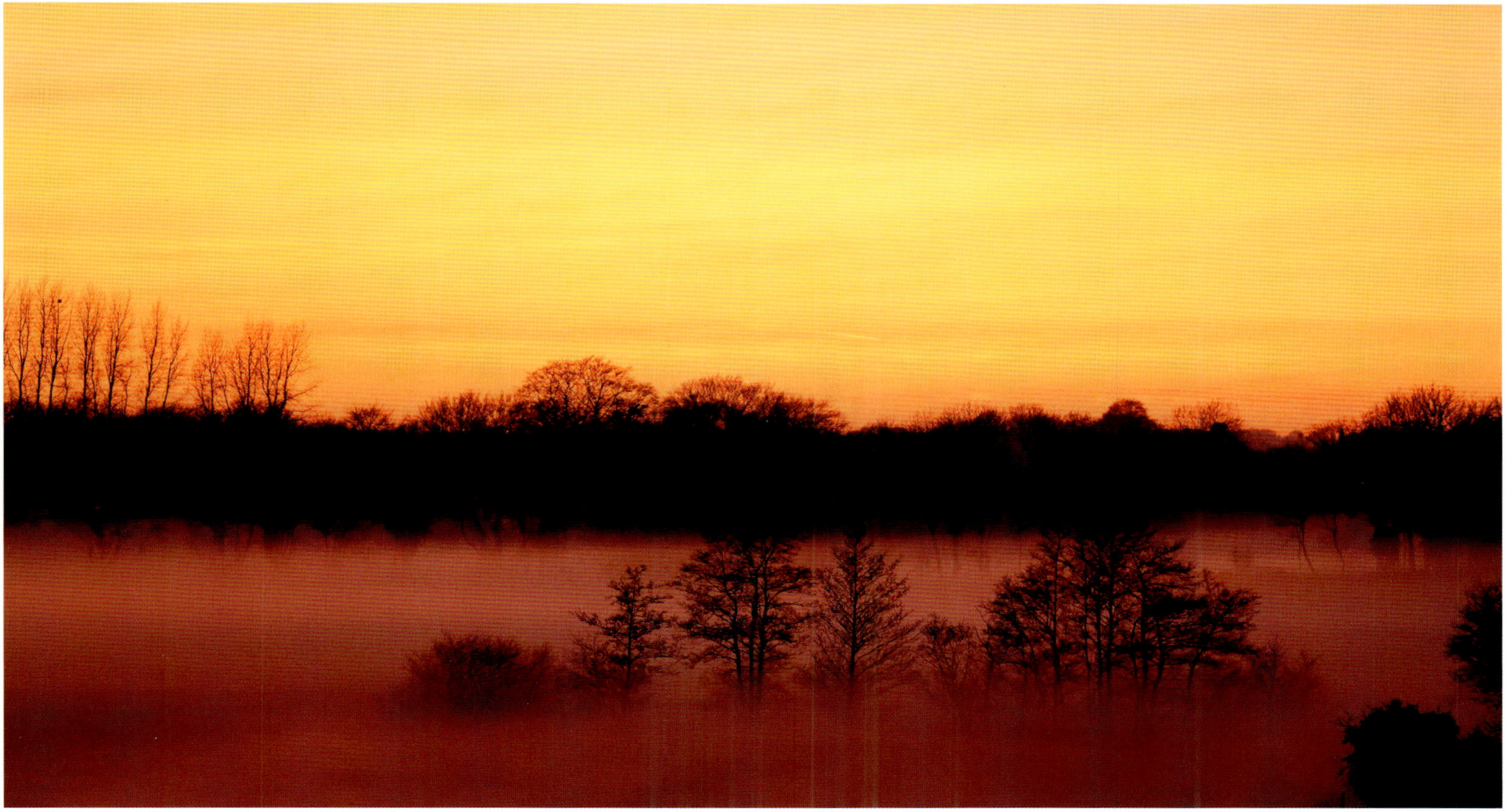

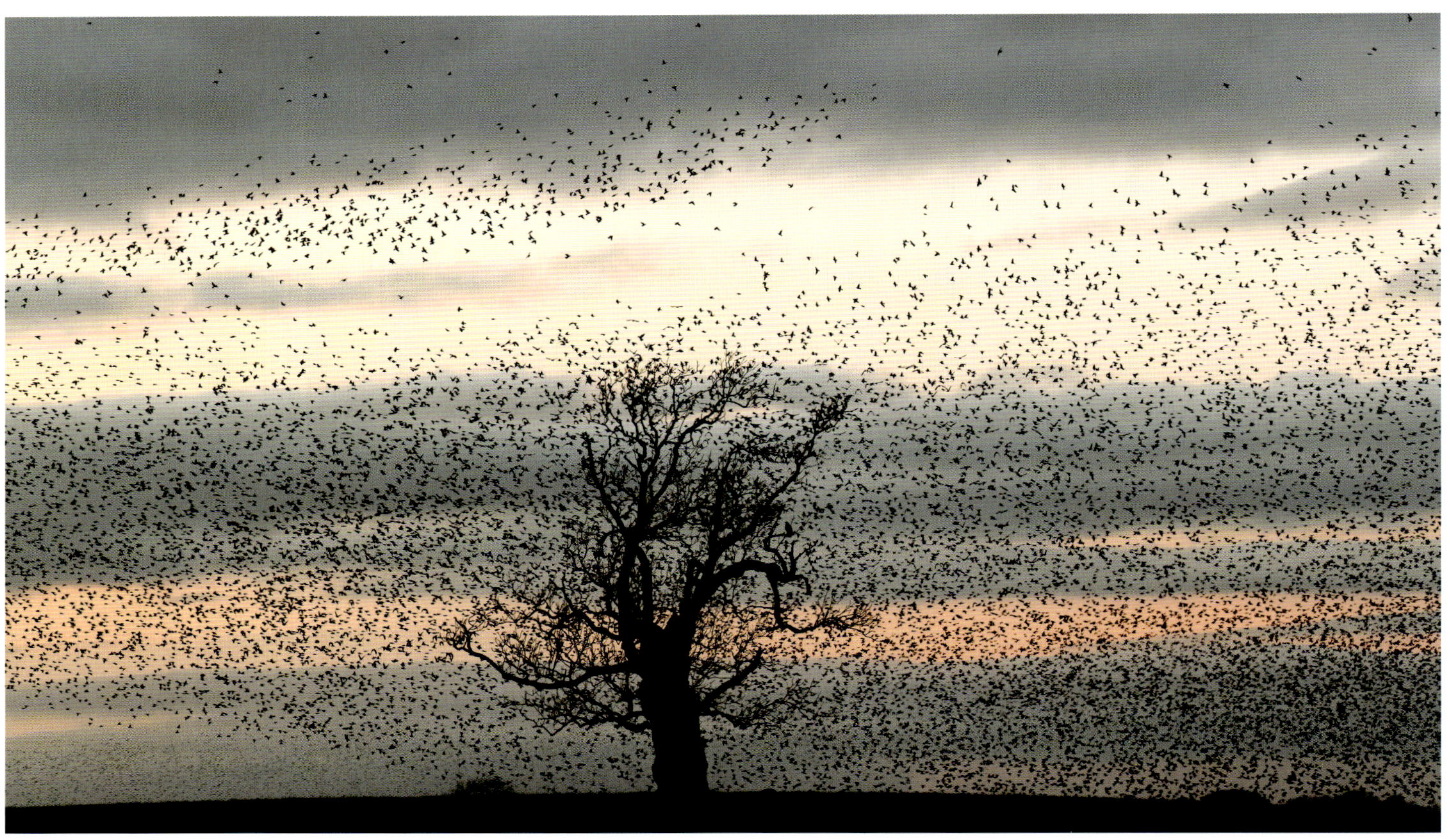

62

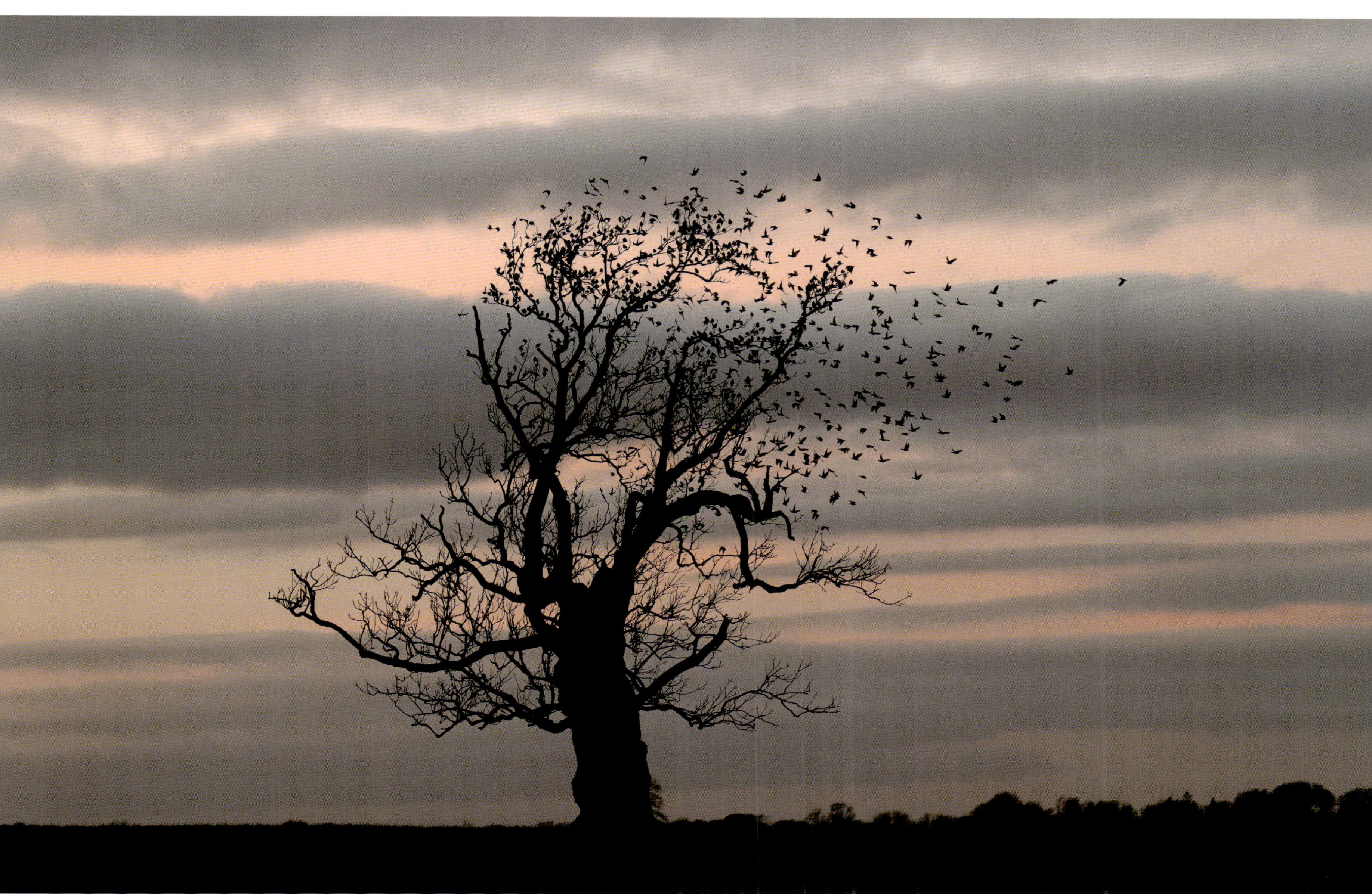

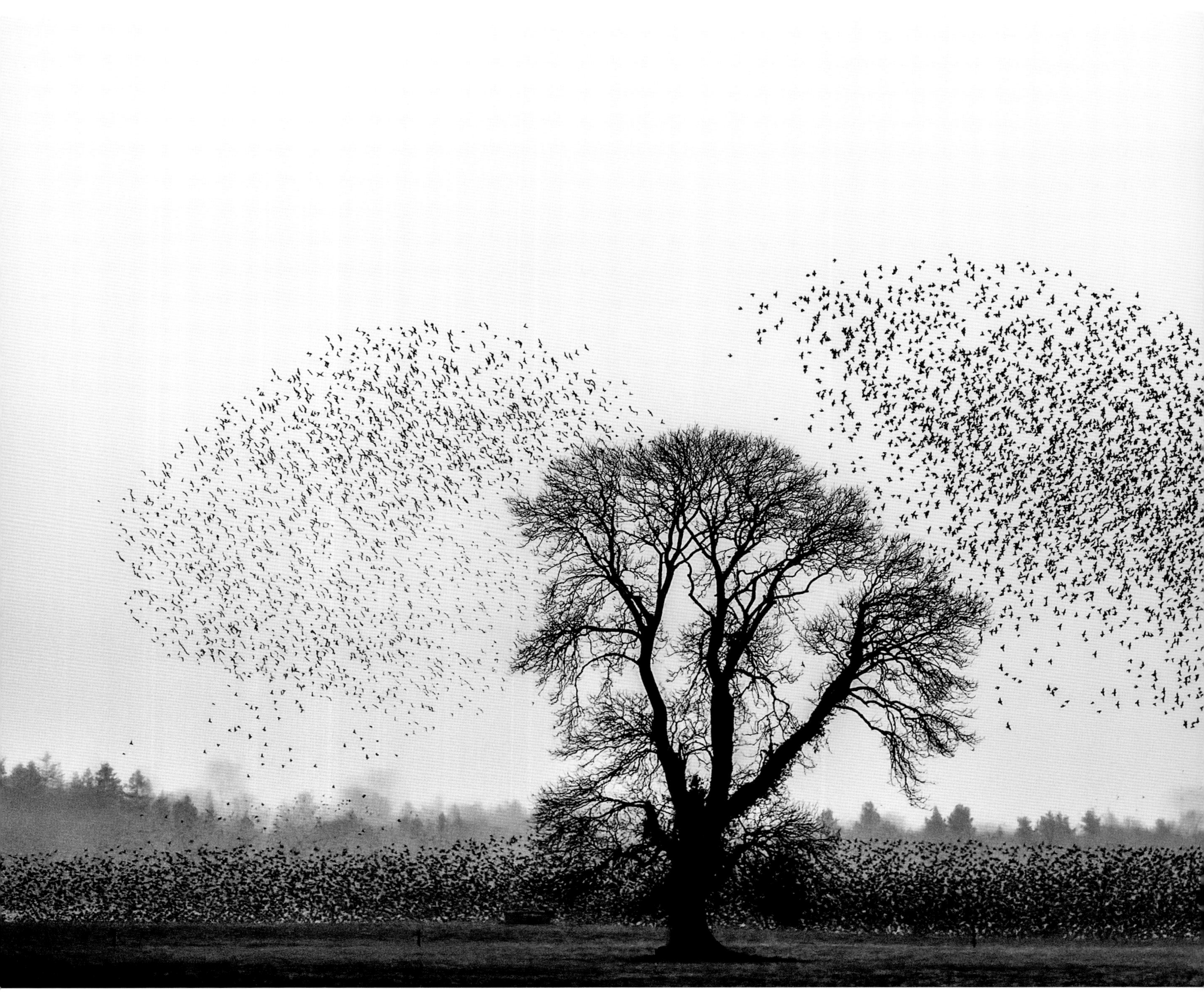

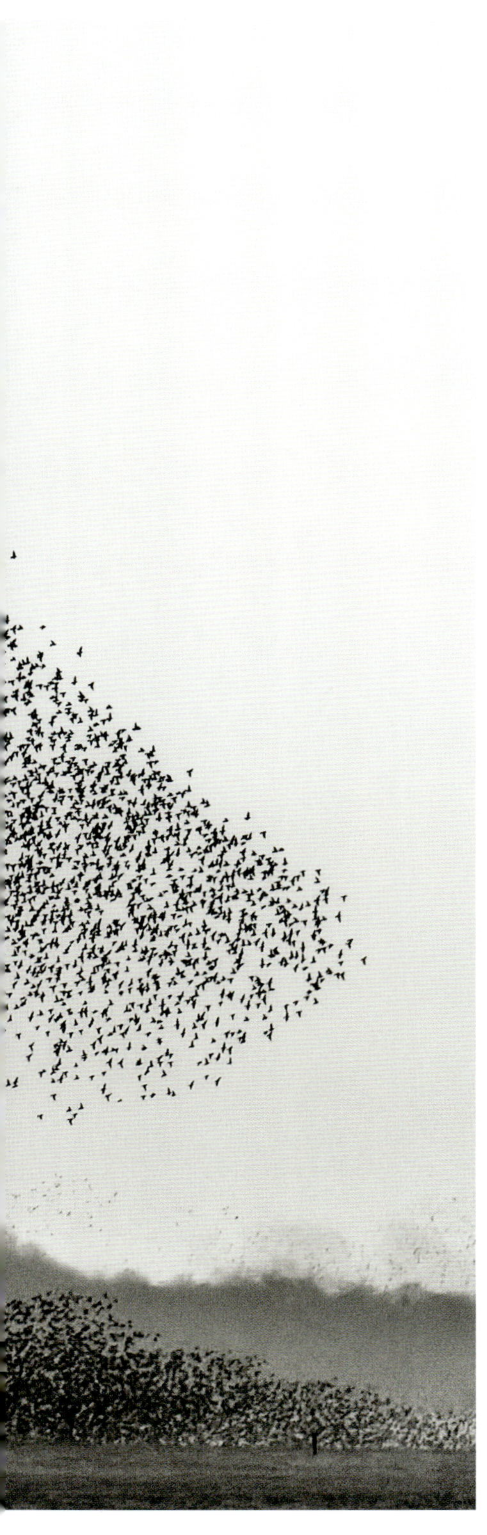

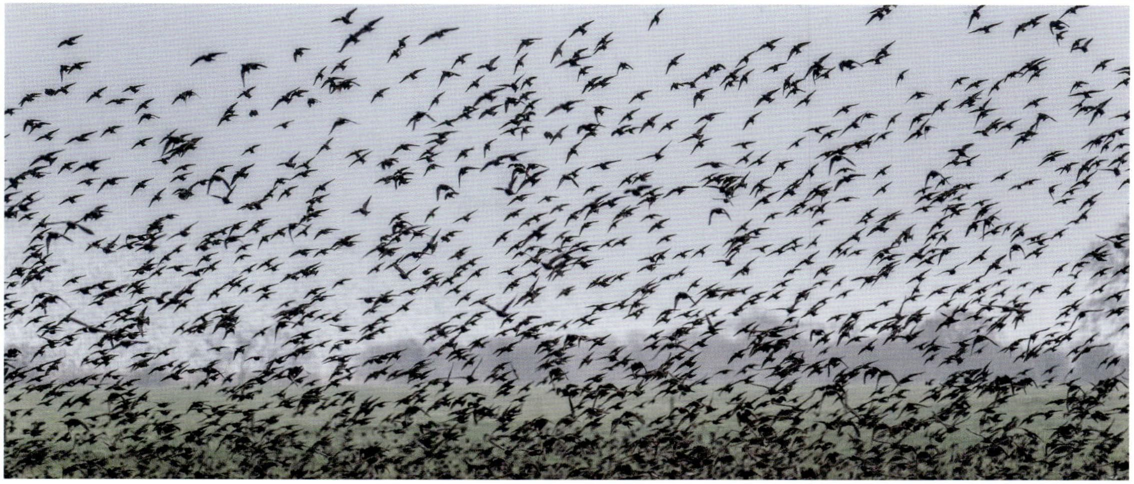

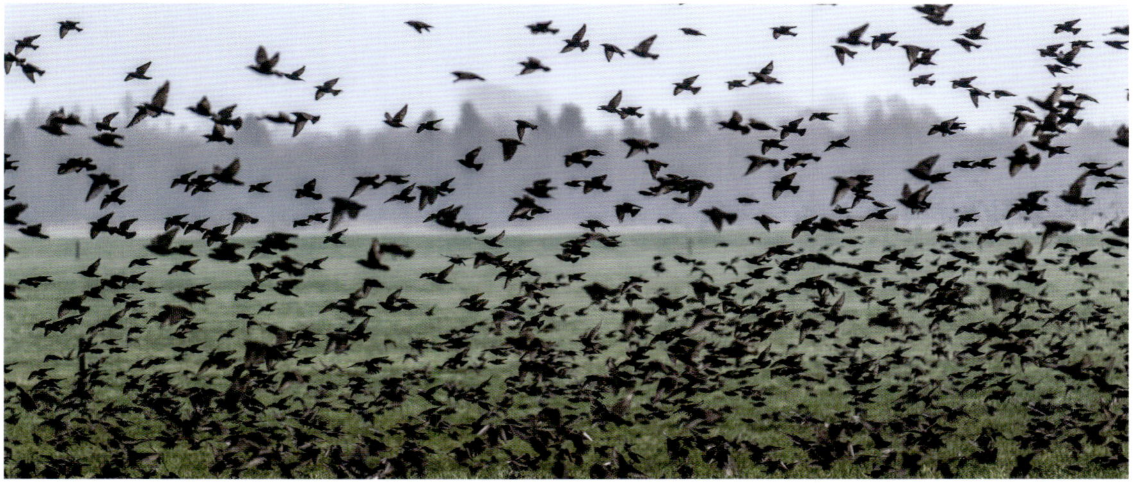

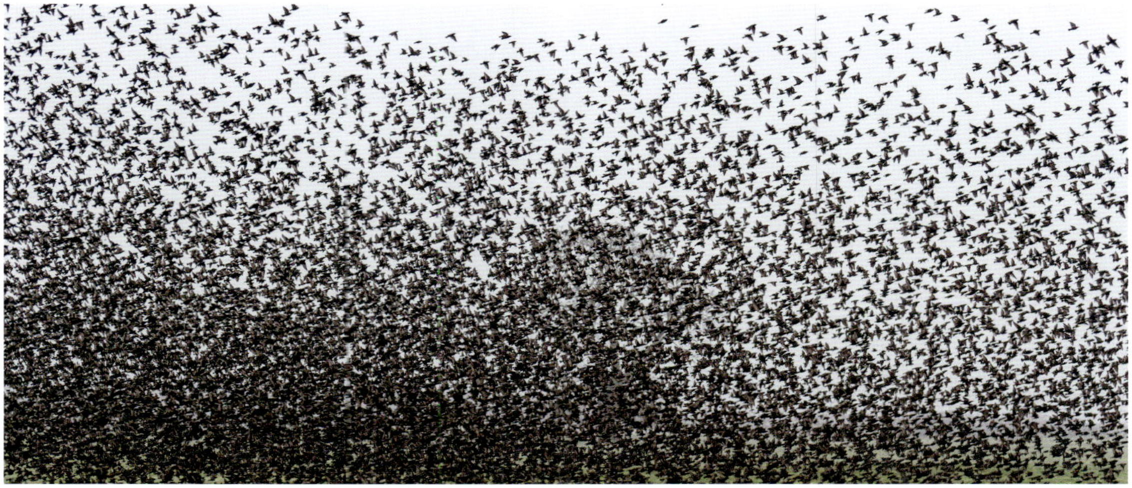

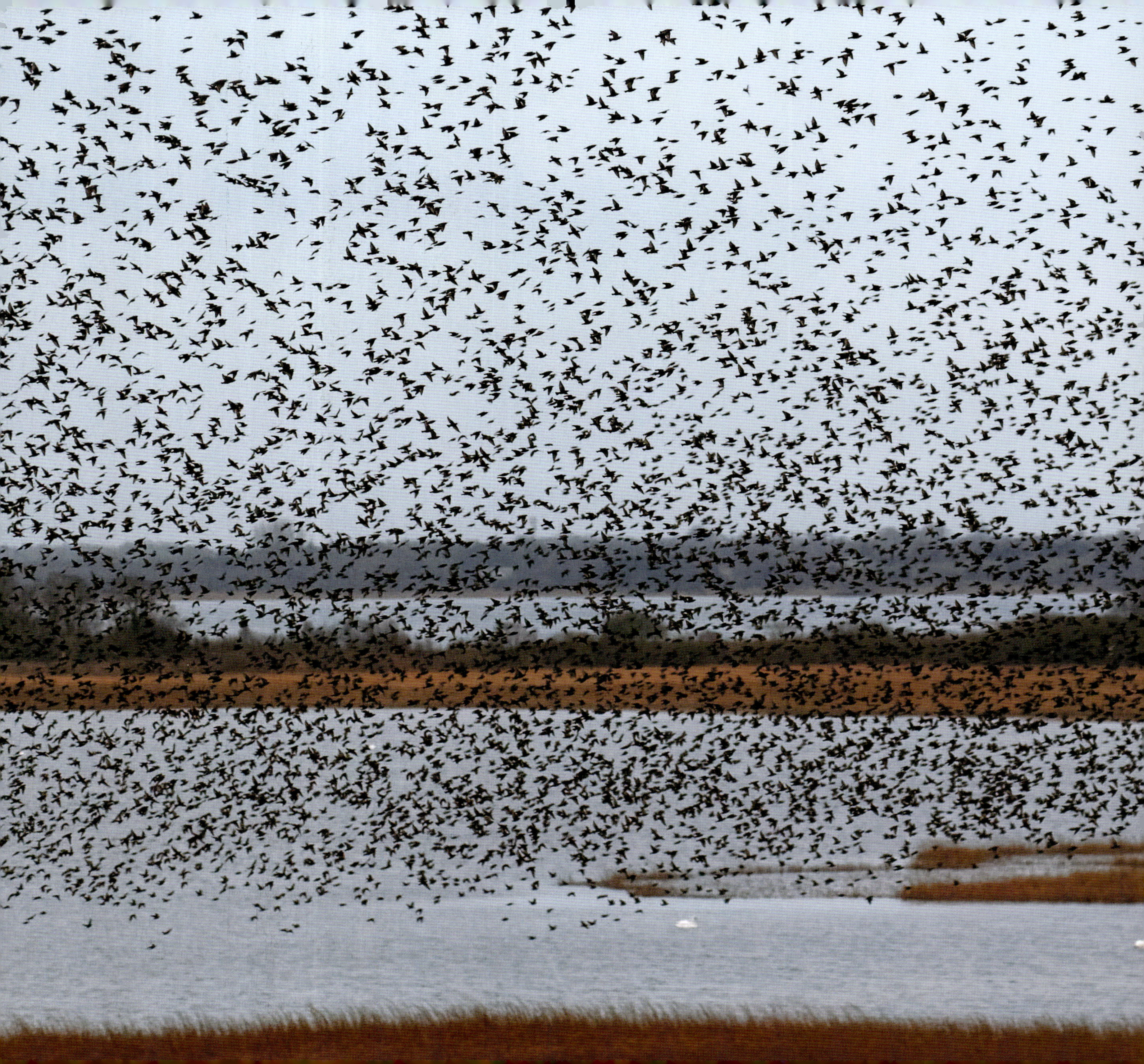

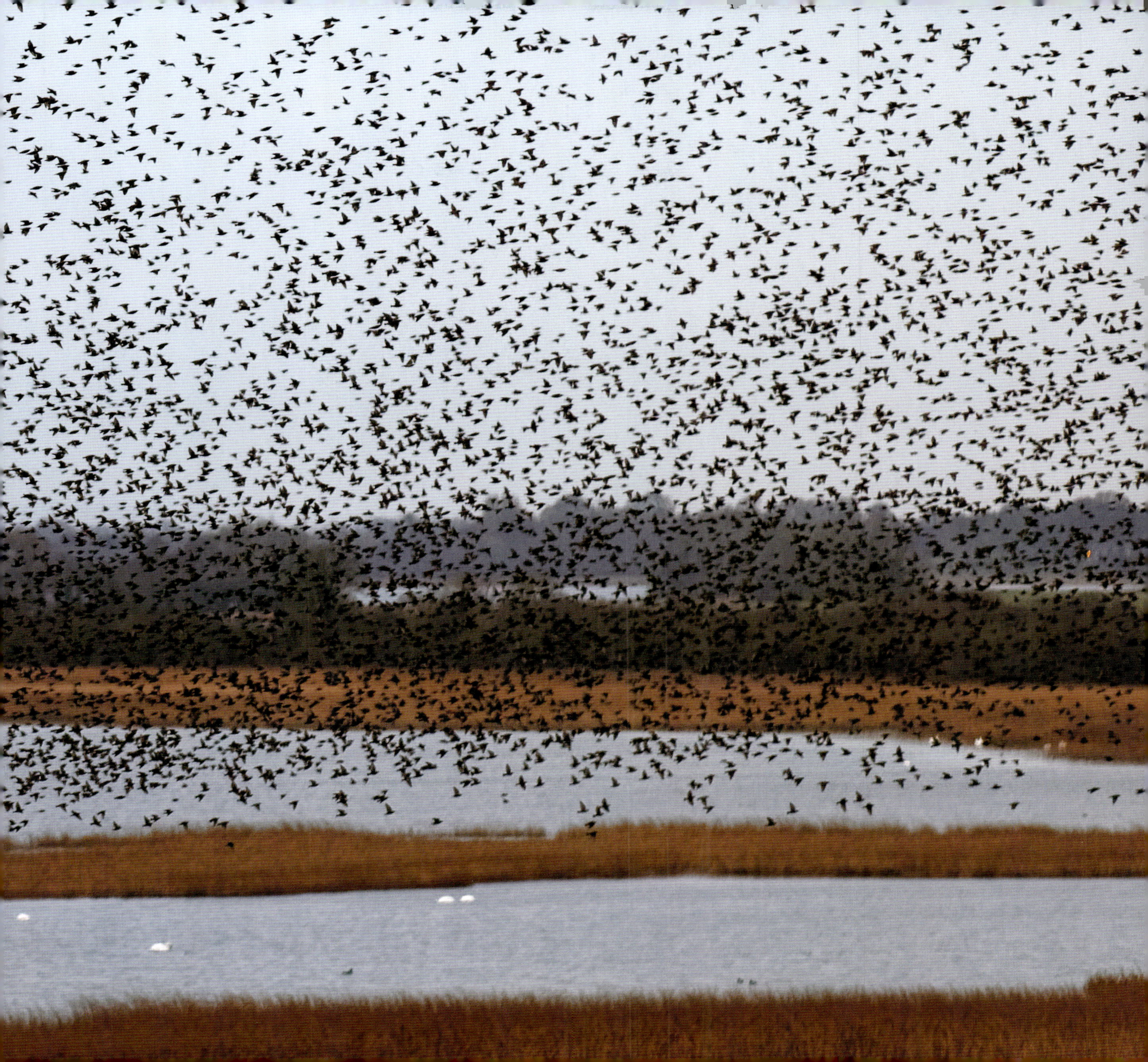

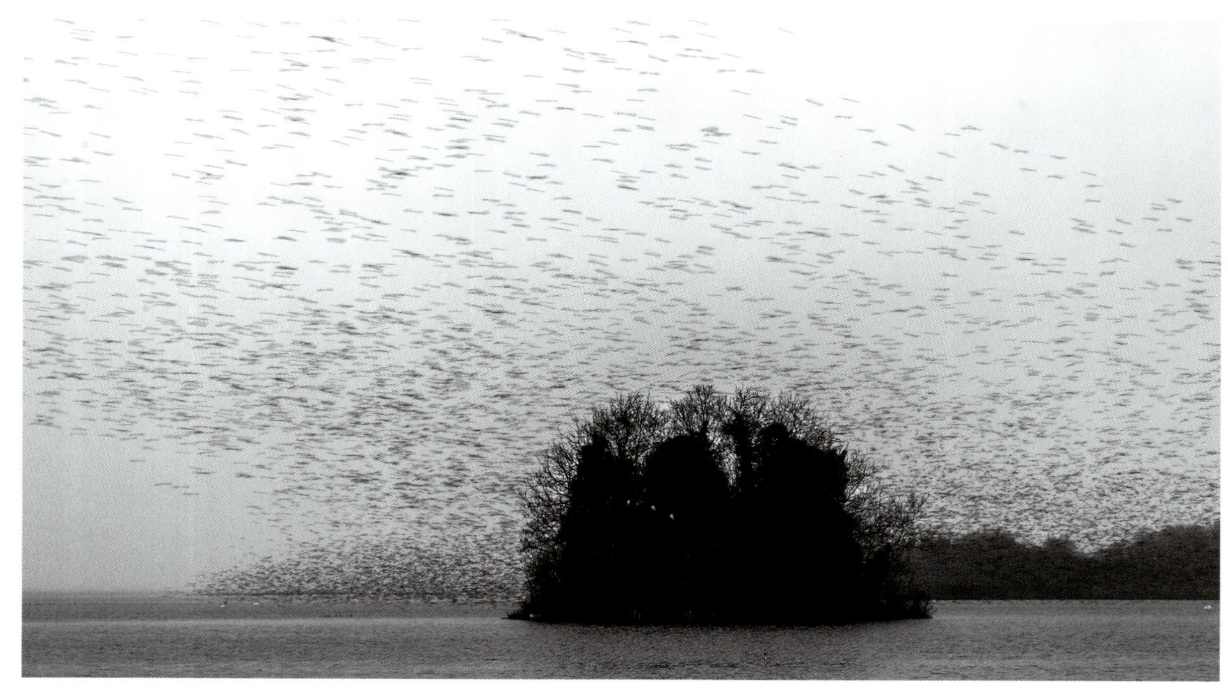

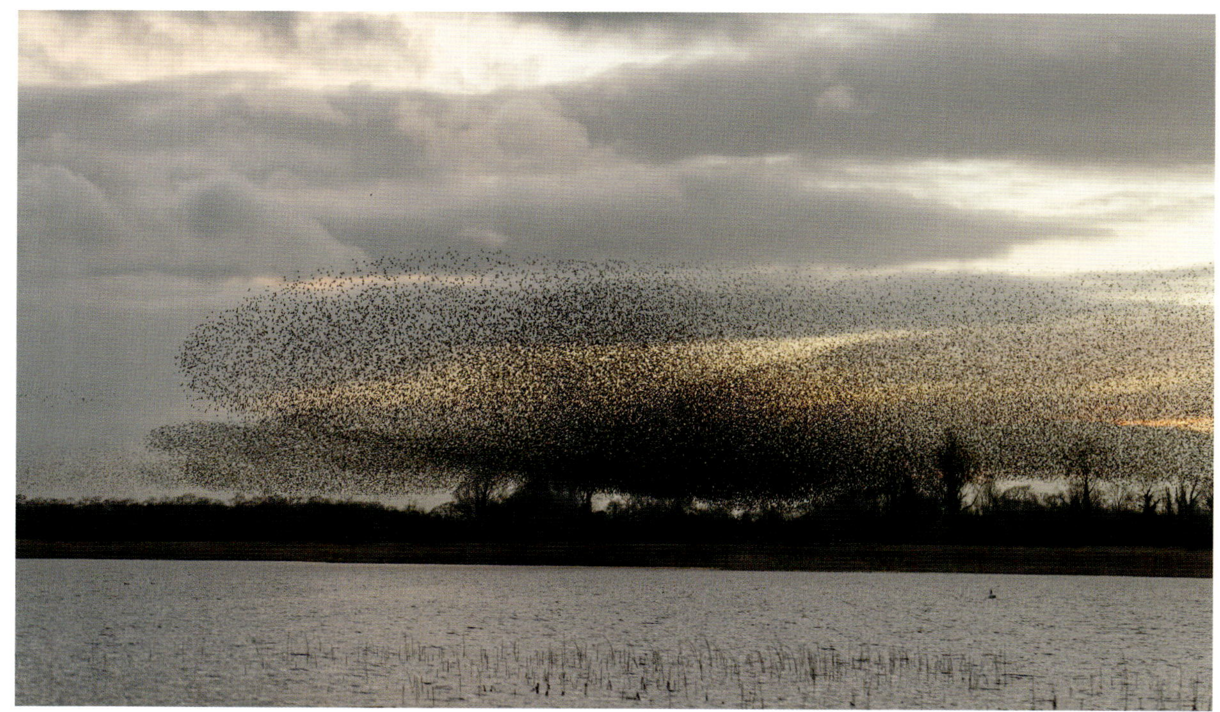

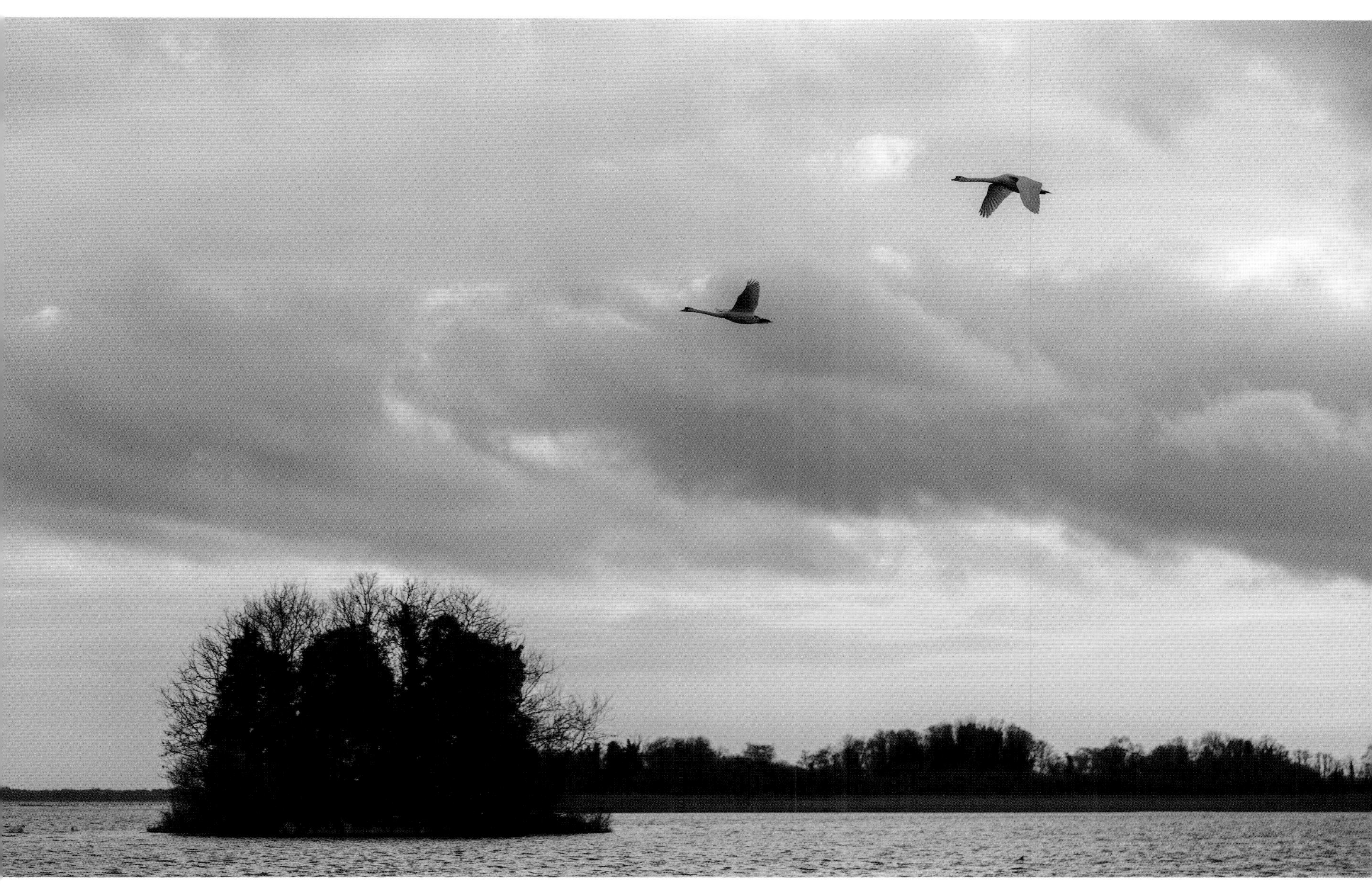

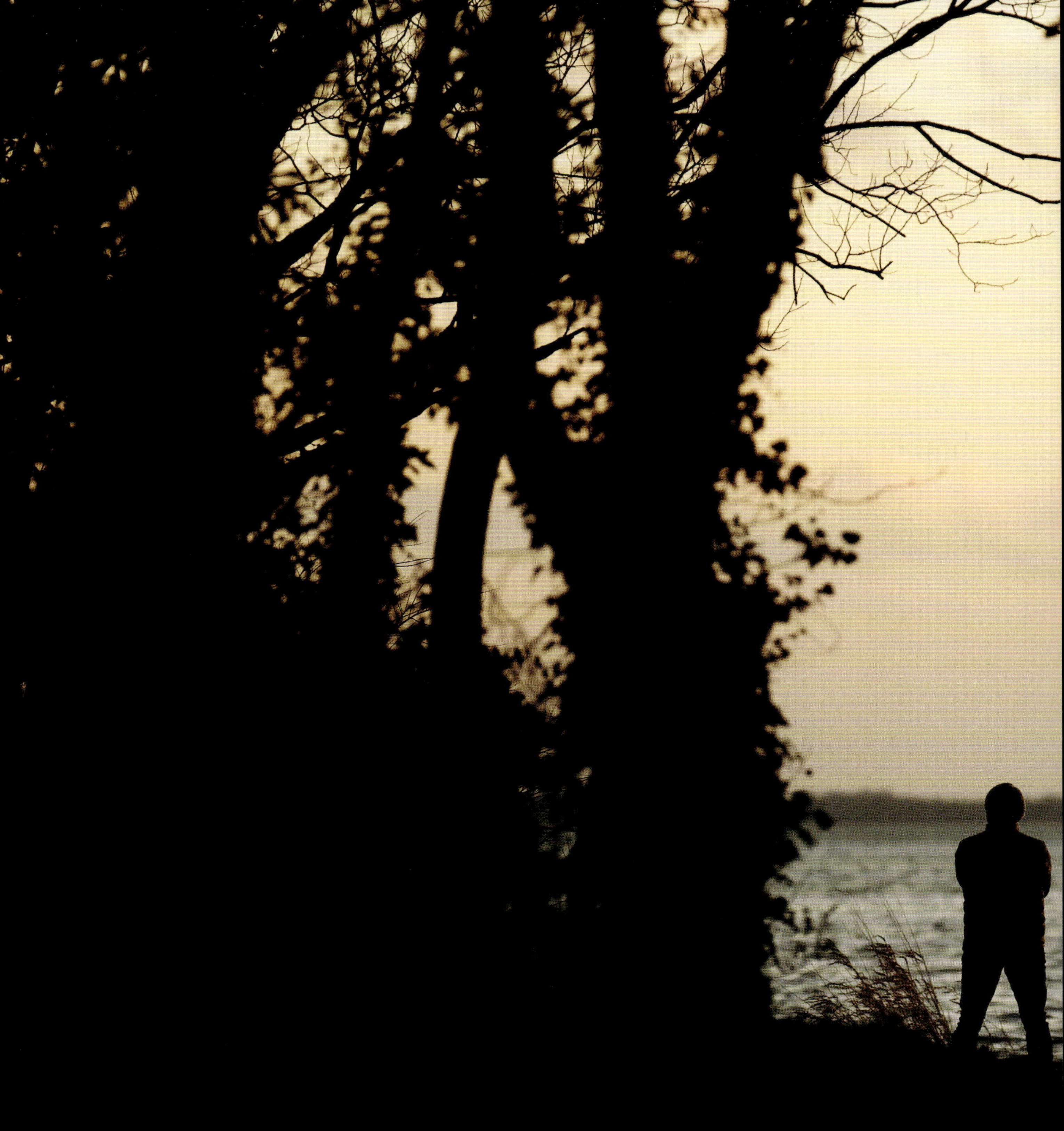

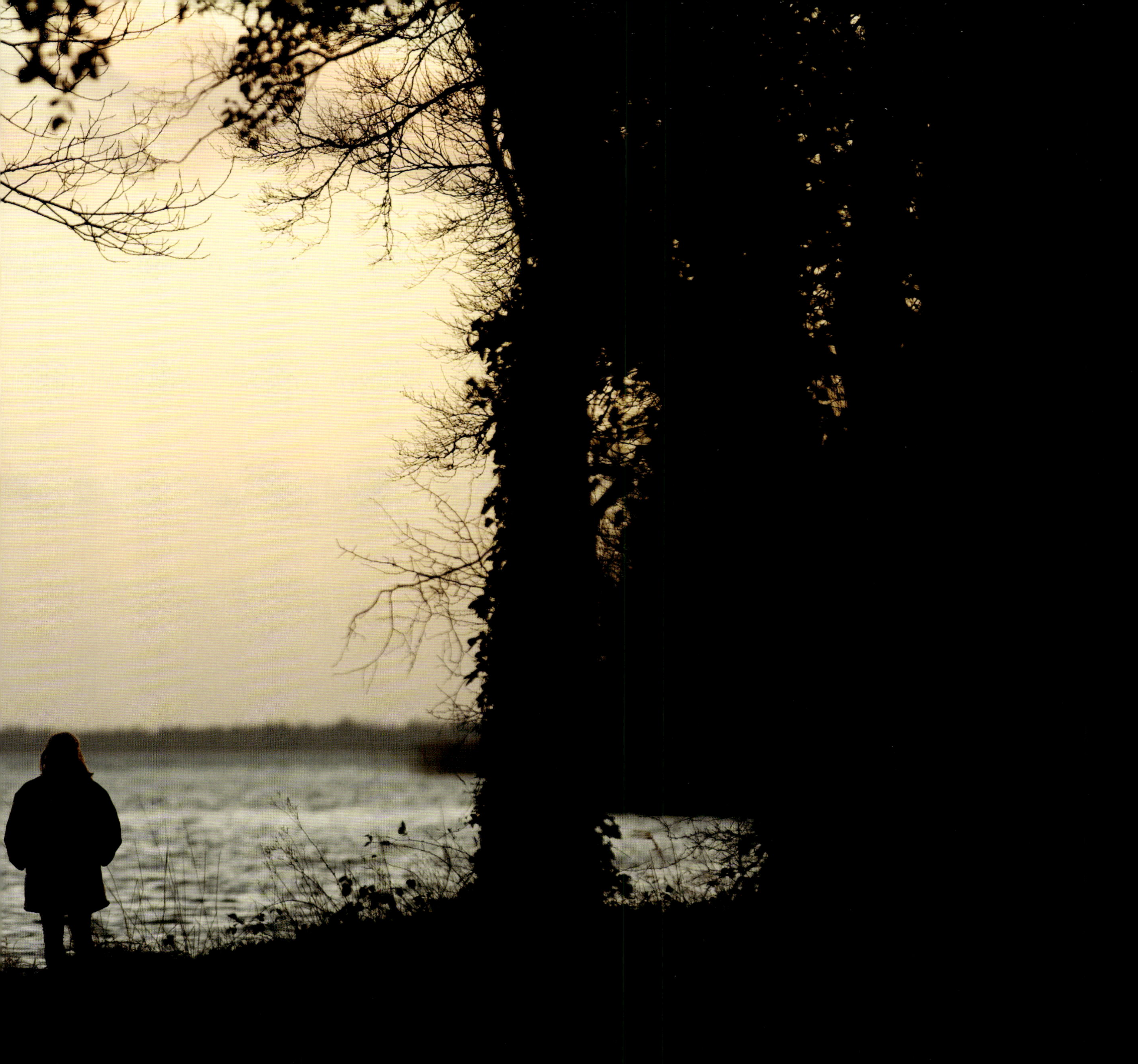

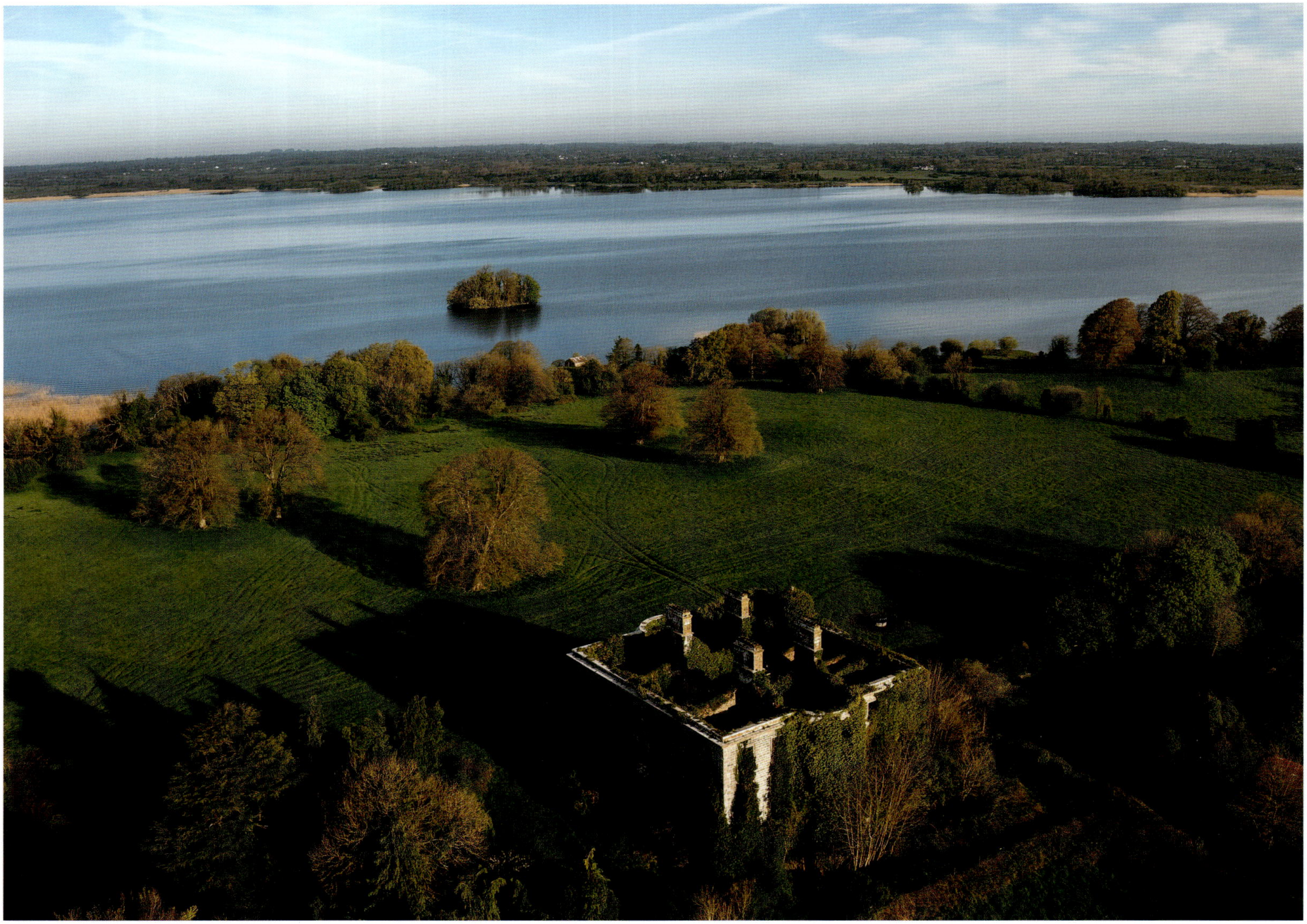

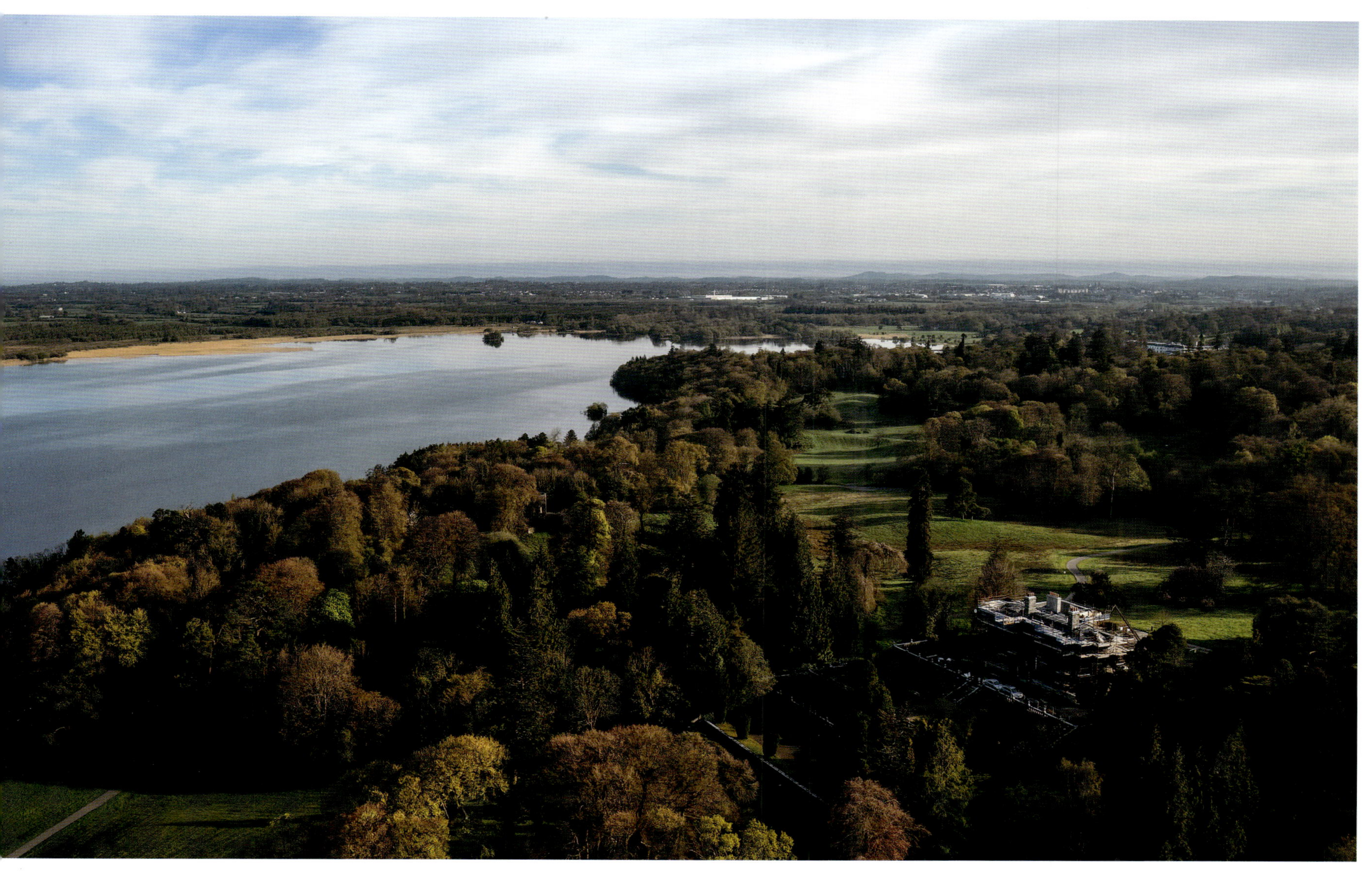

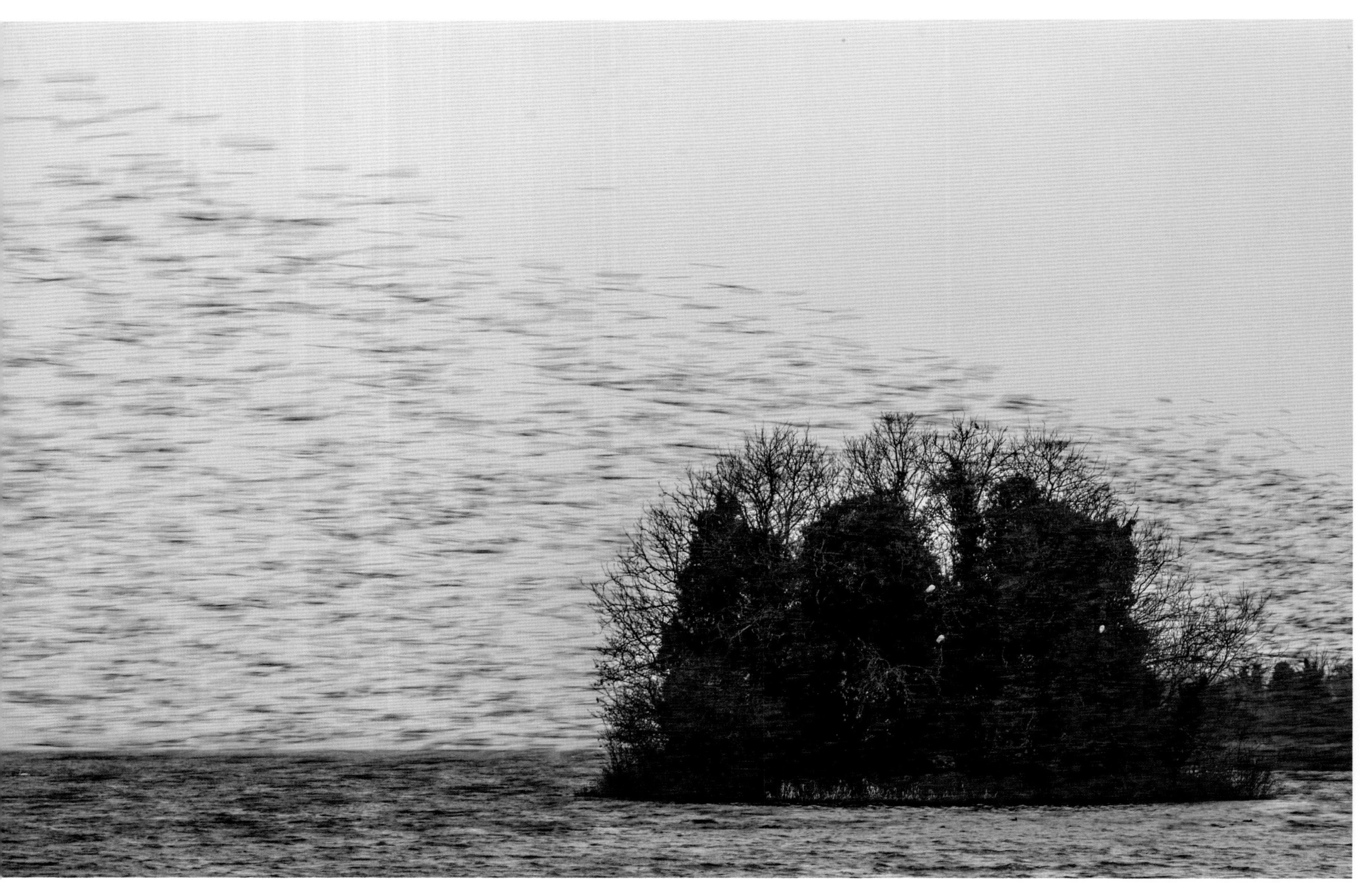

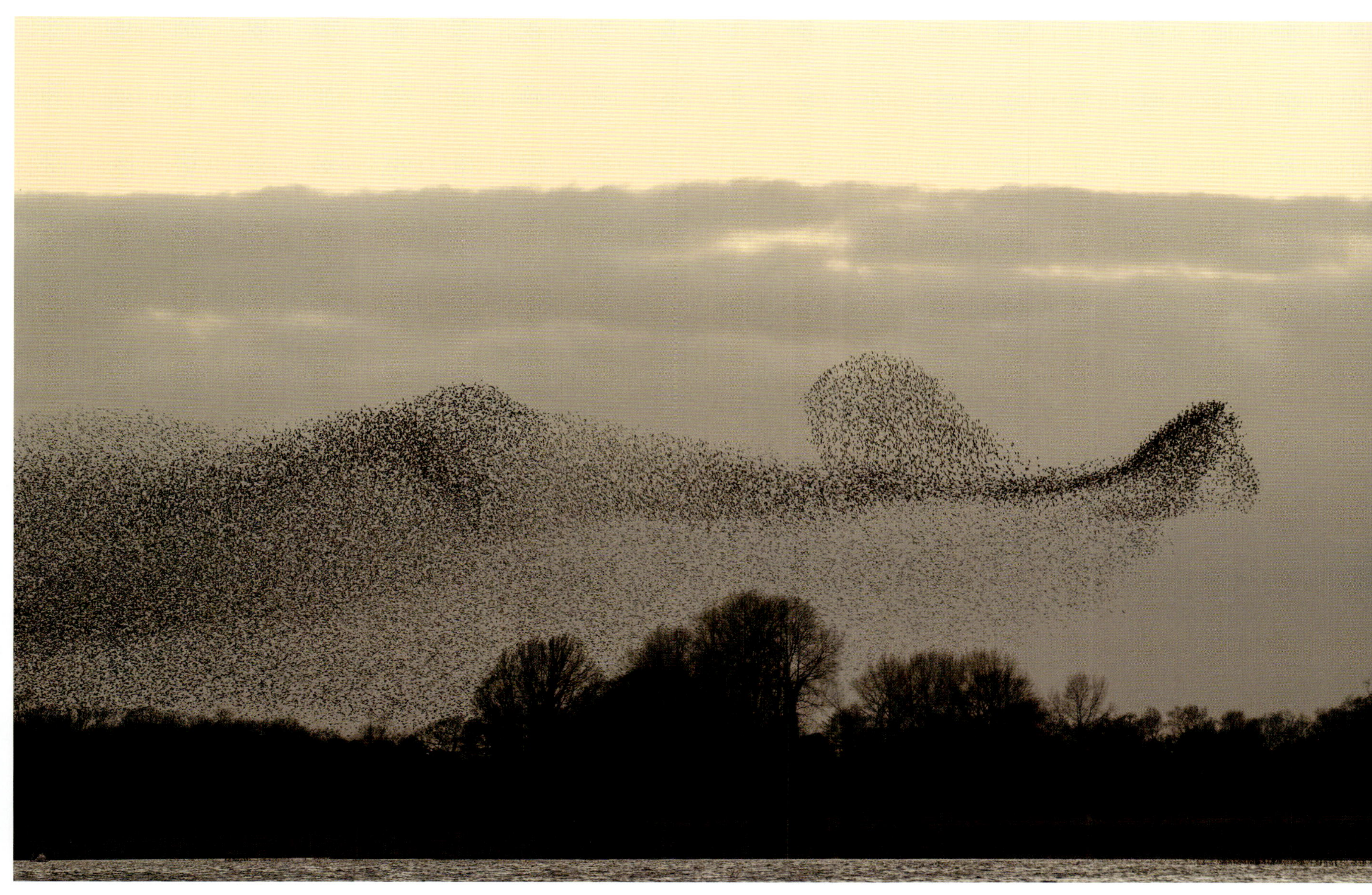

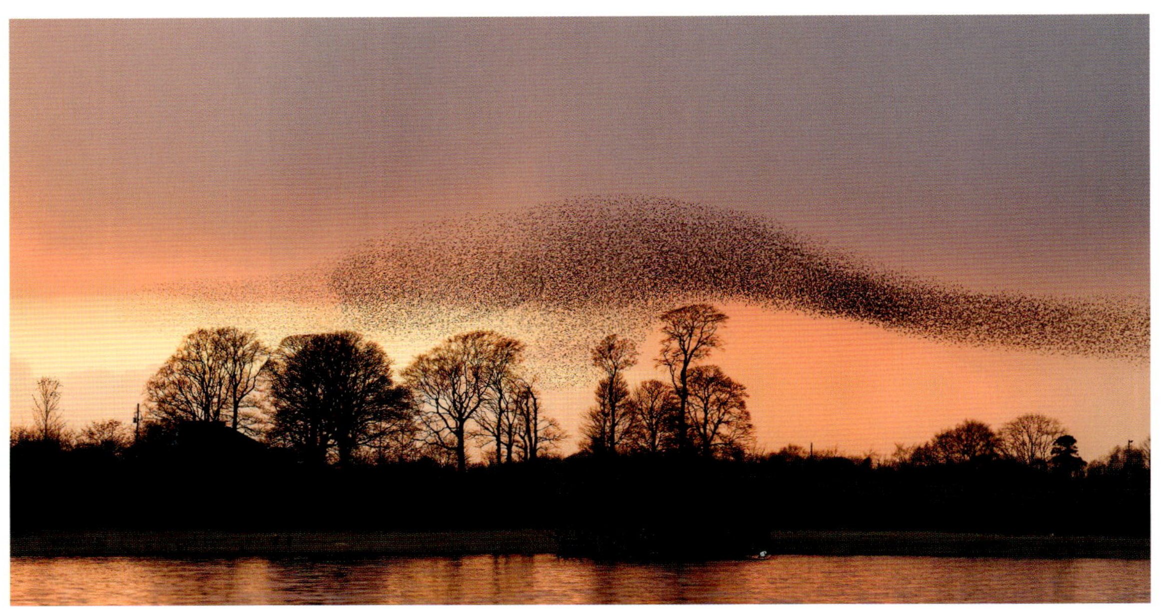

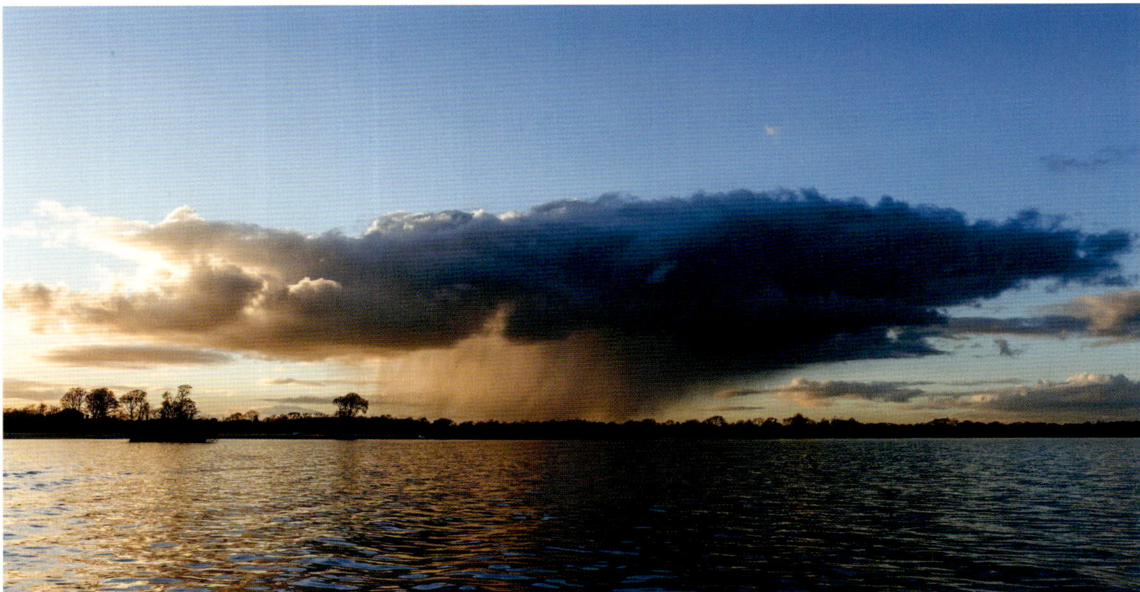

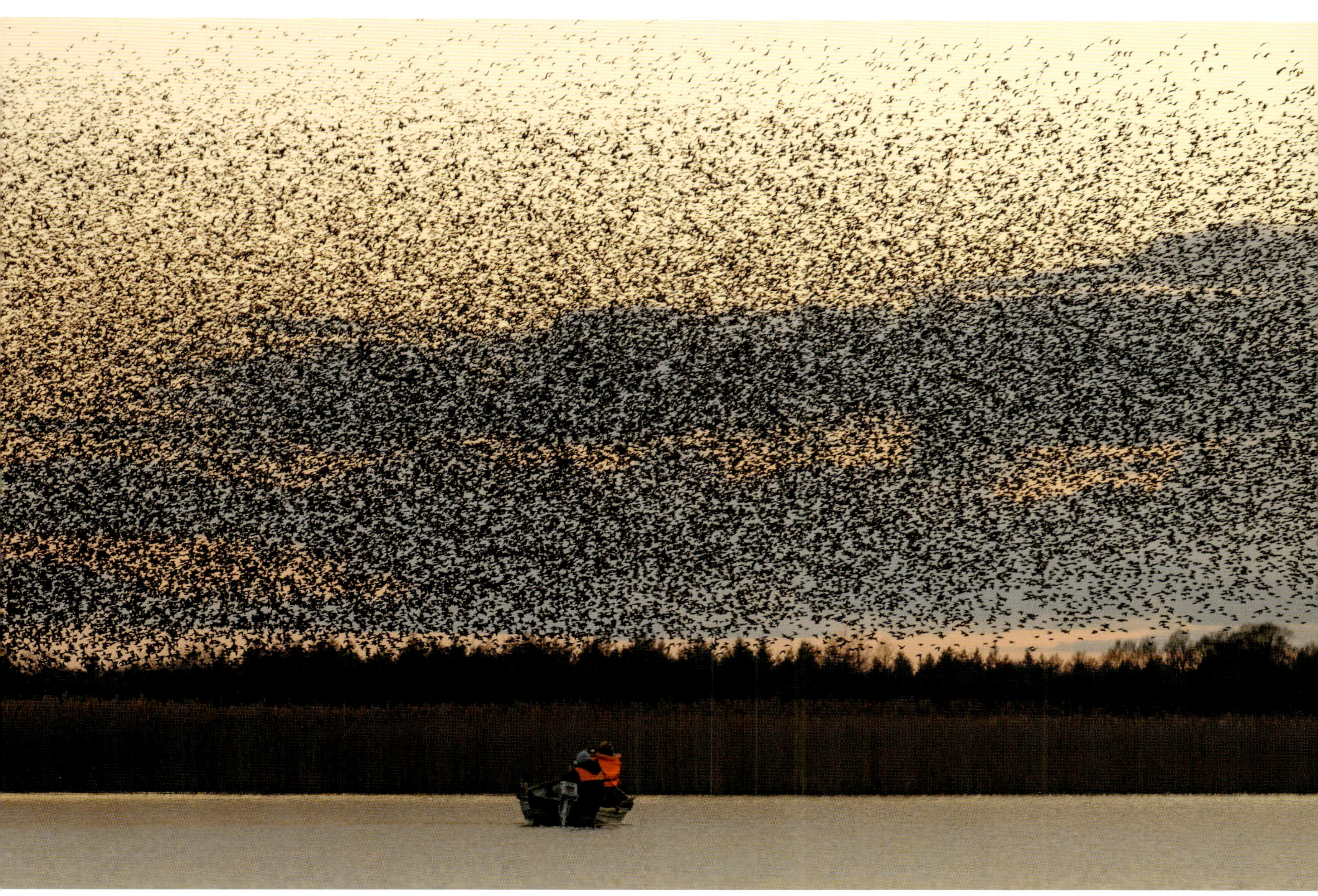

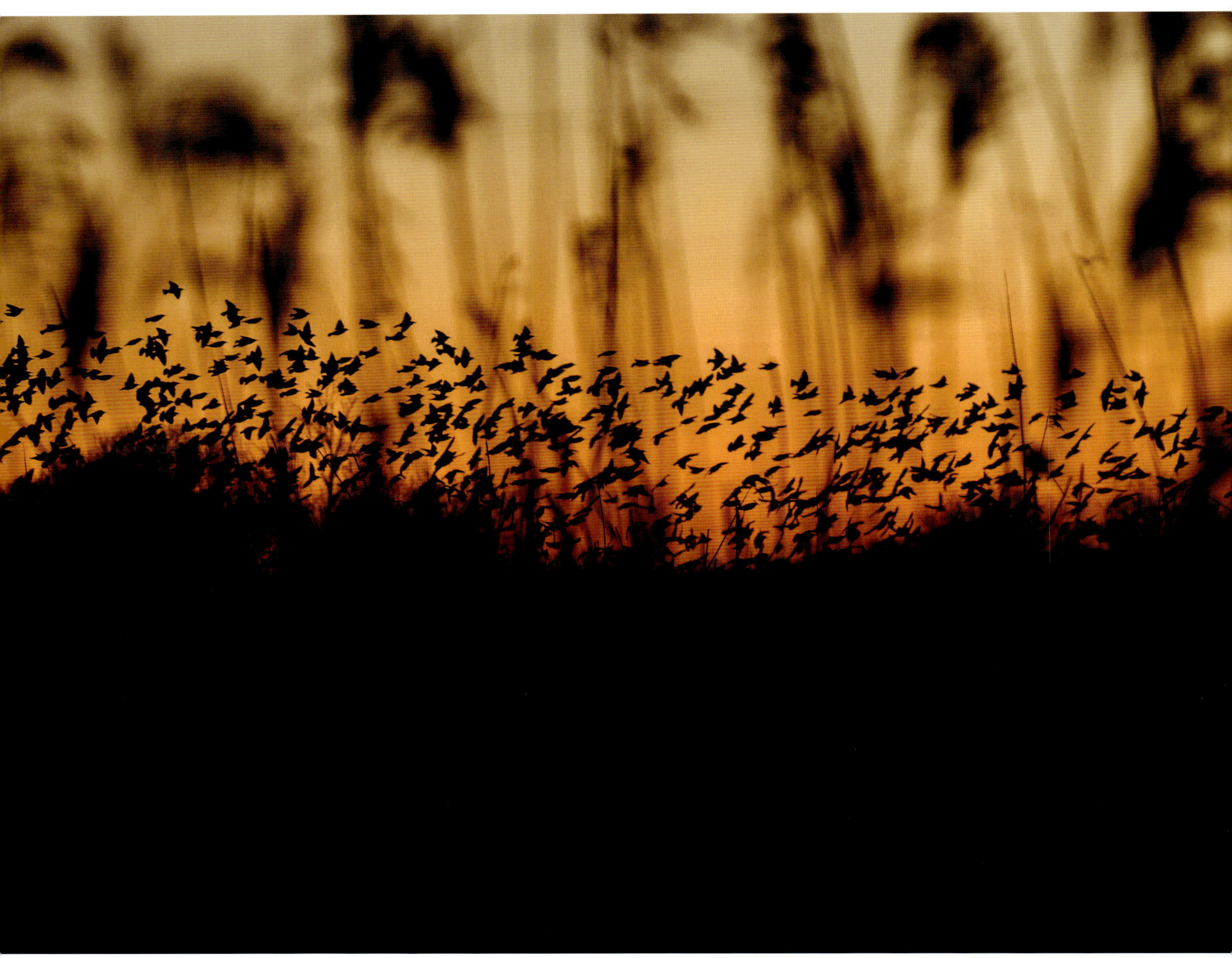

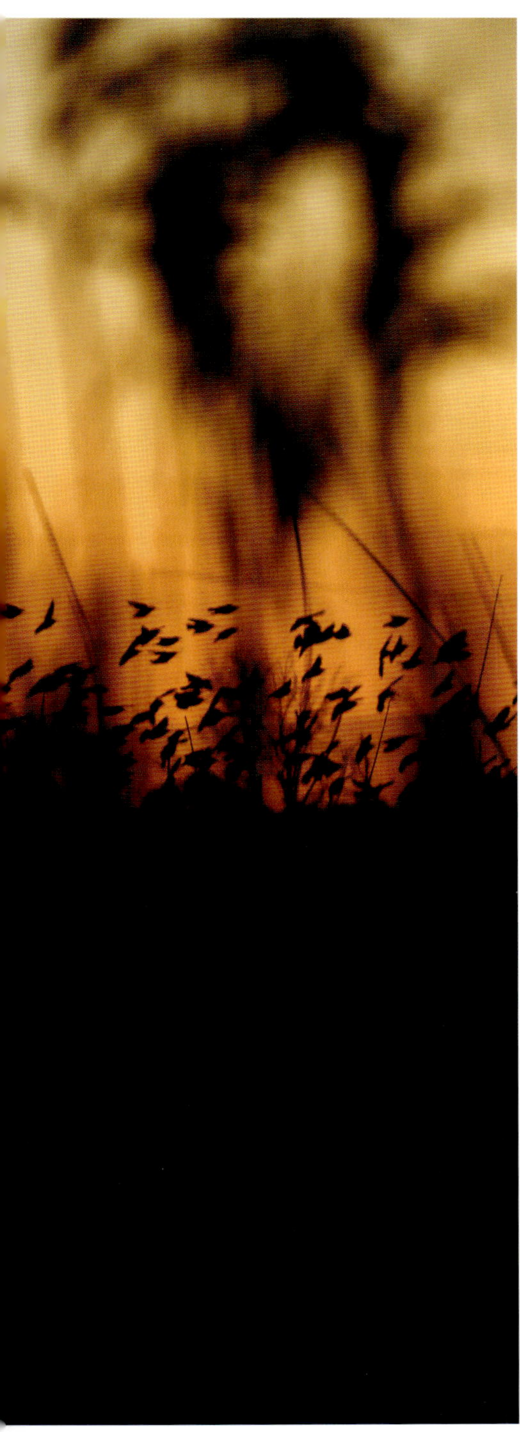

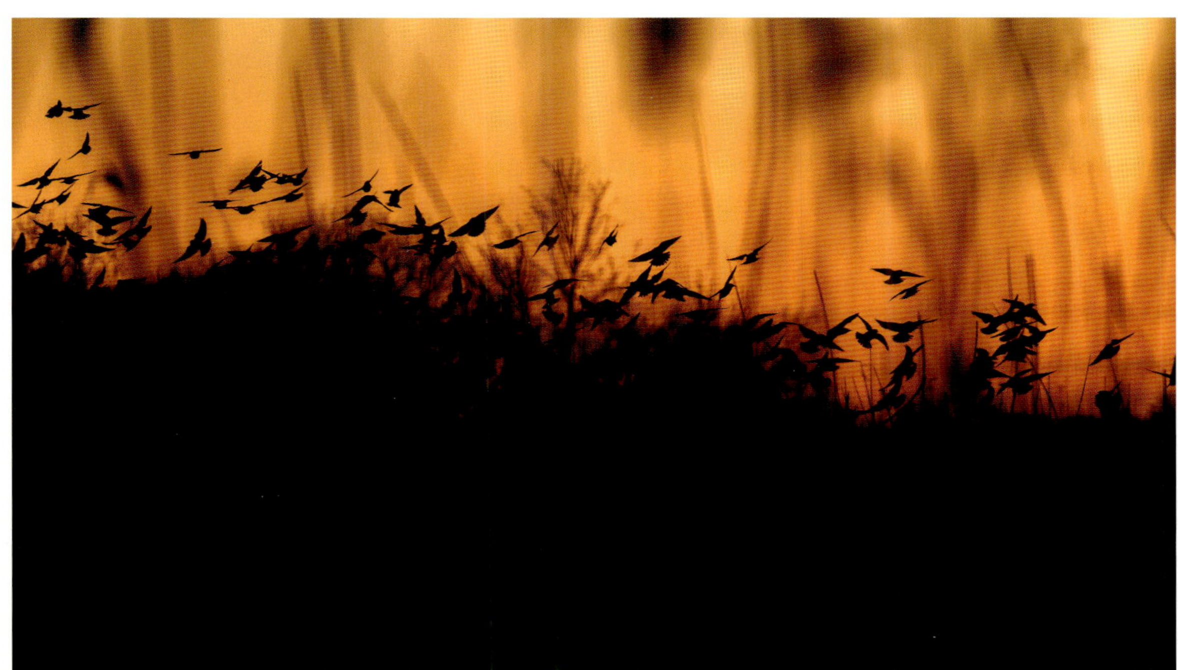

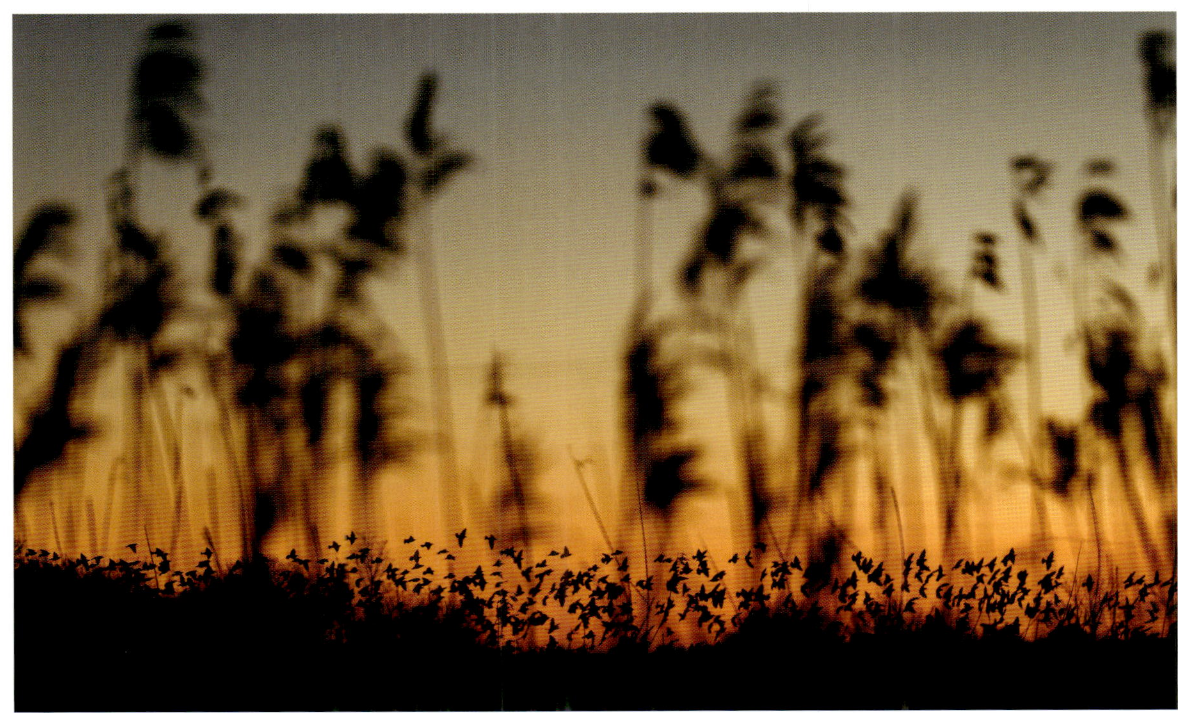

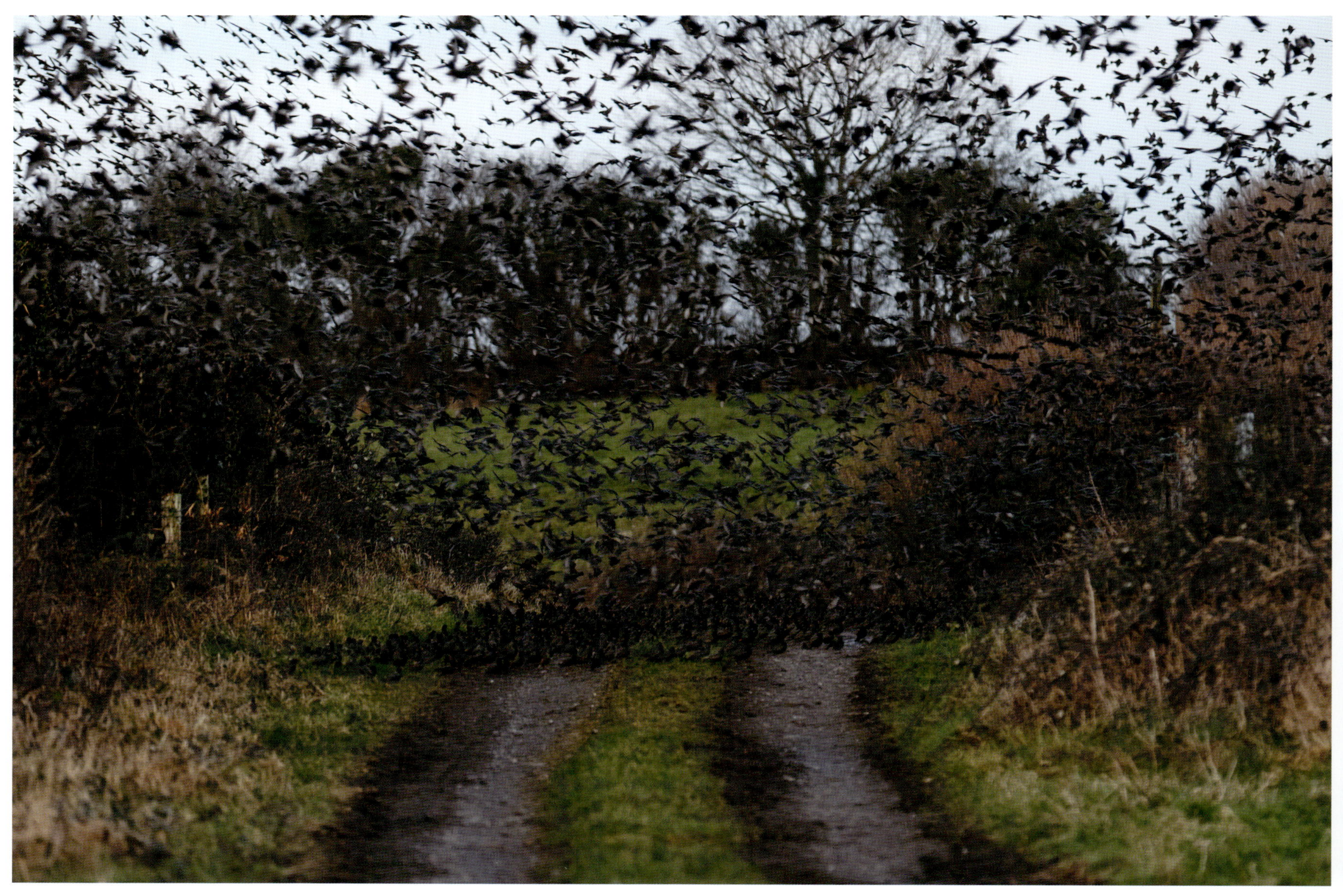

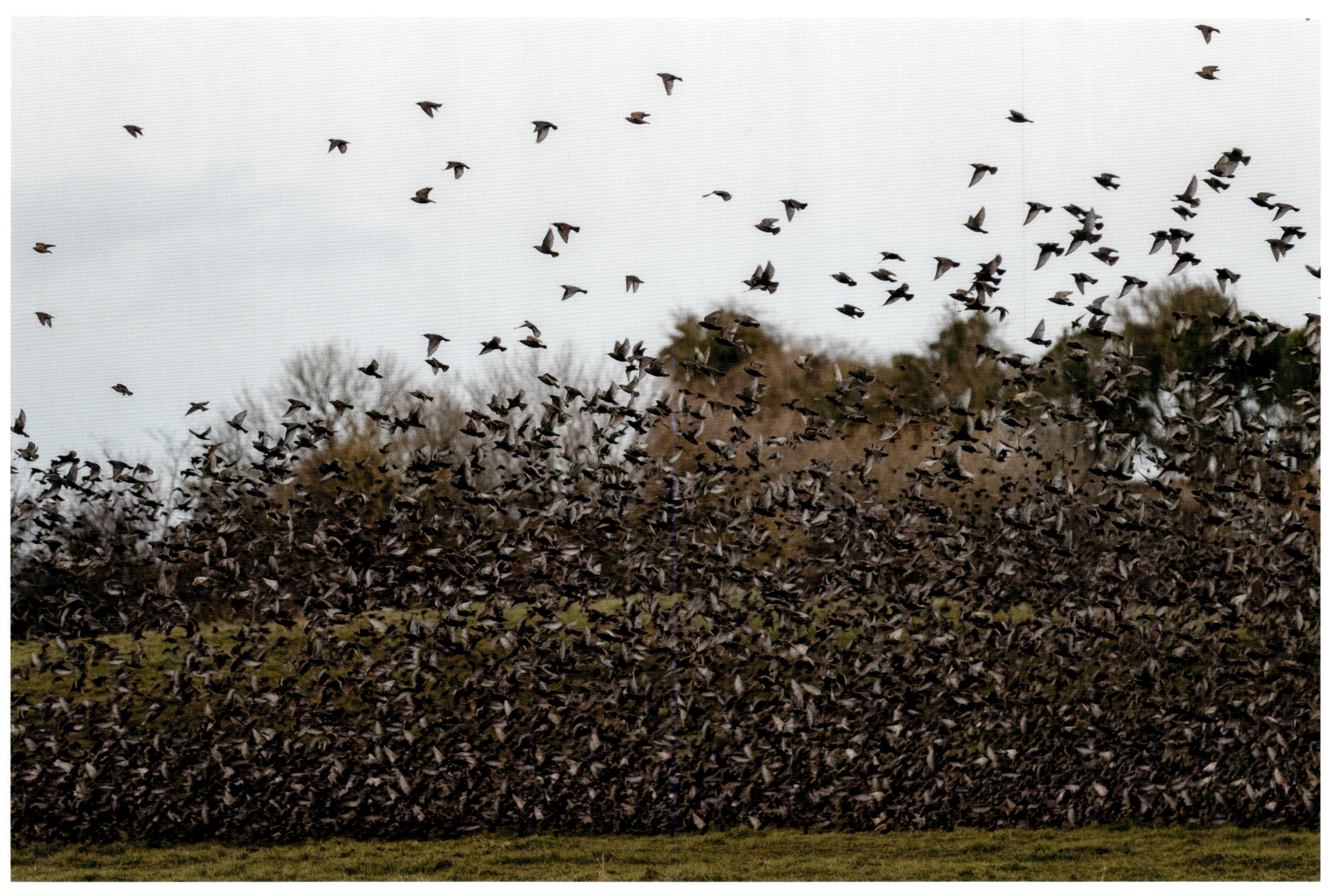

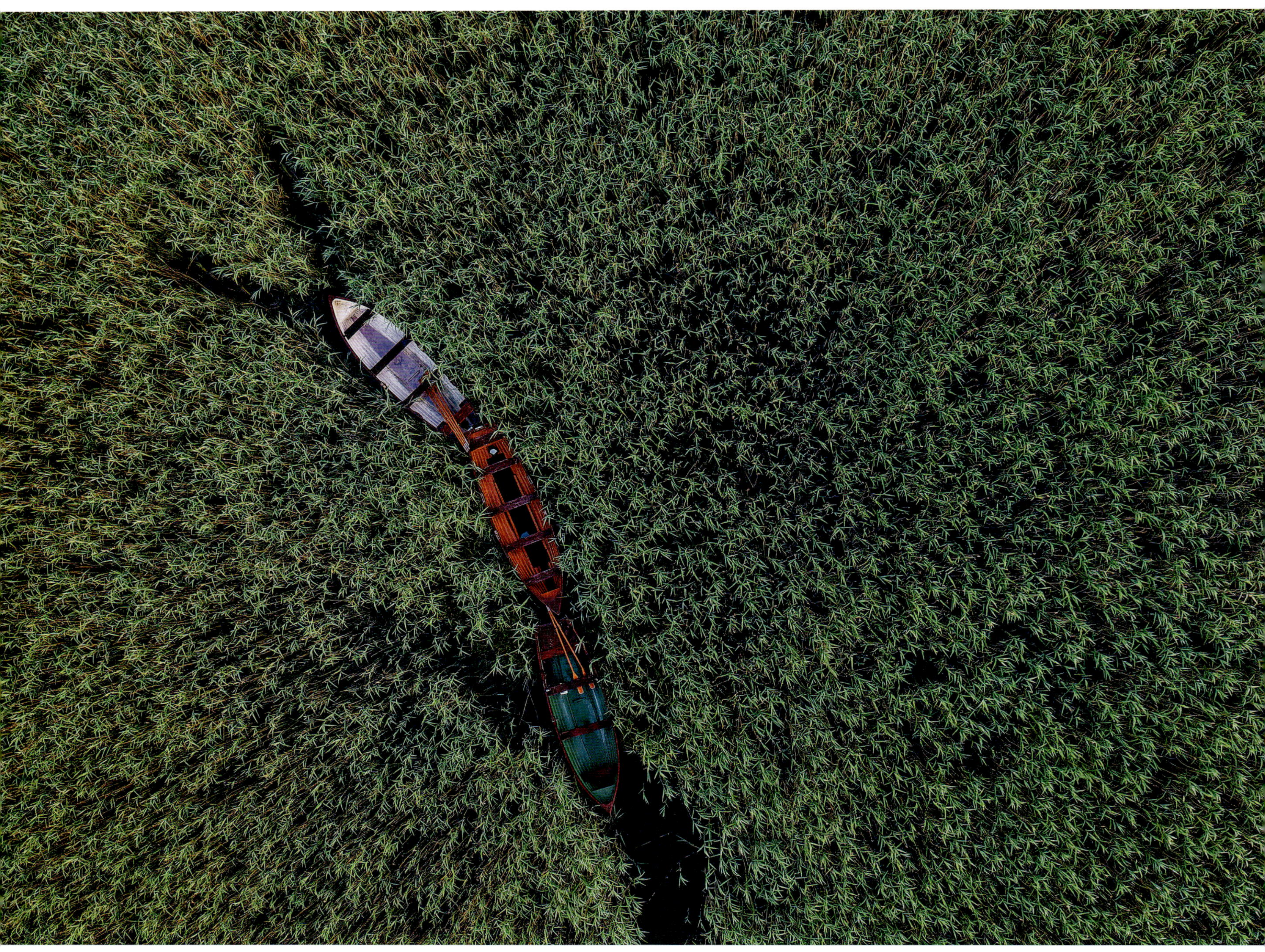

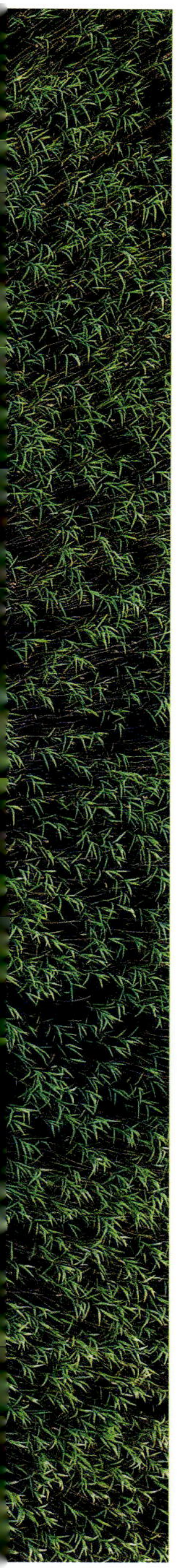

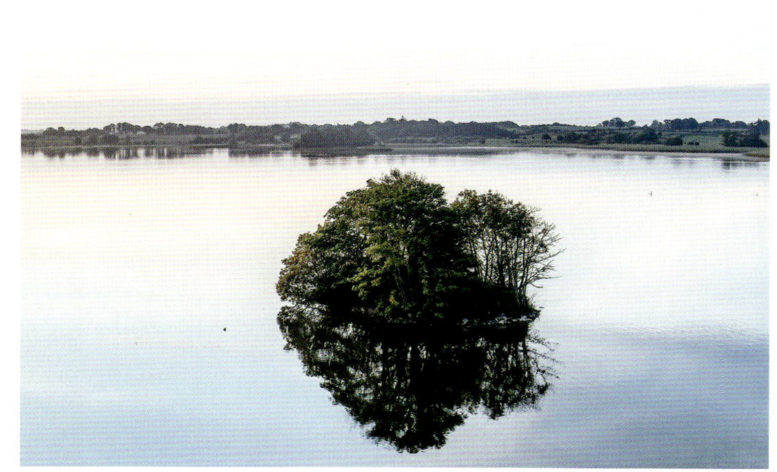

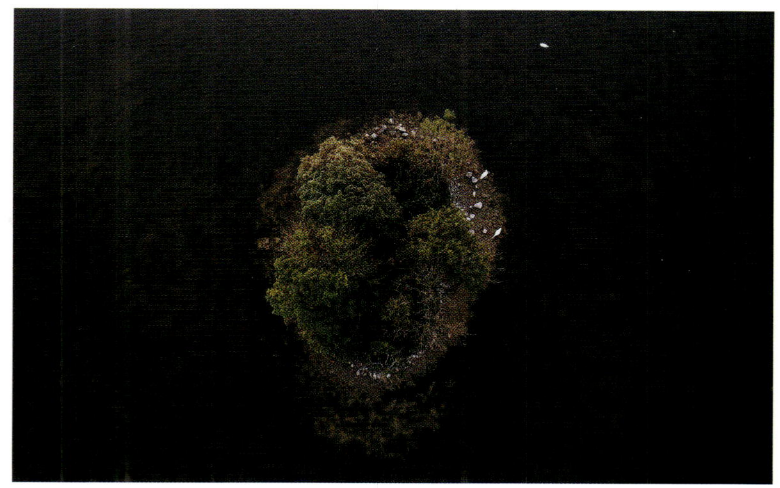

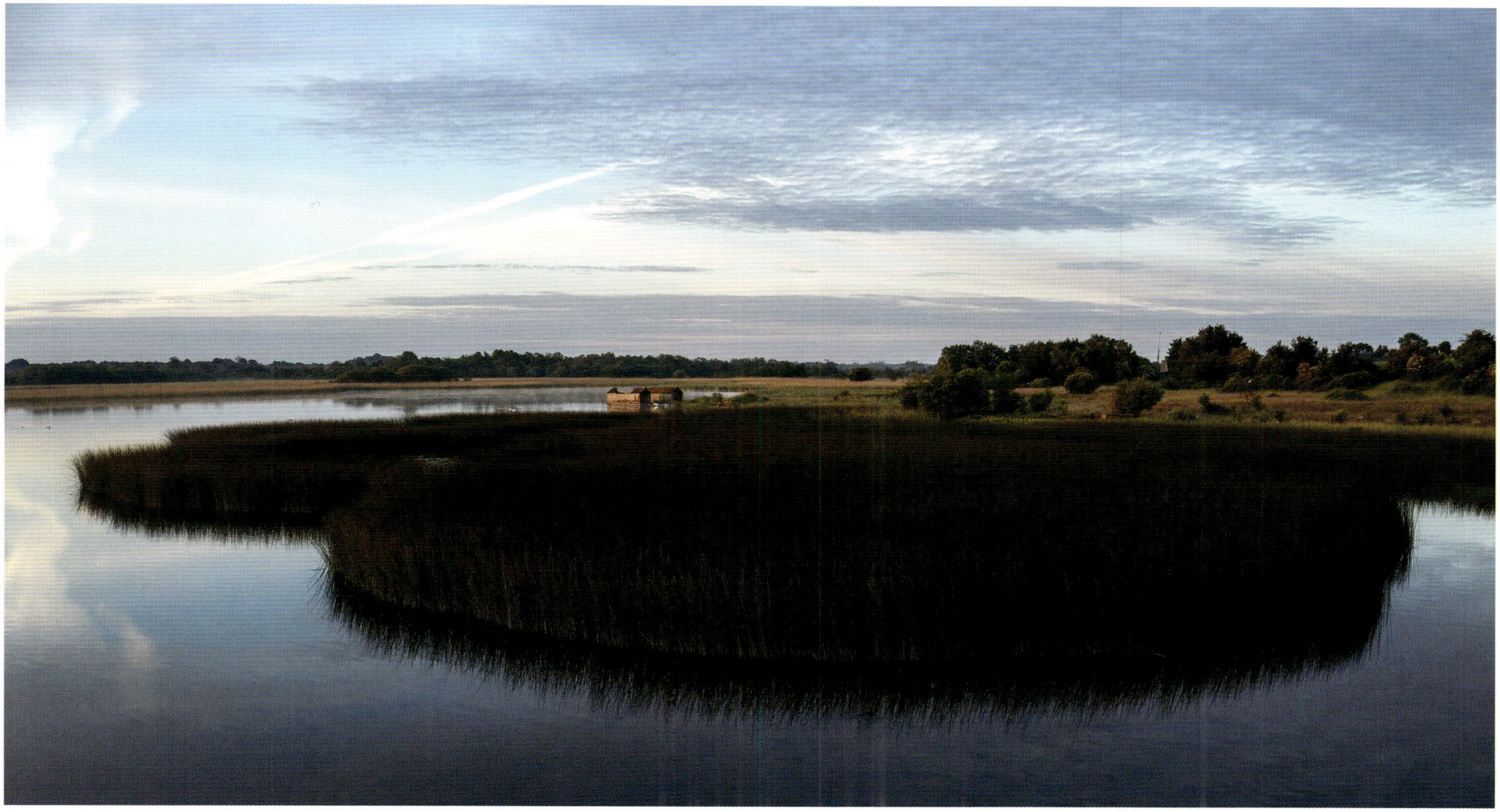

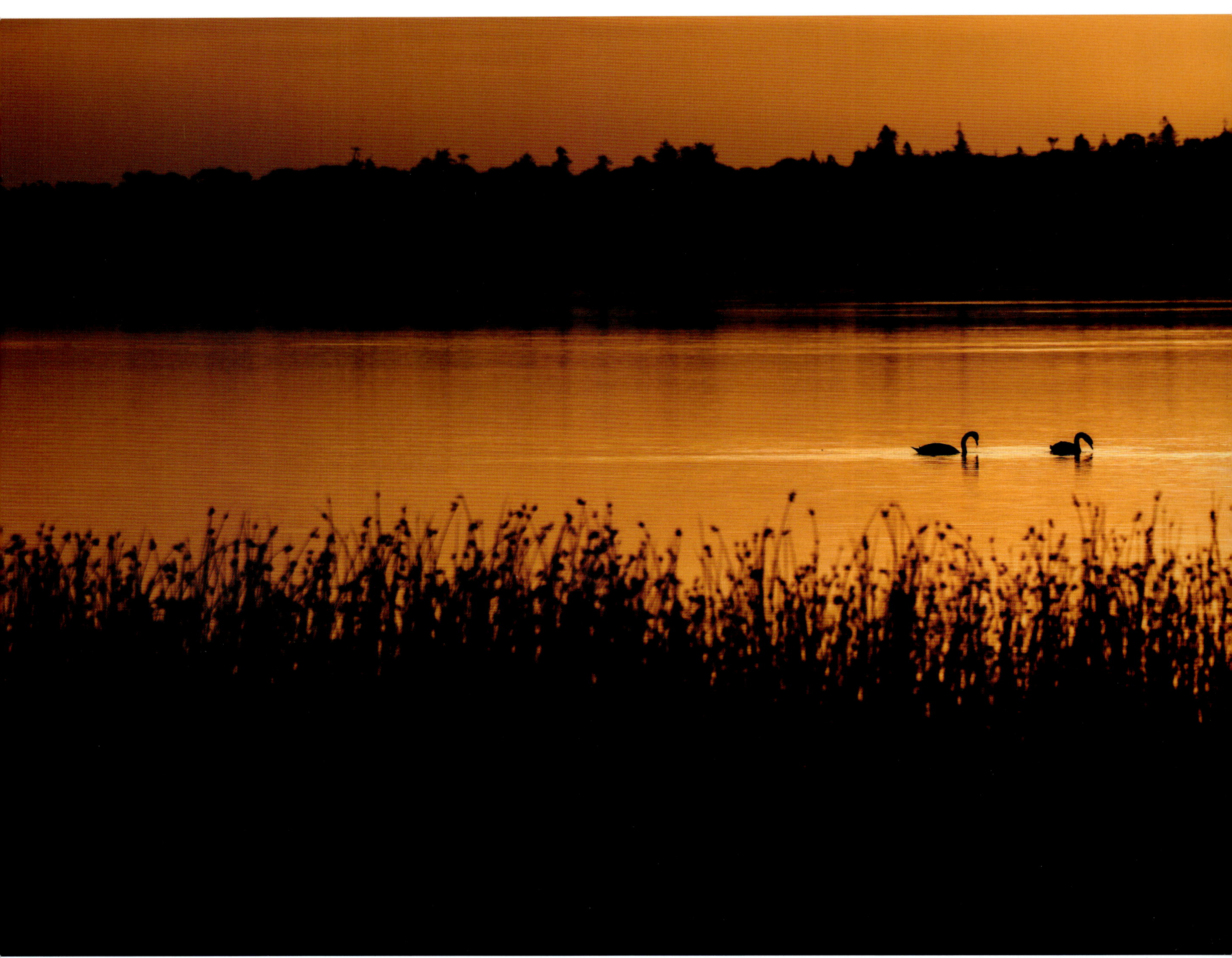

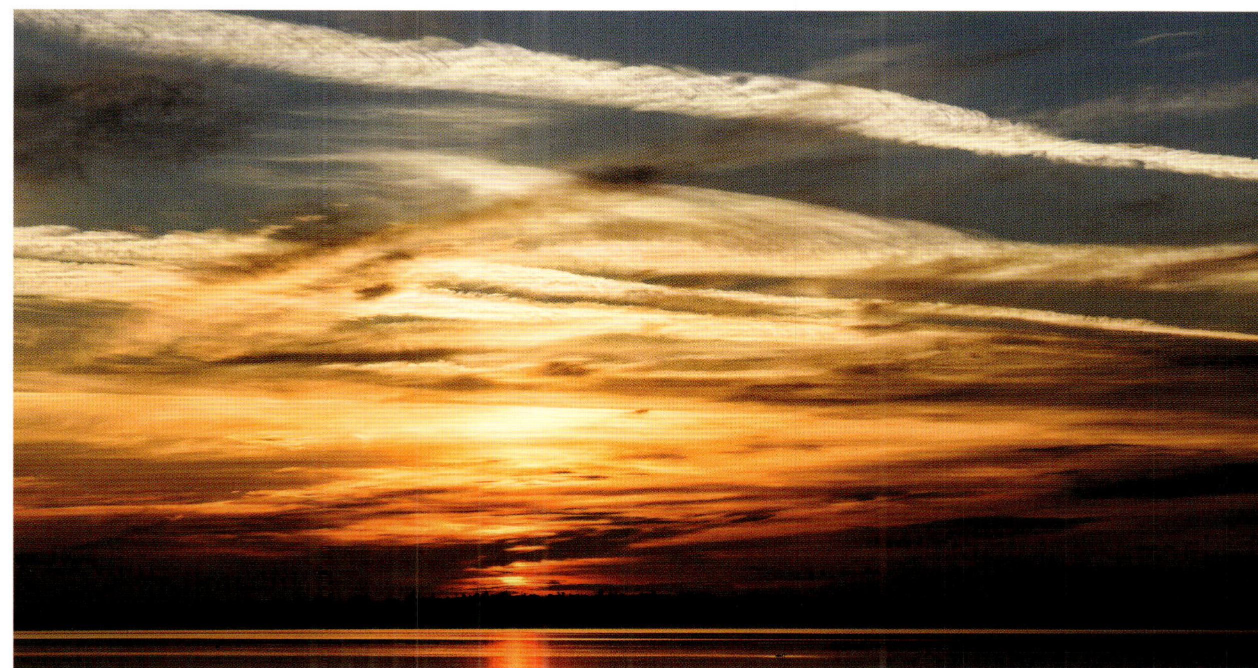
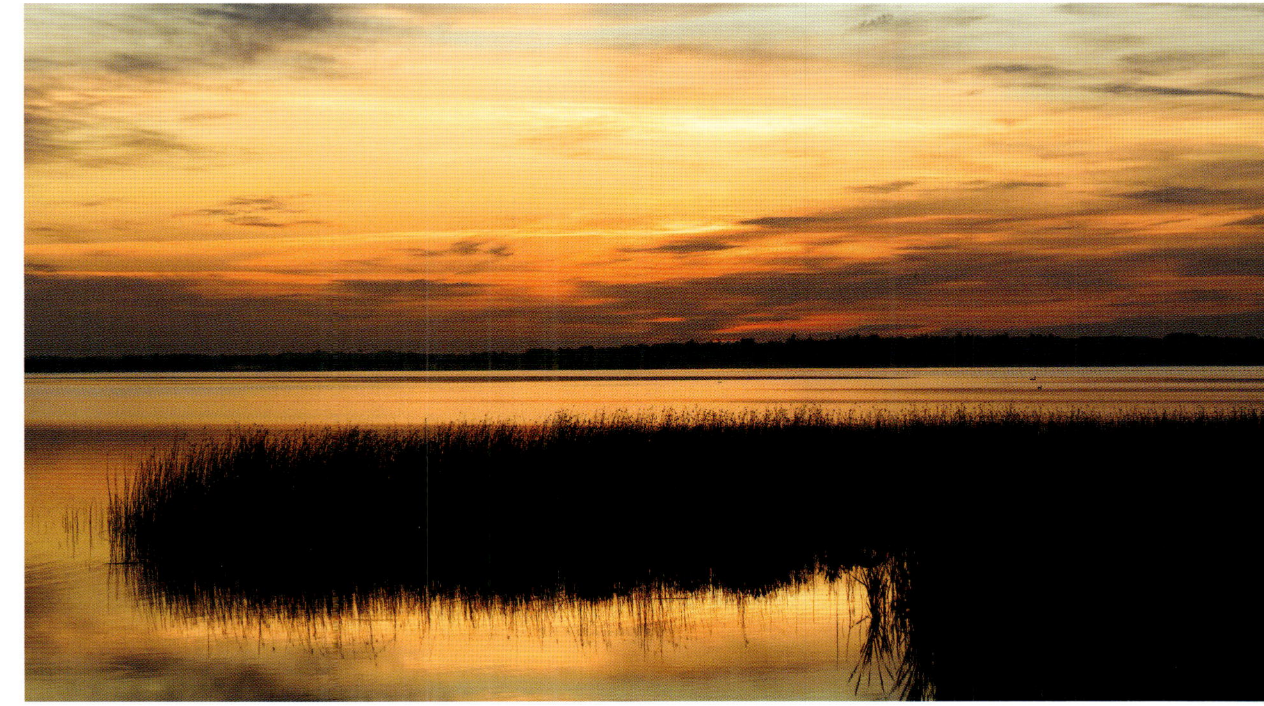

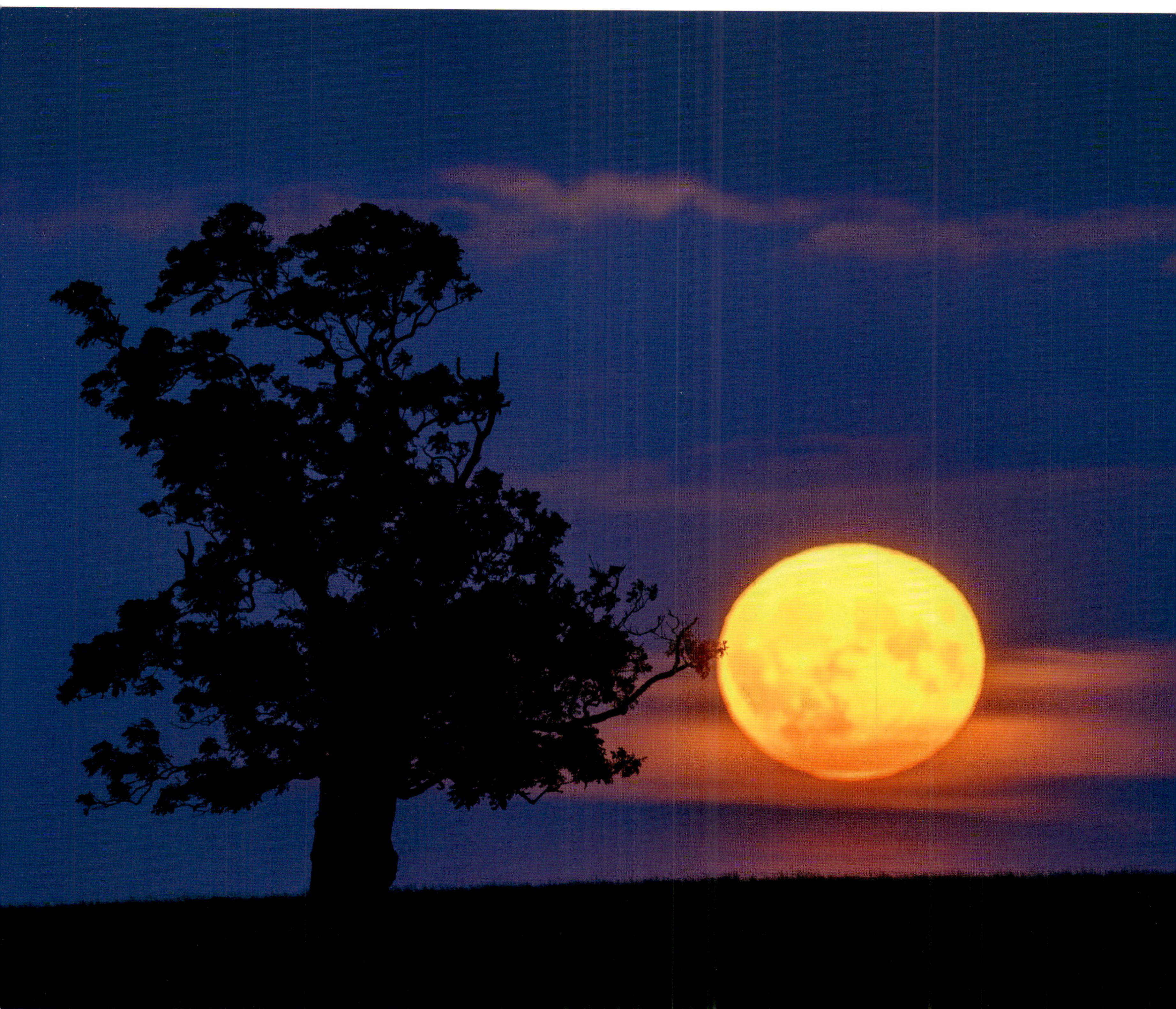

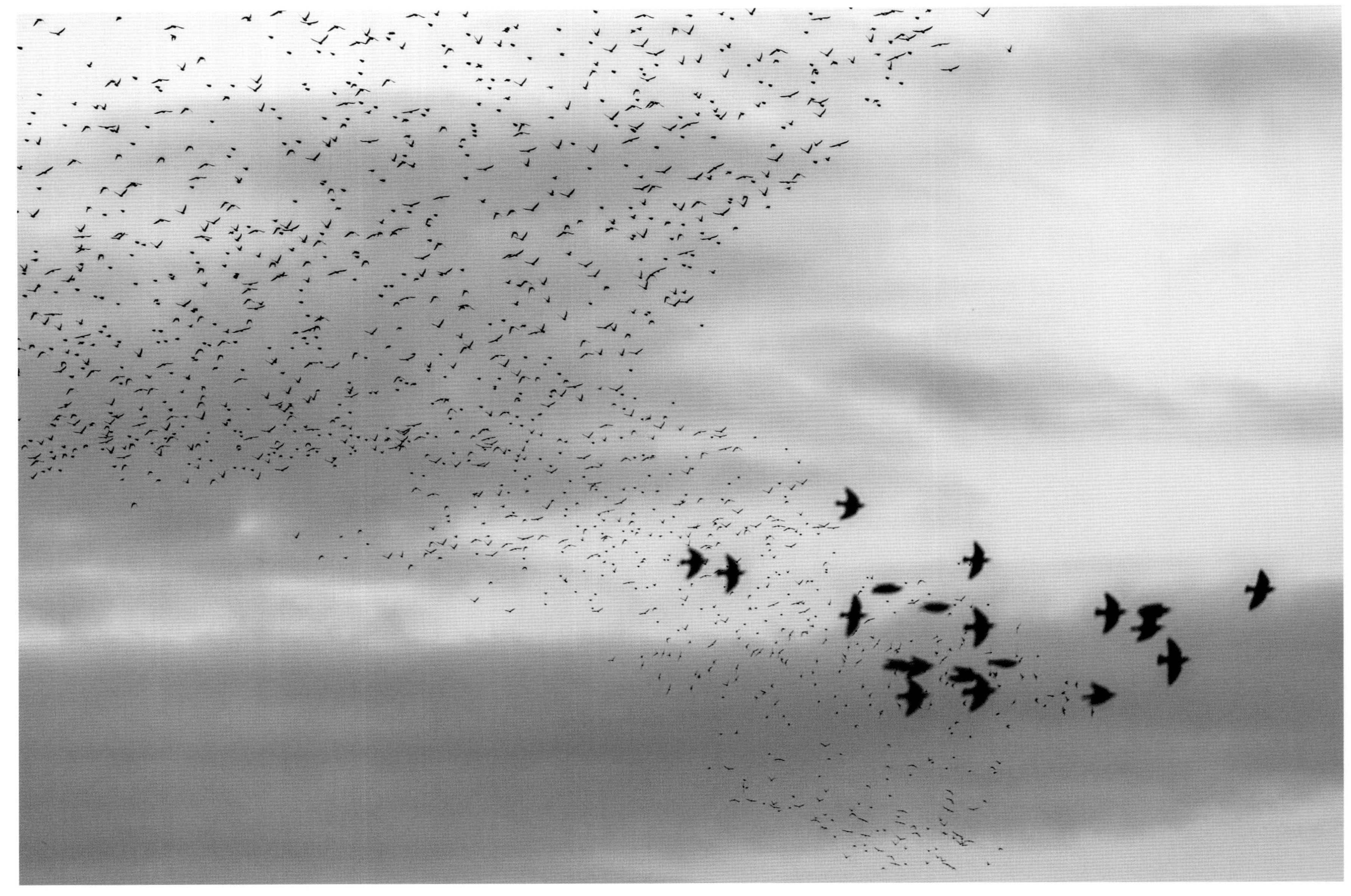

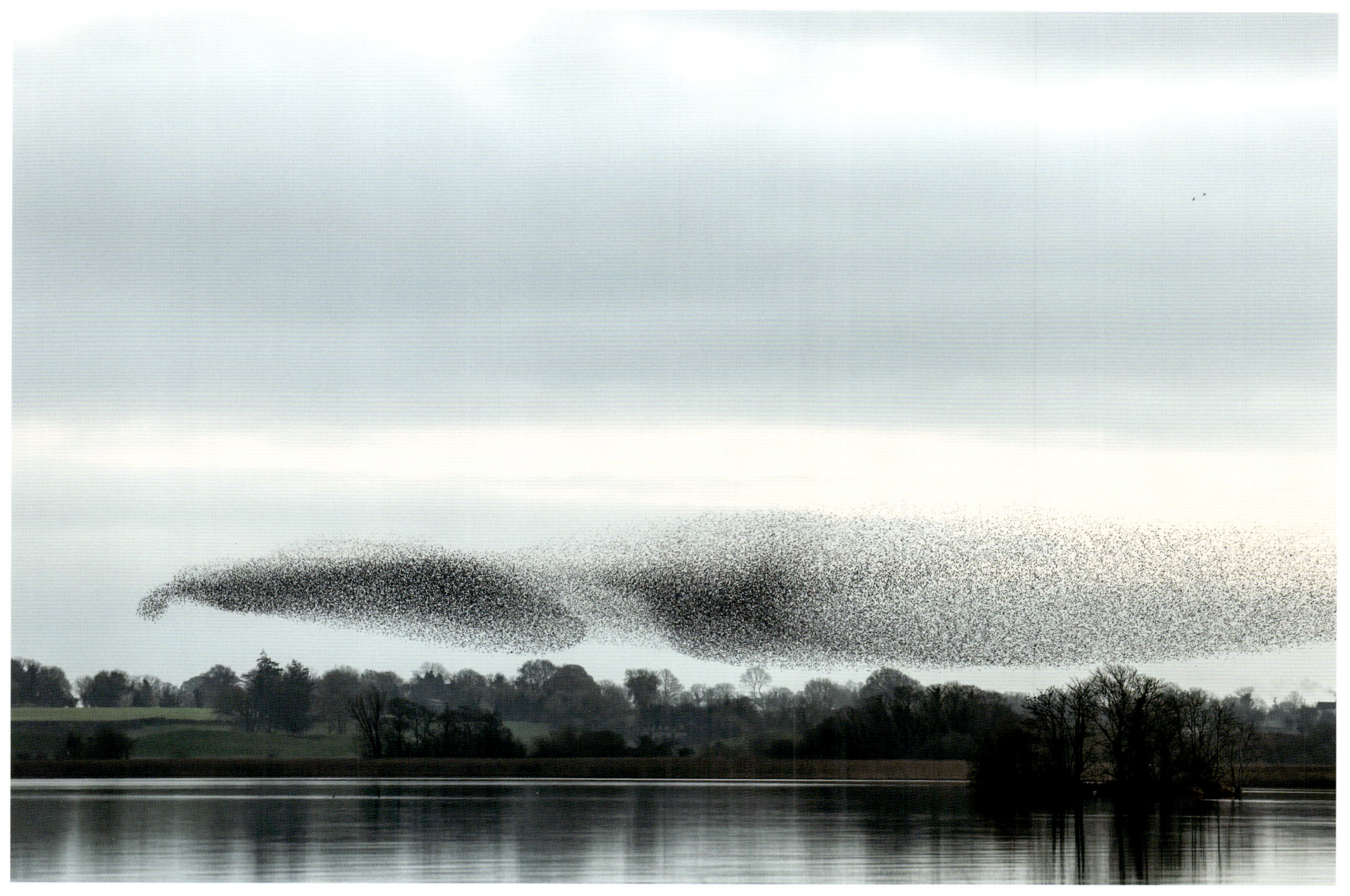

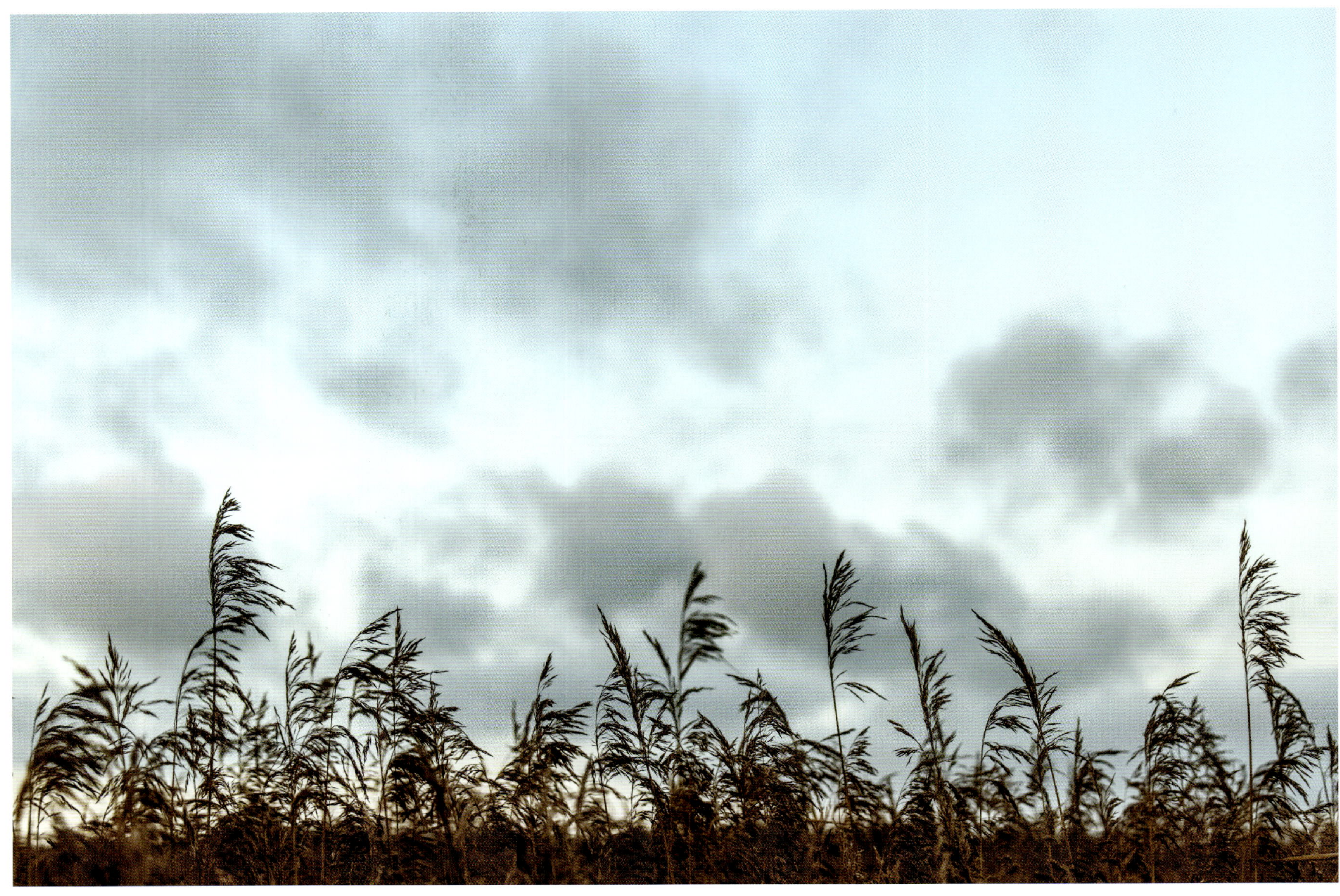

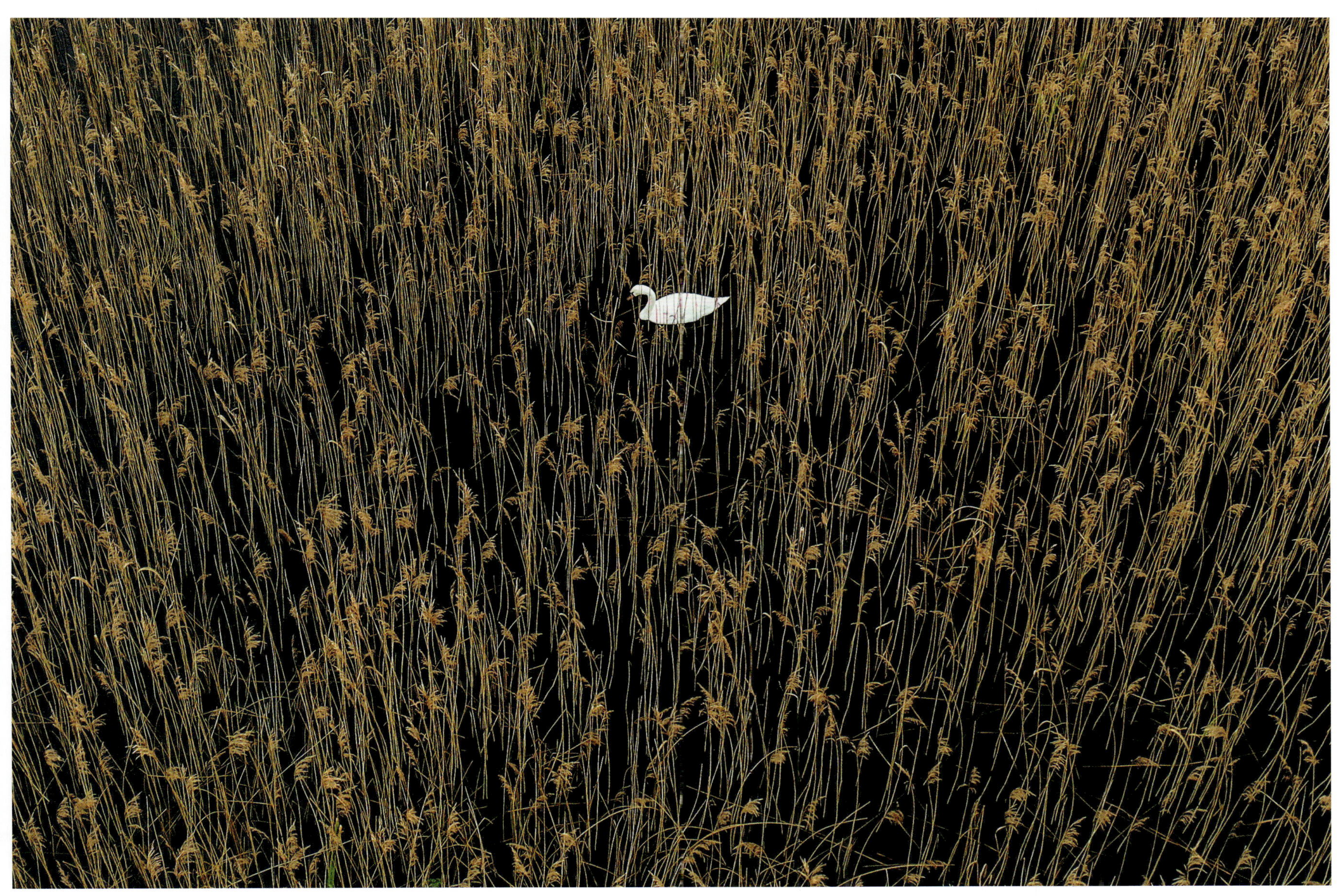

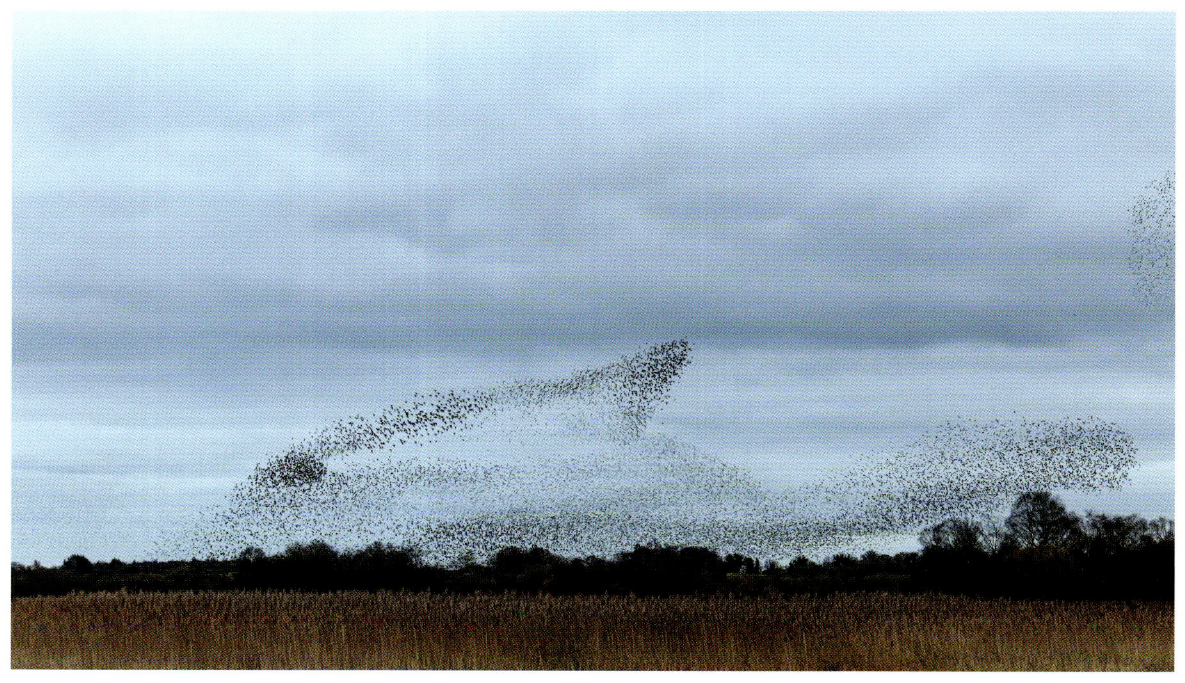

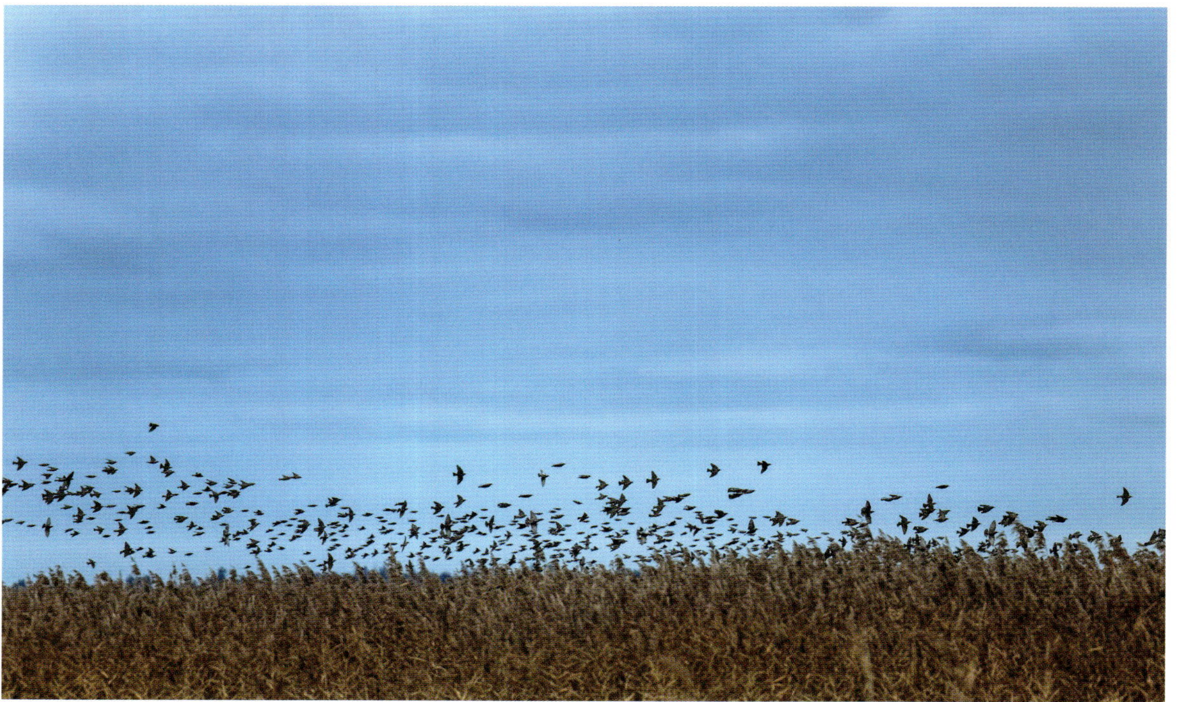

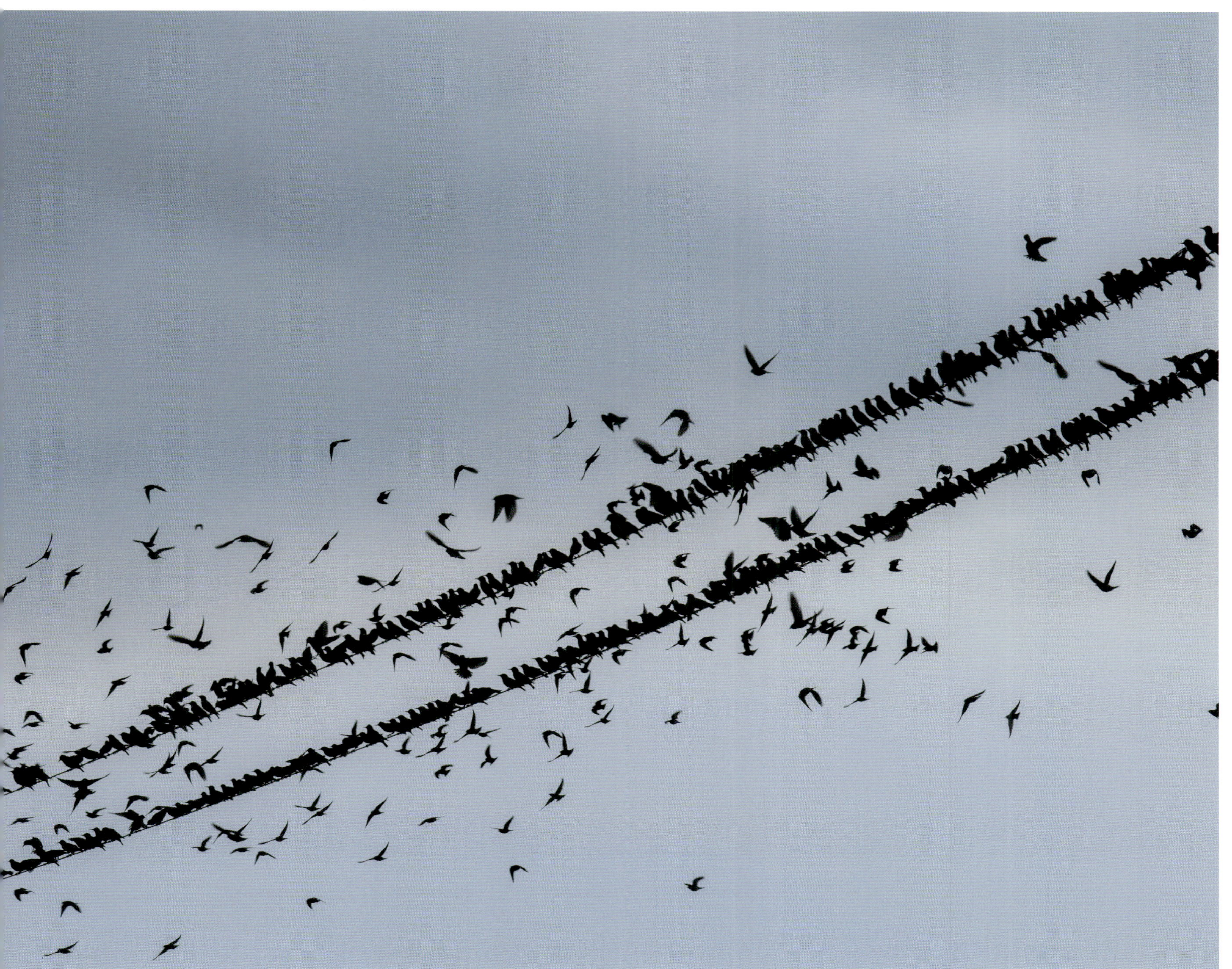

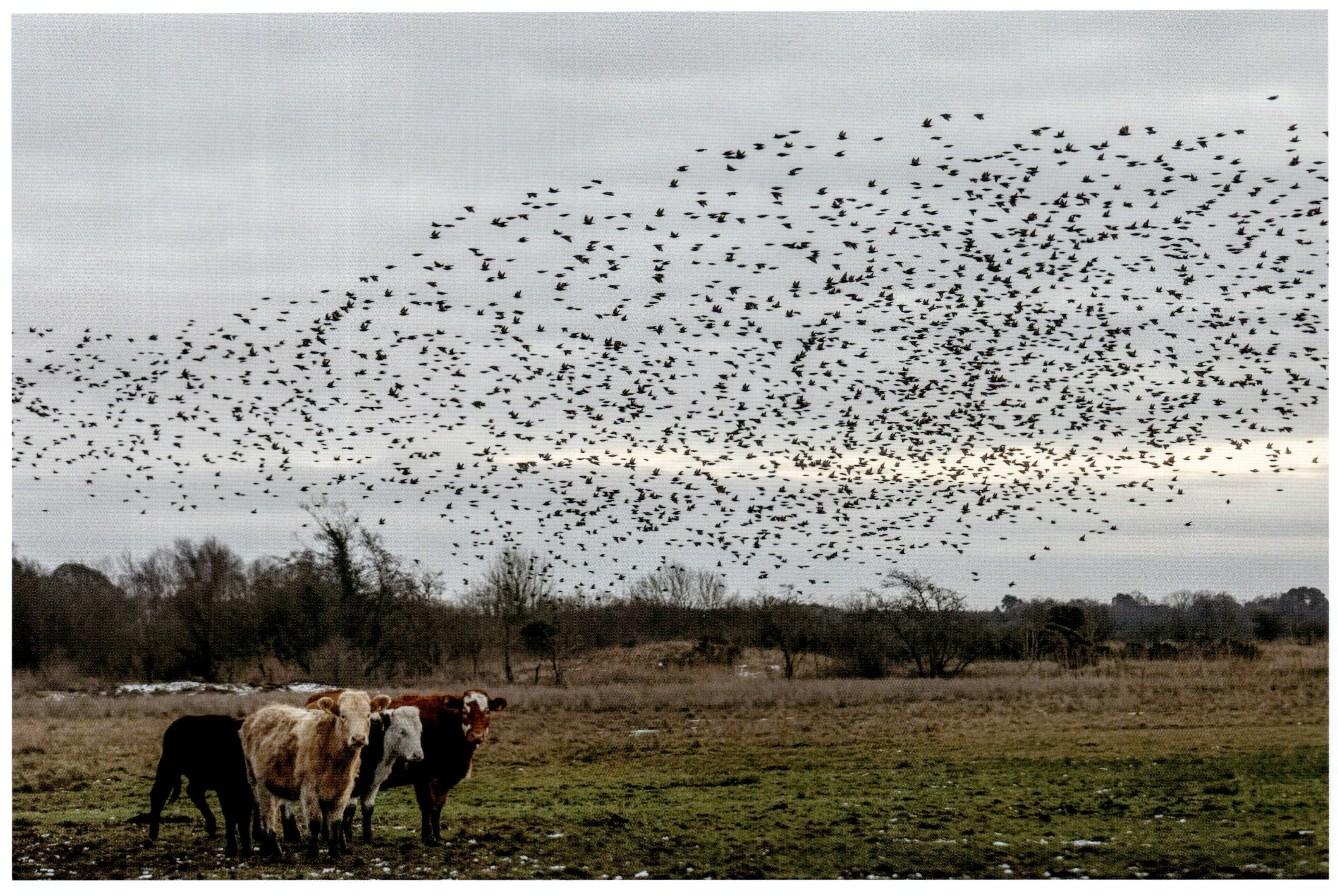

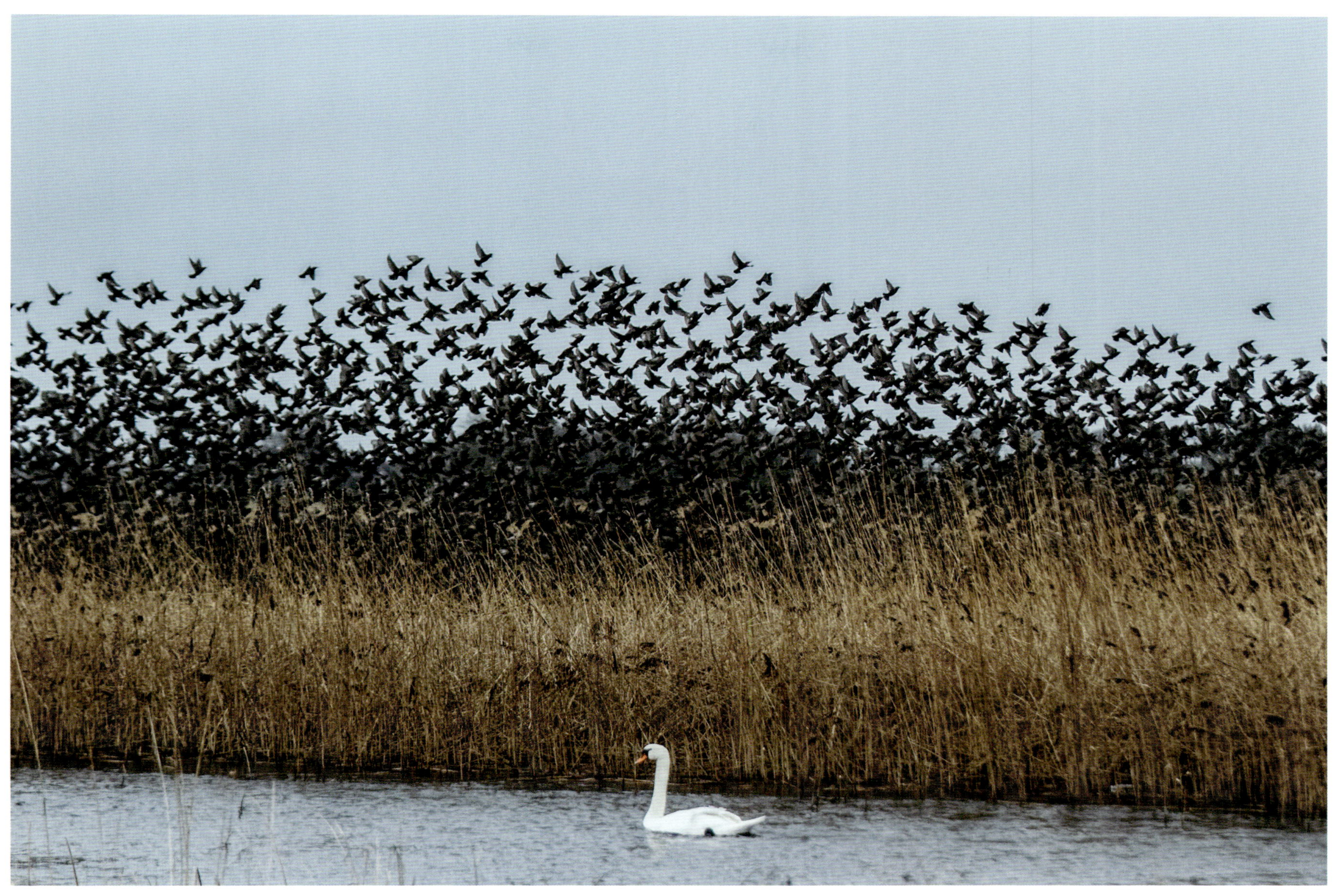

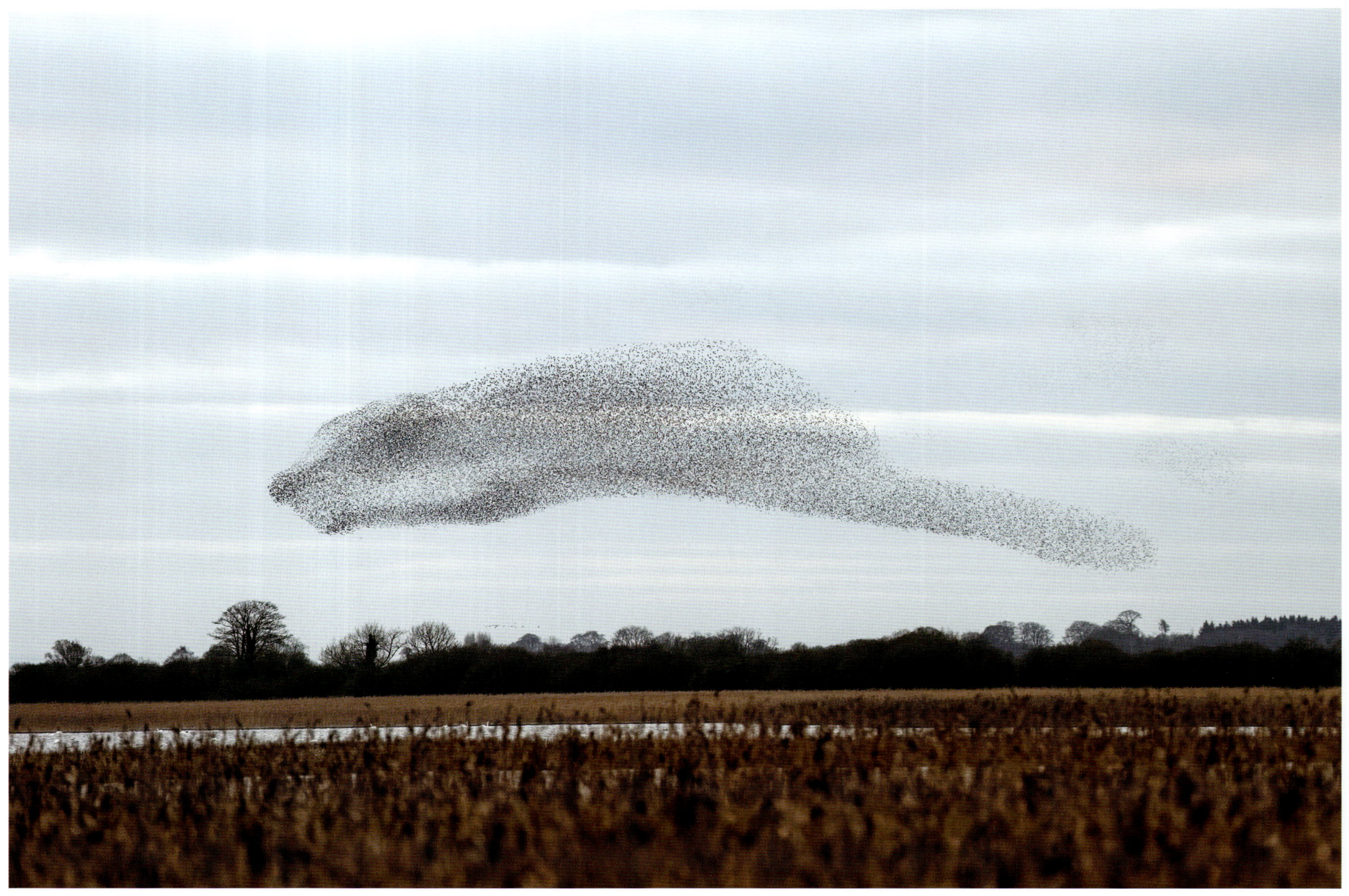

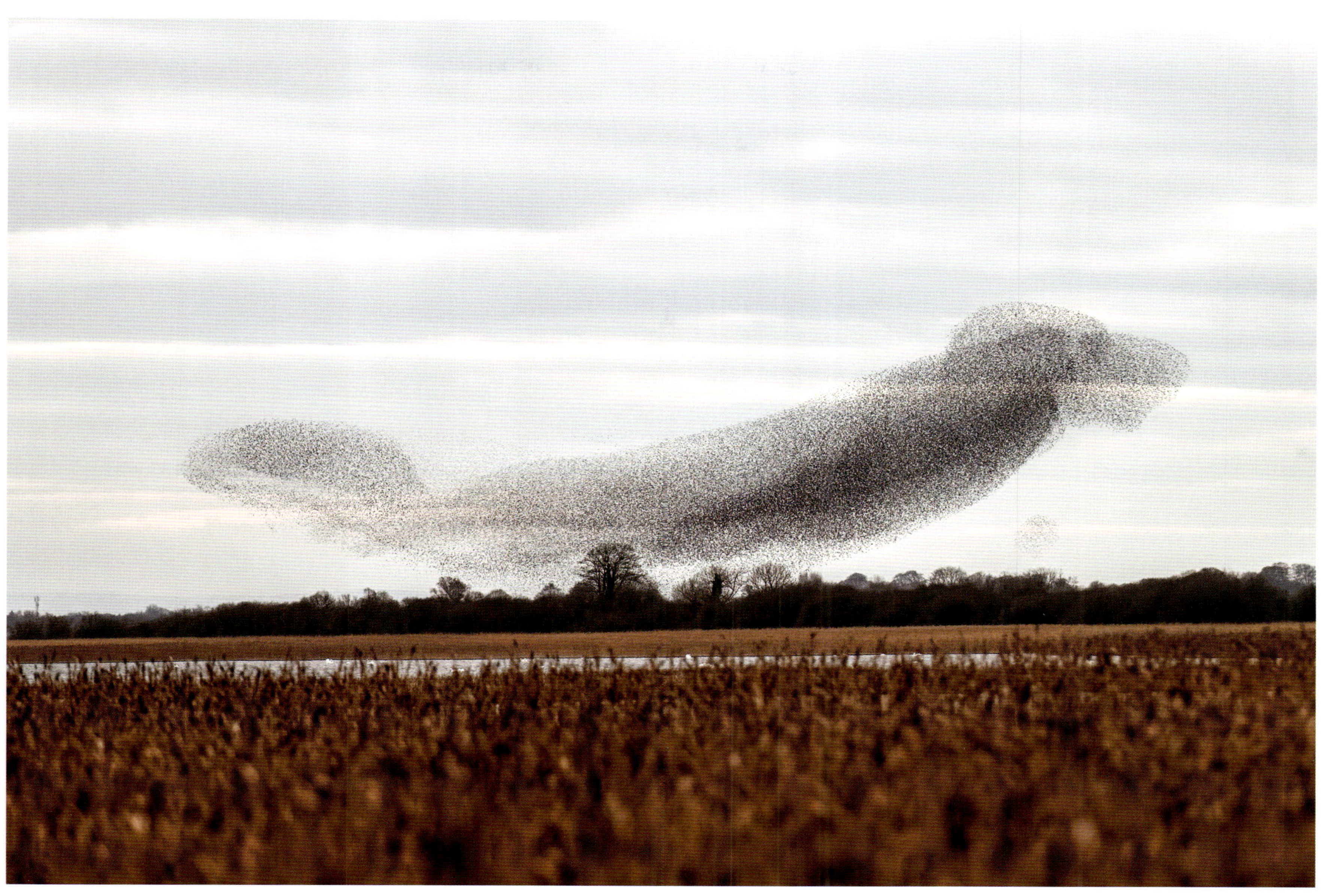

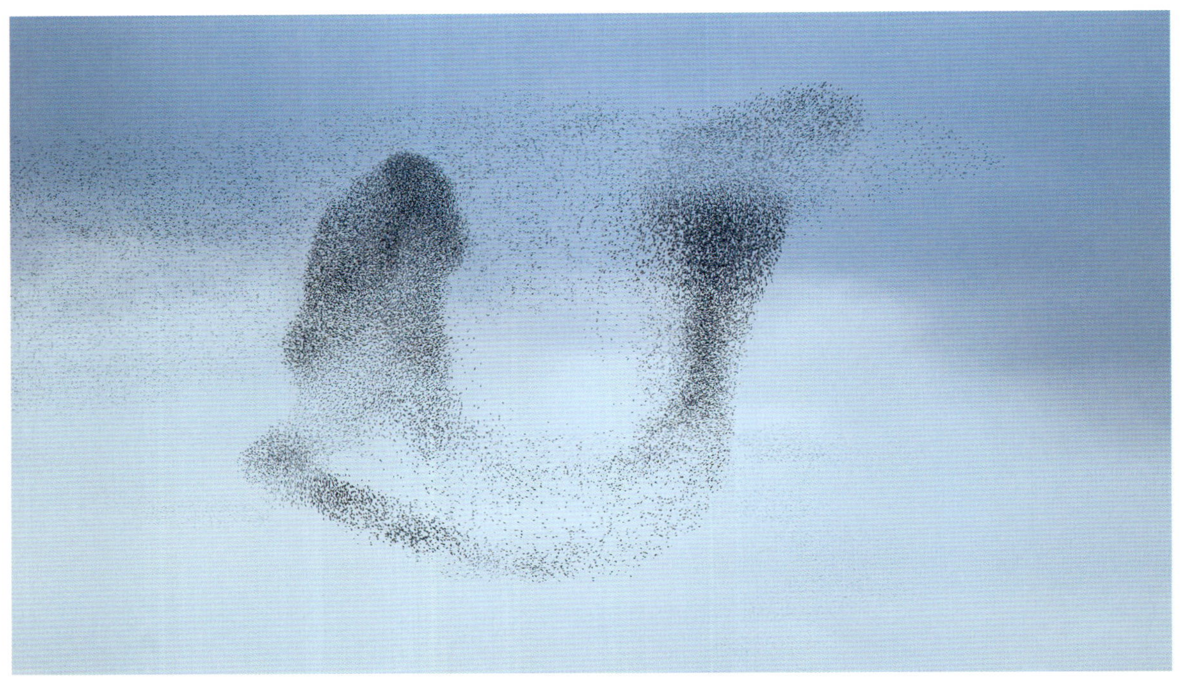

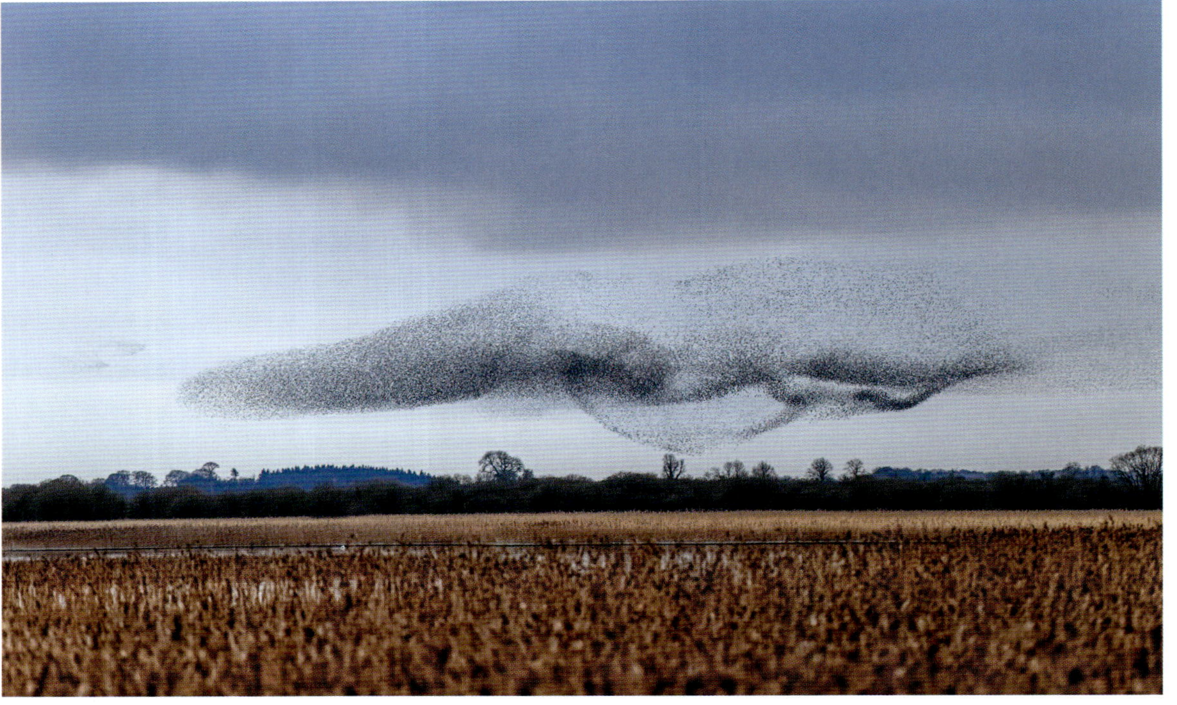

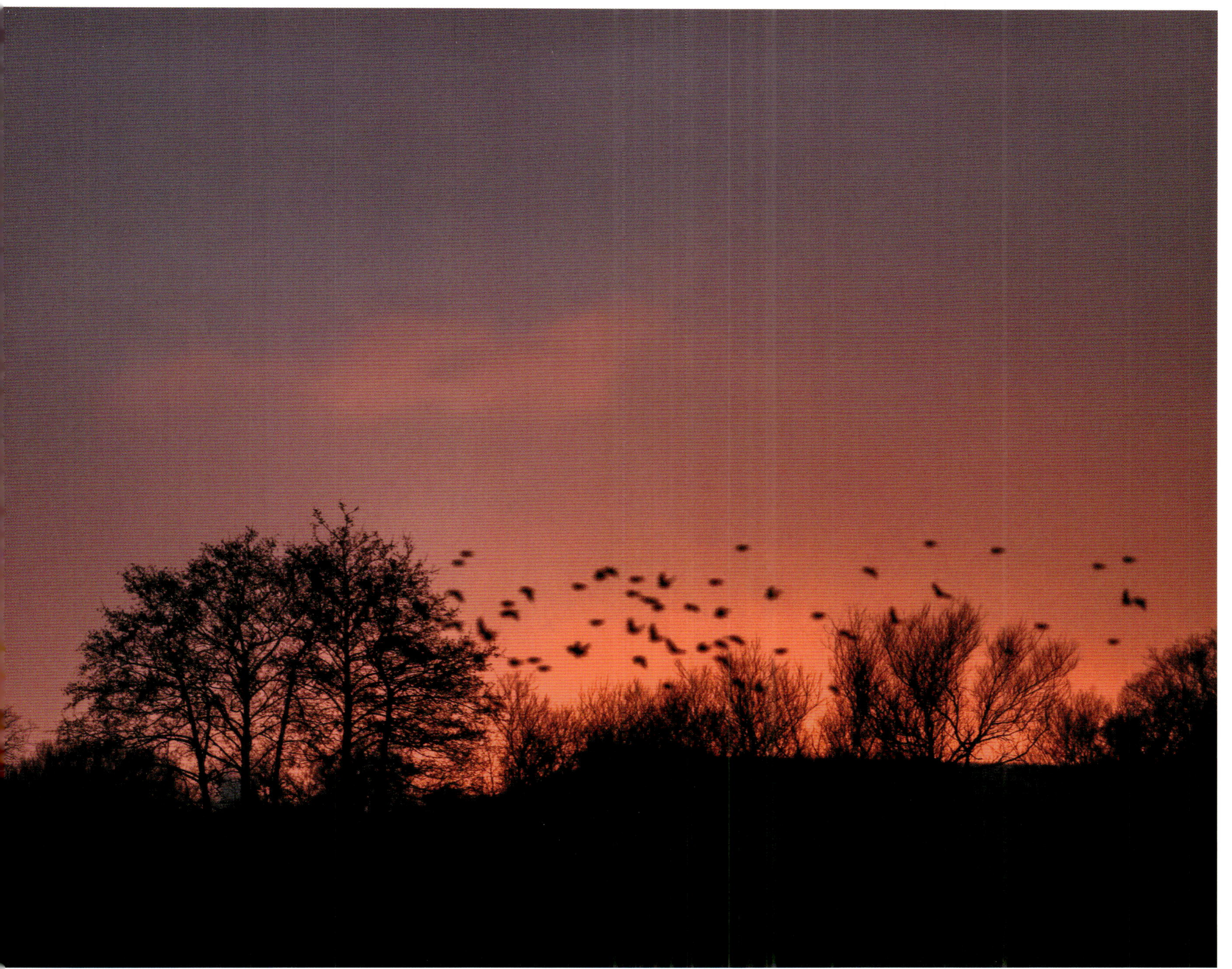

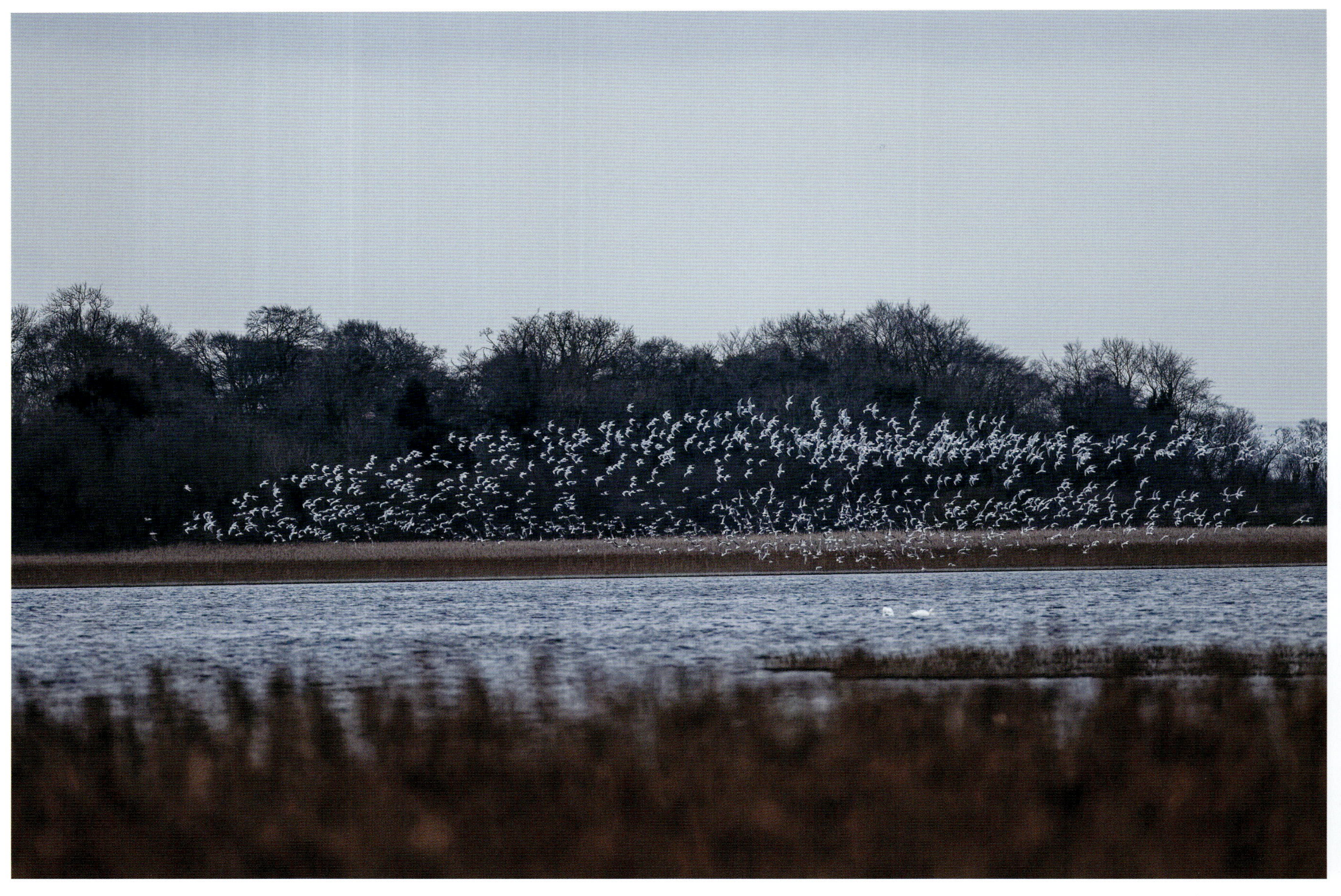

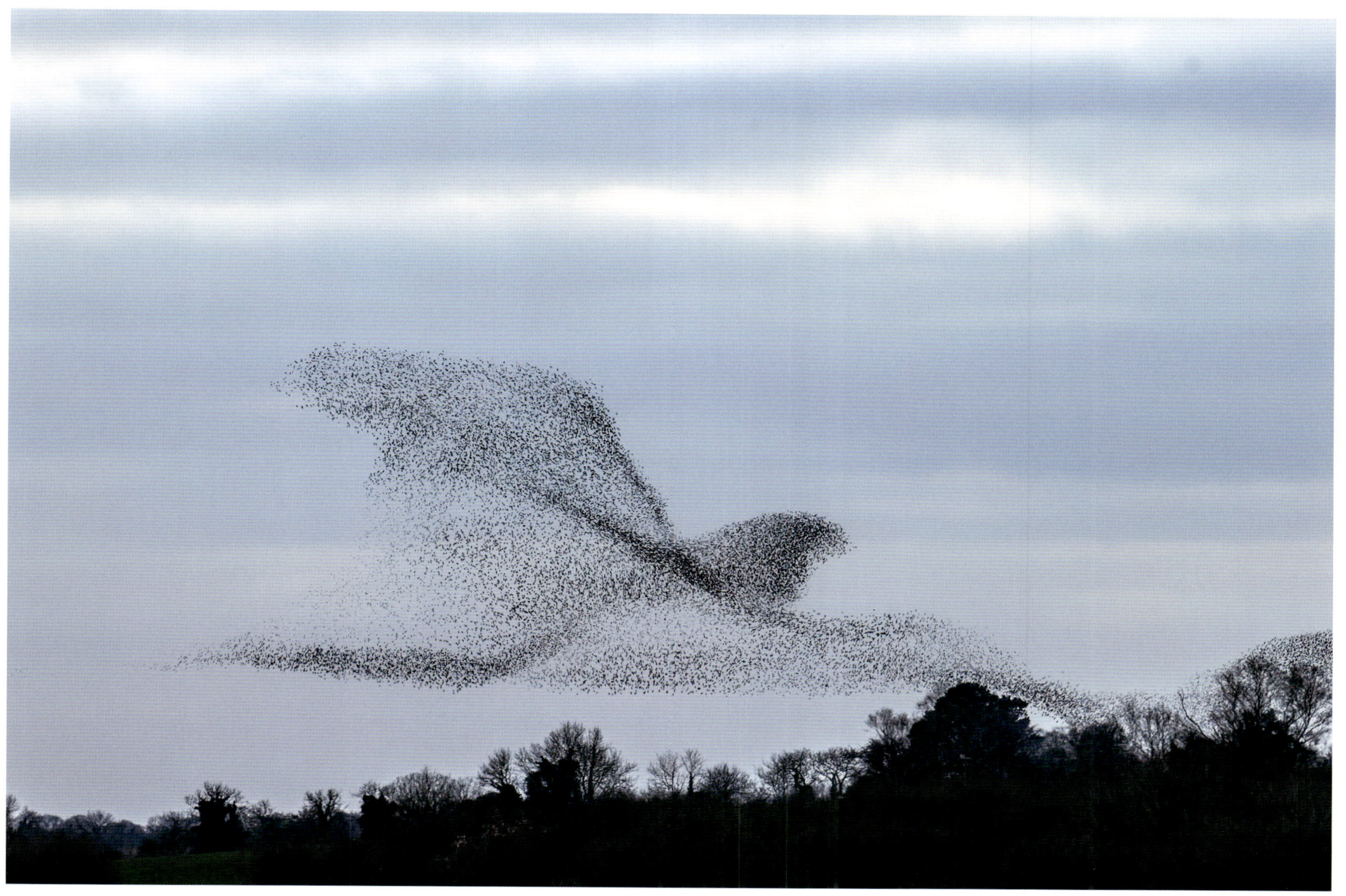

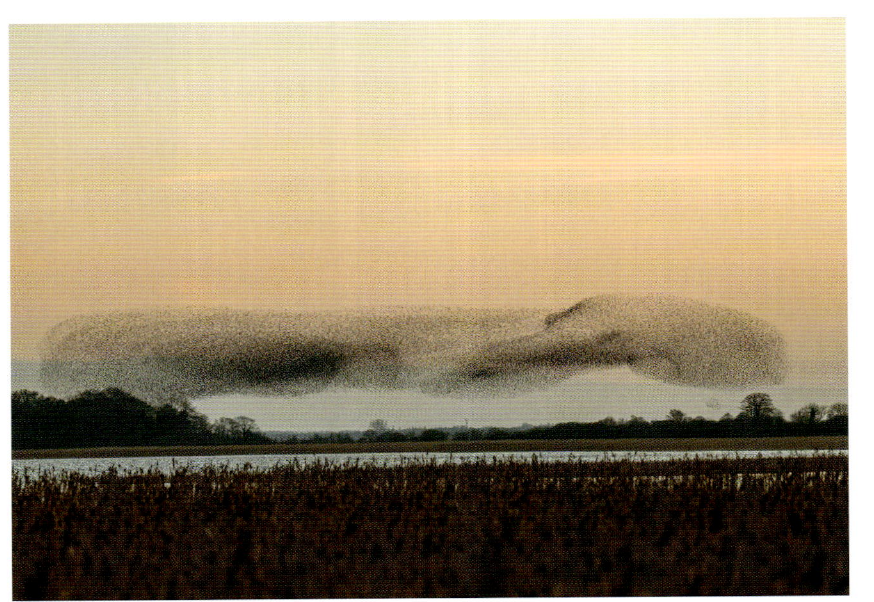
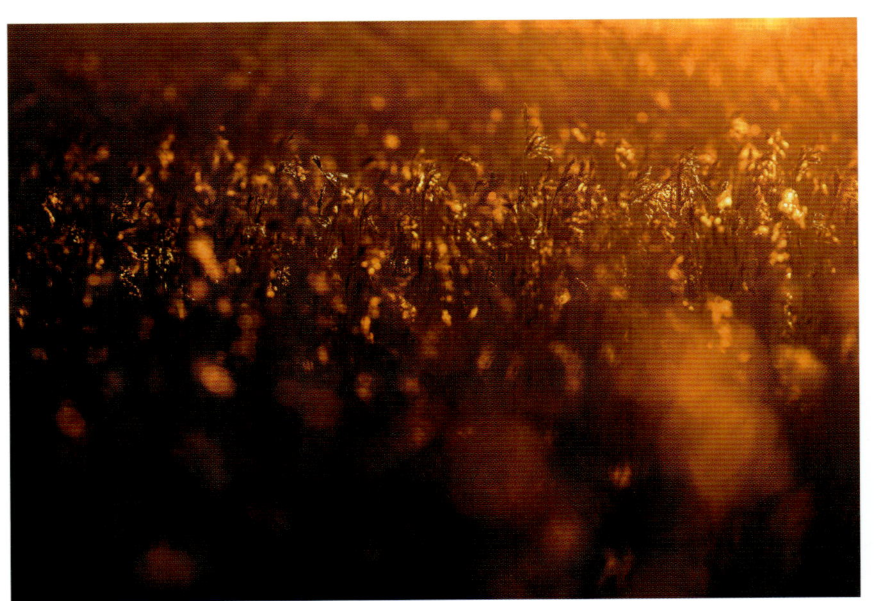

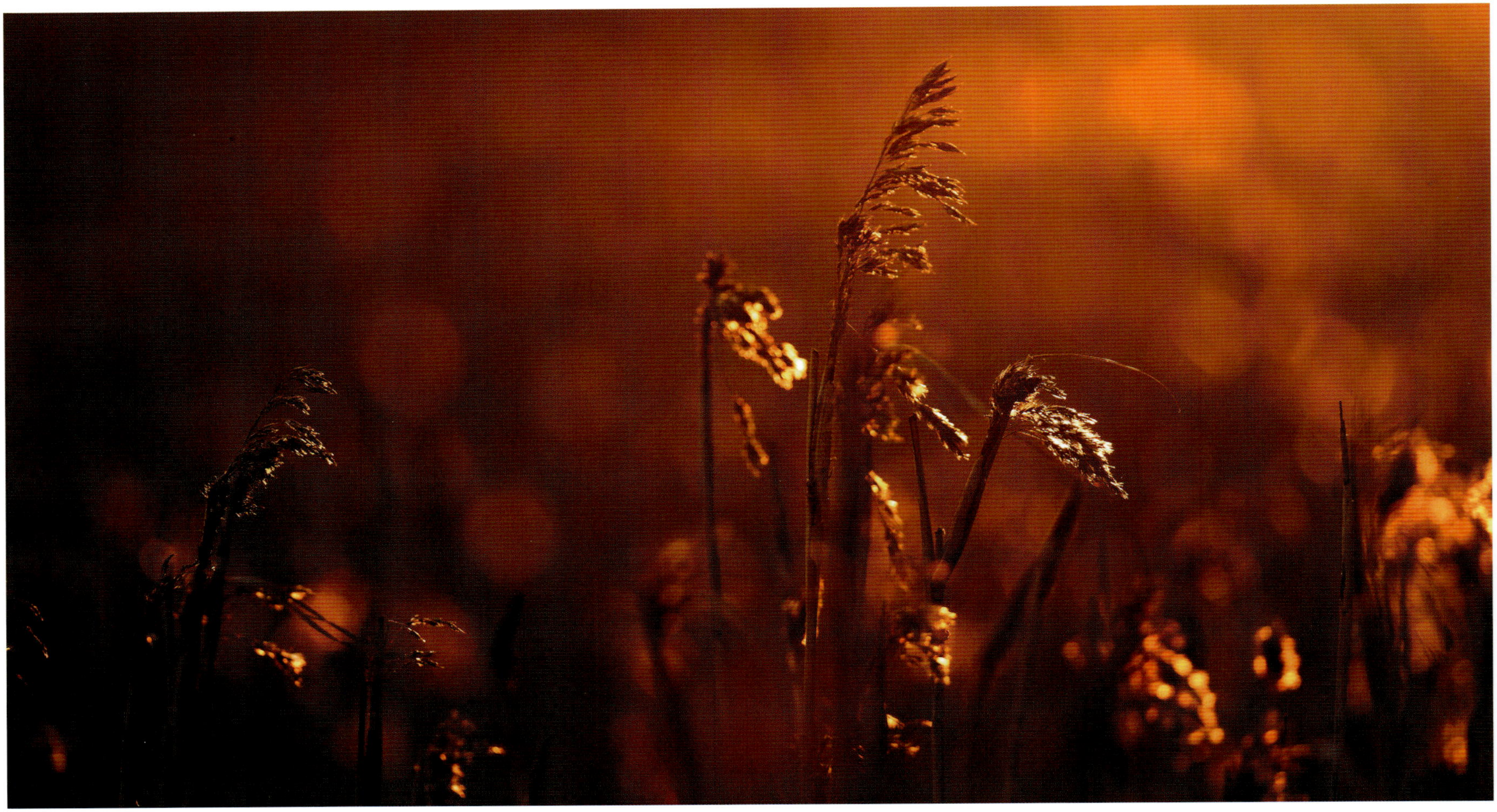

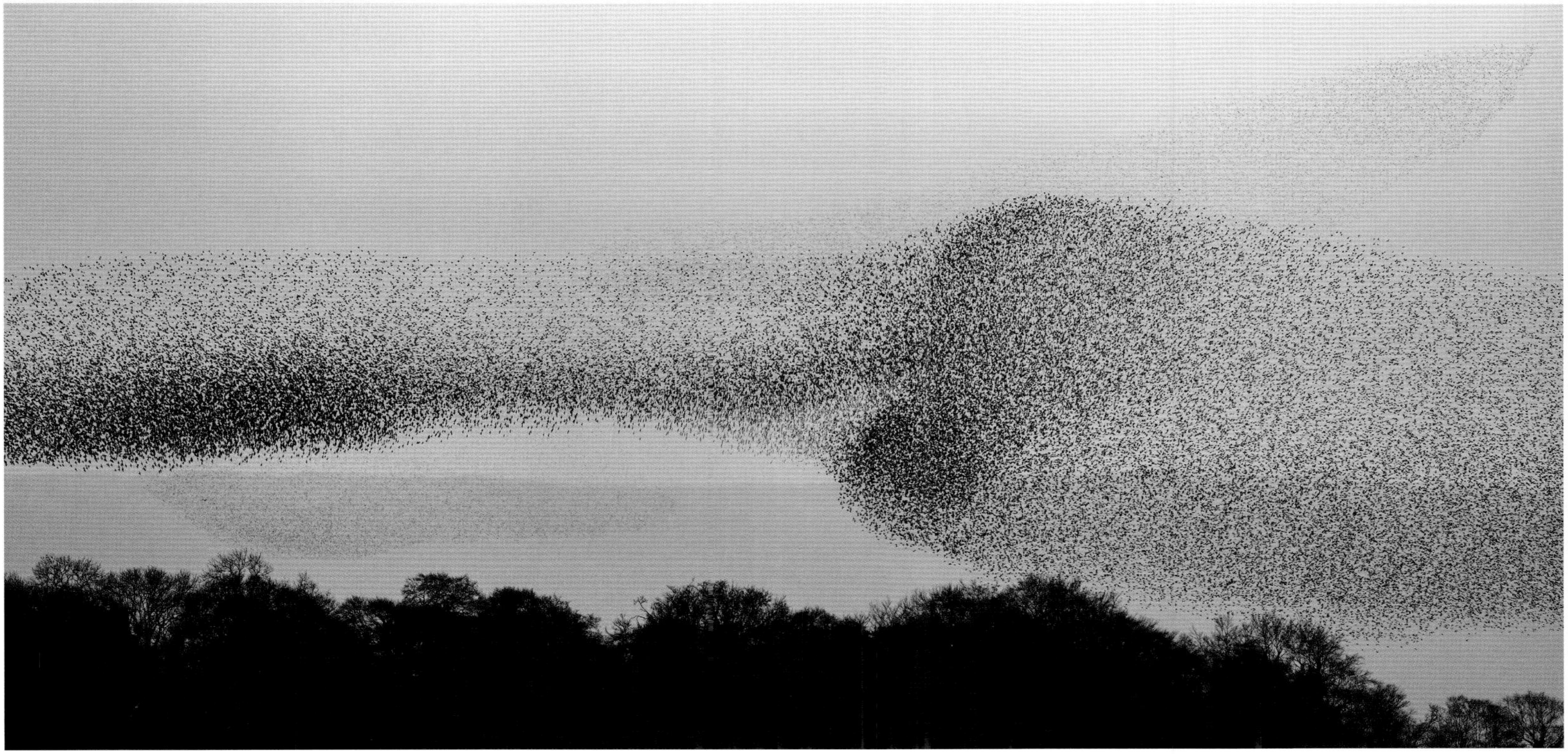

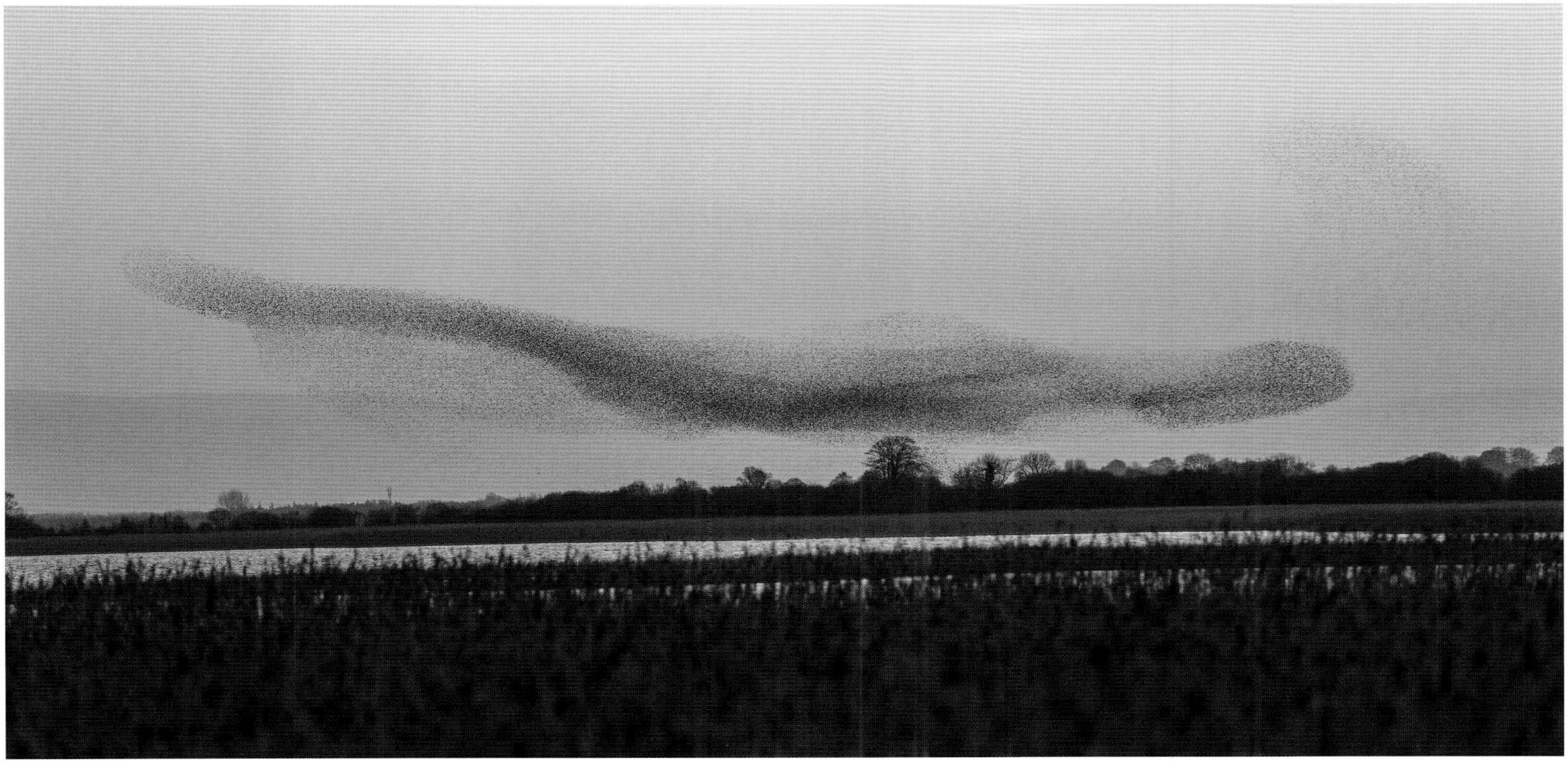

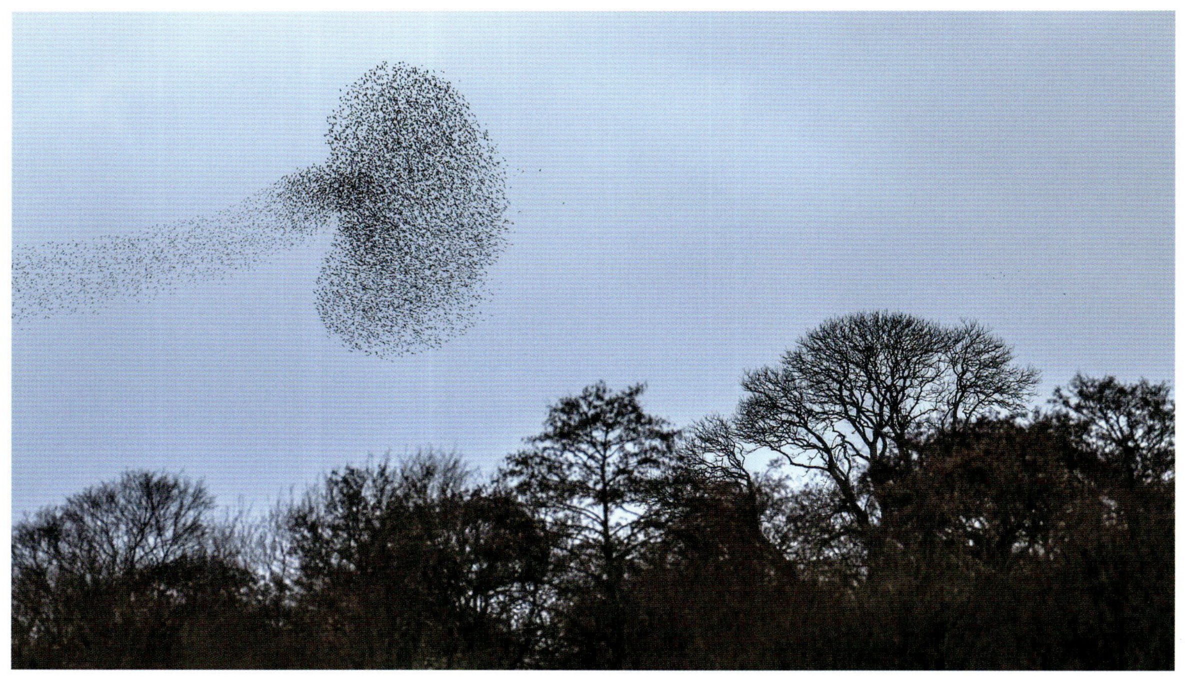

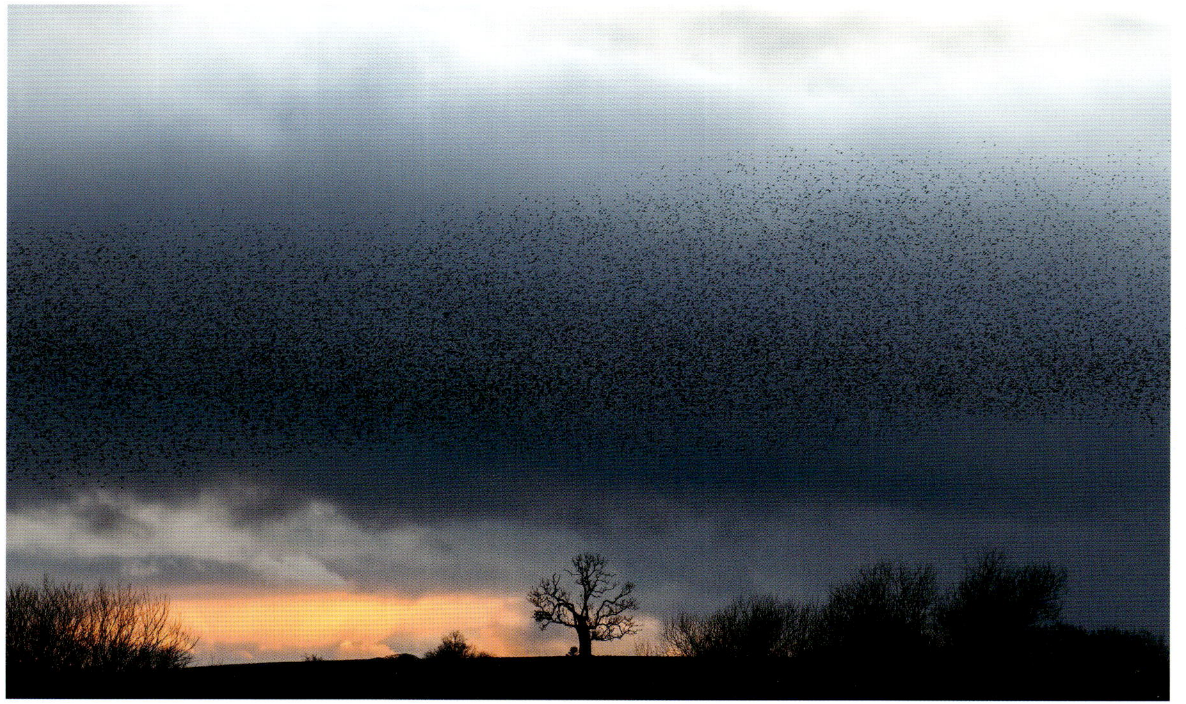

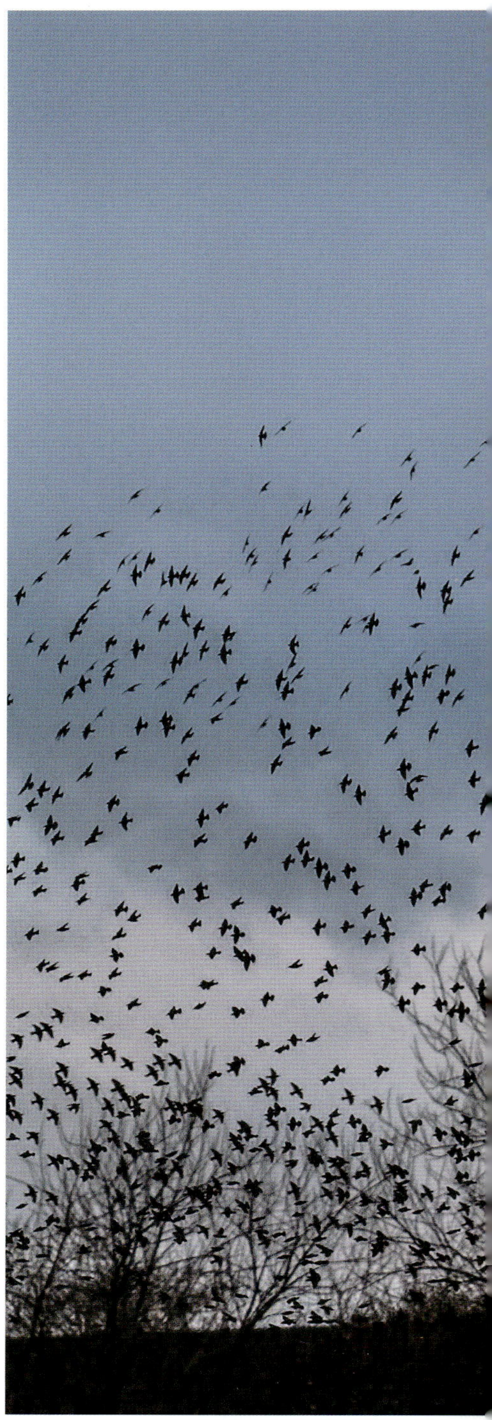

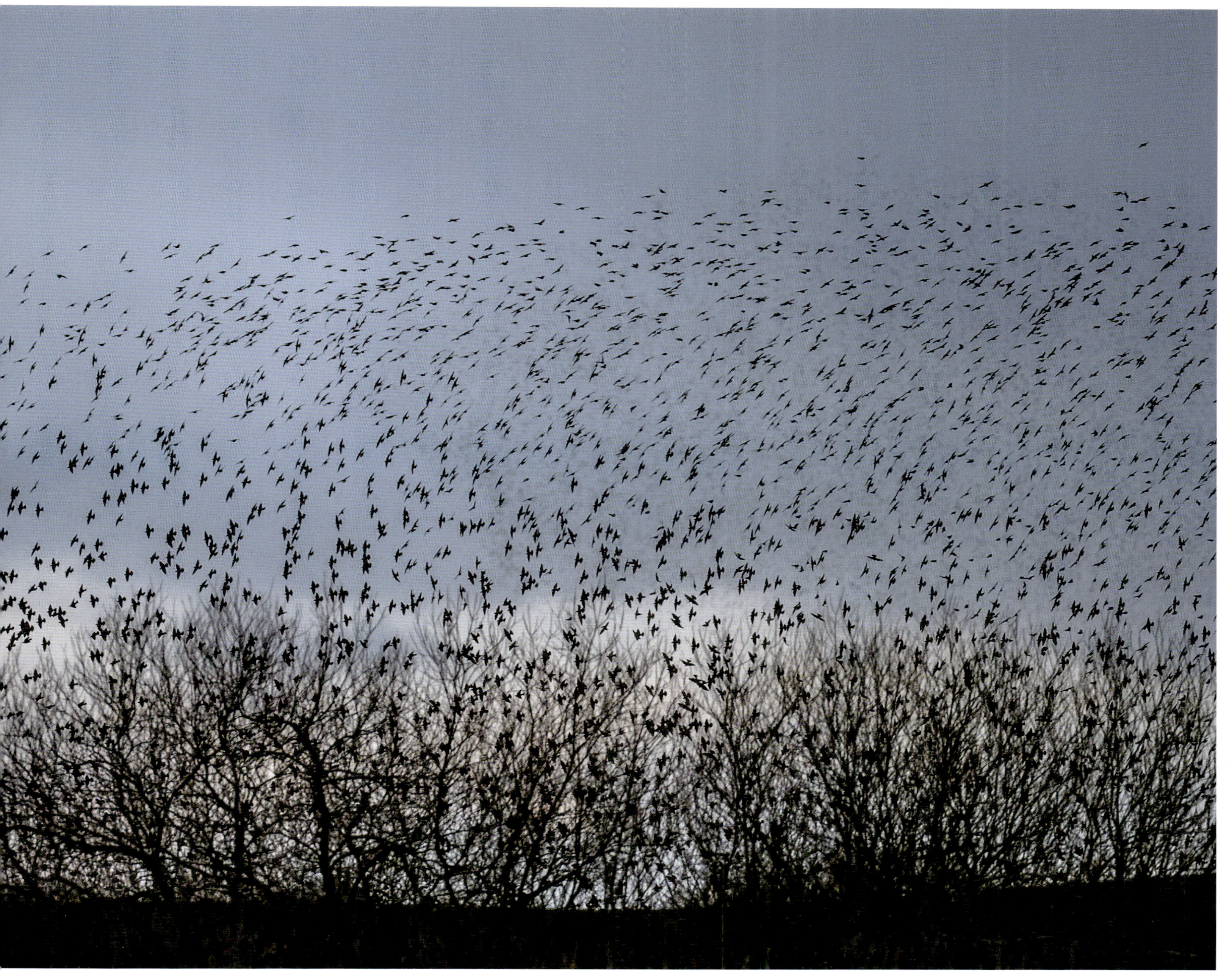

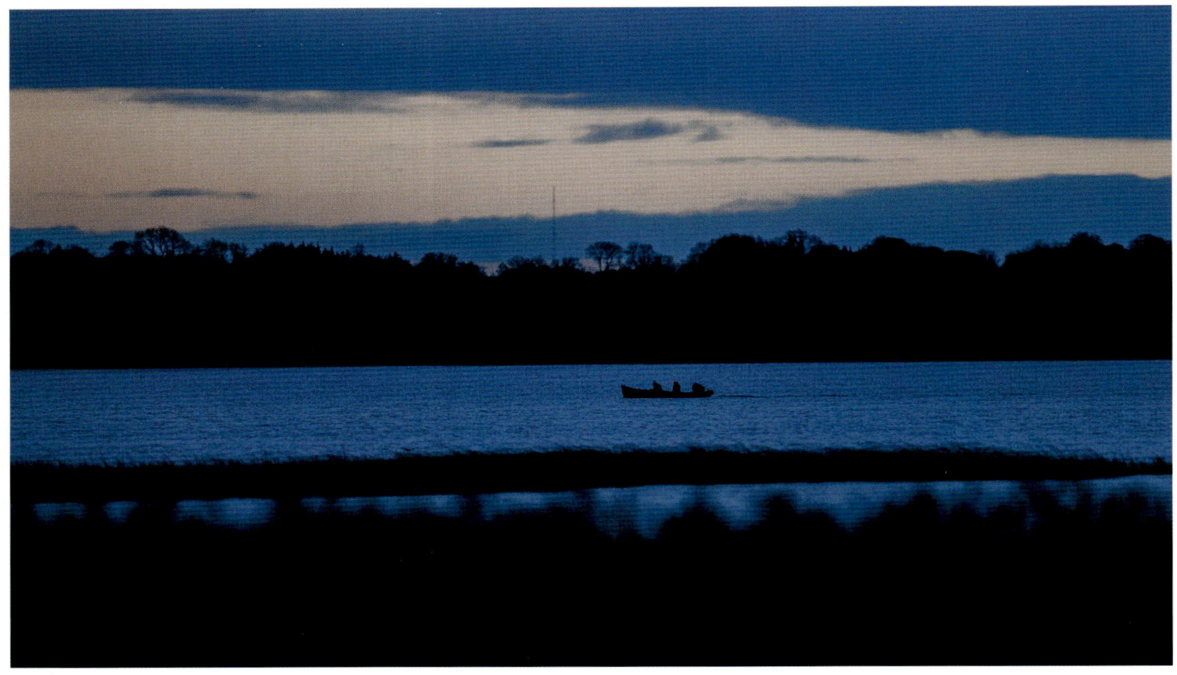

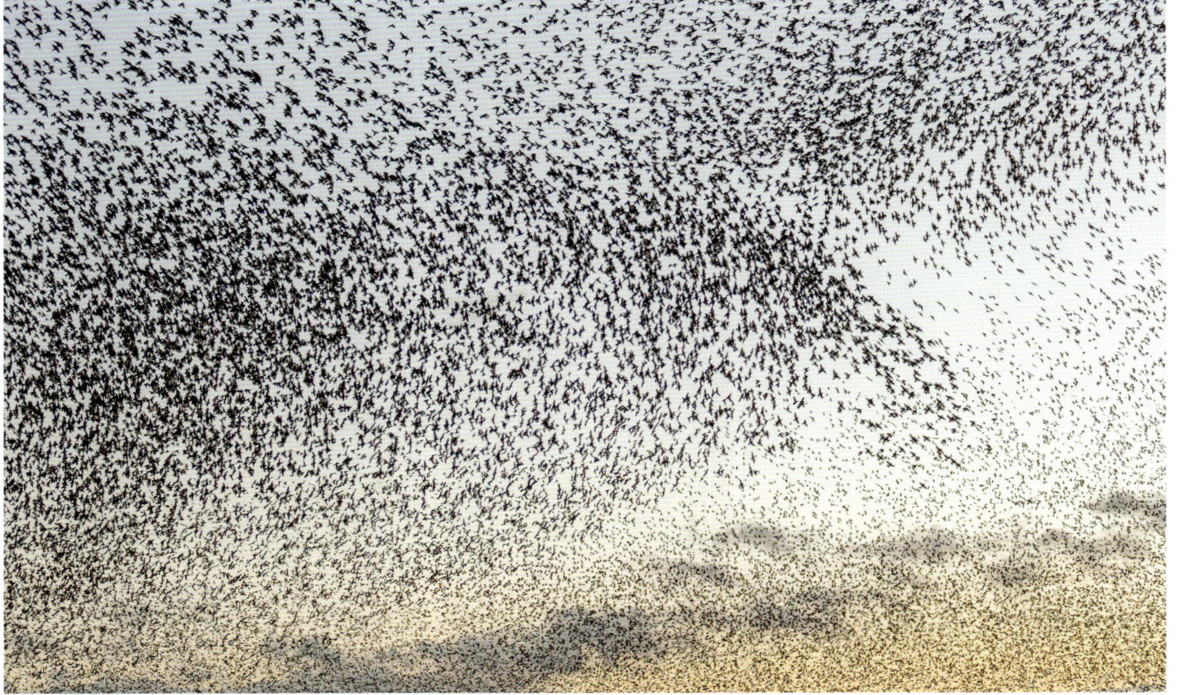

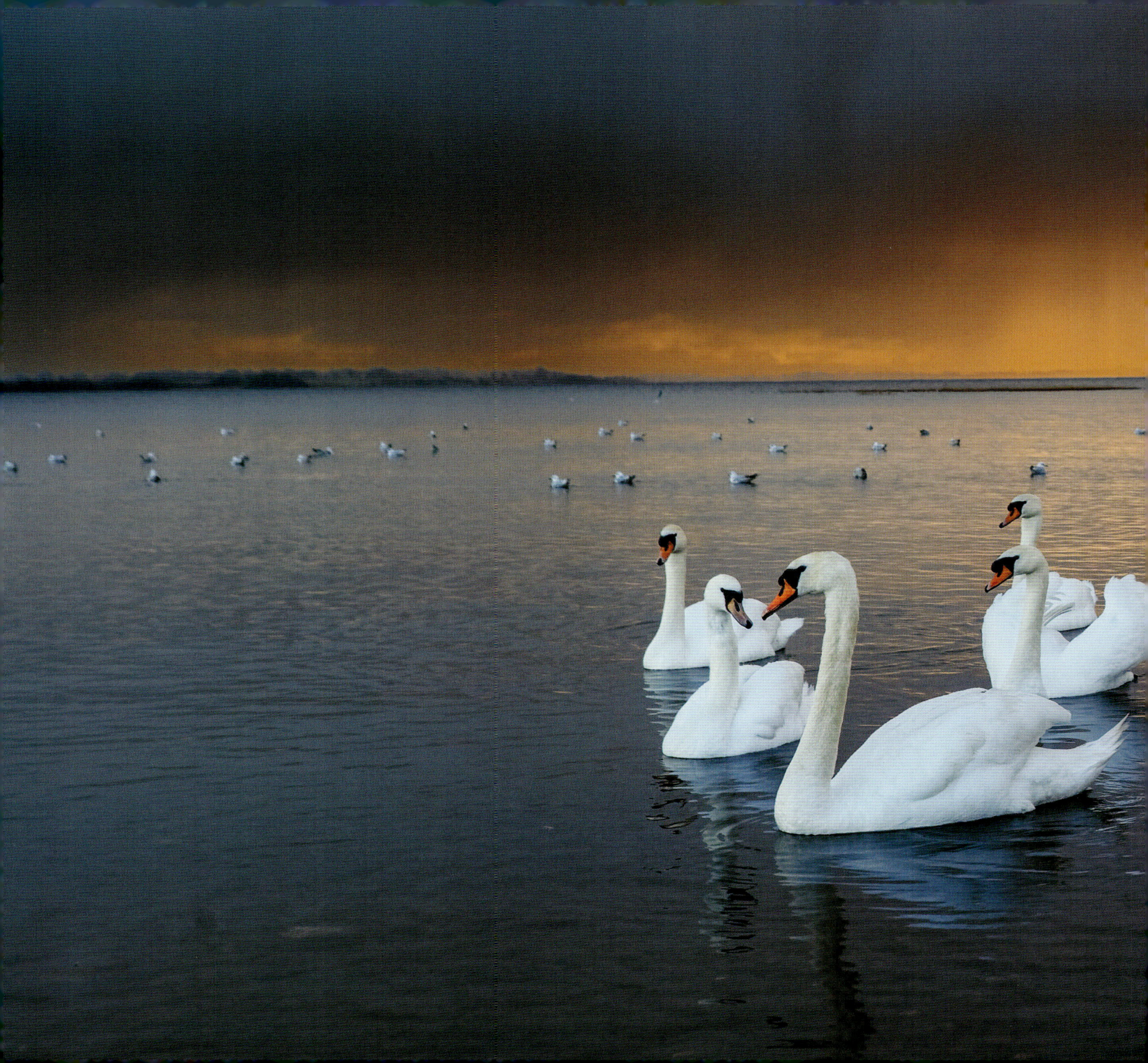

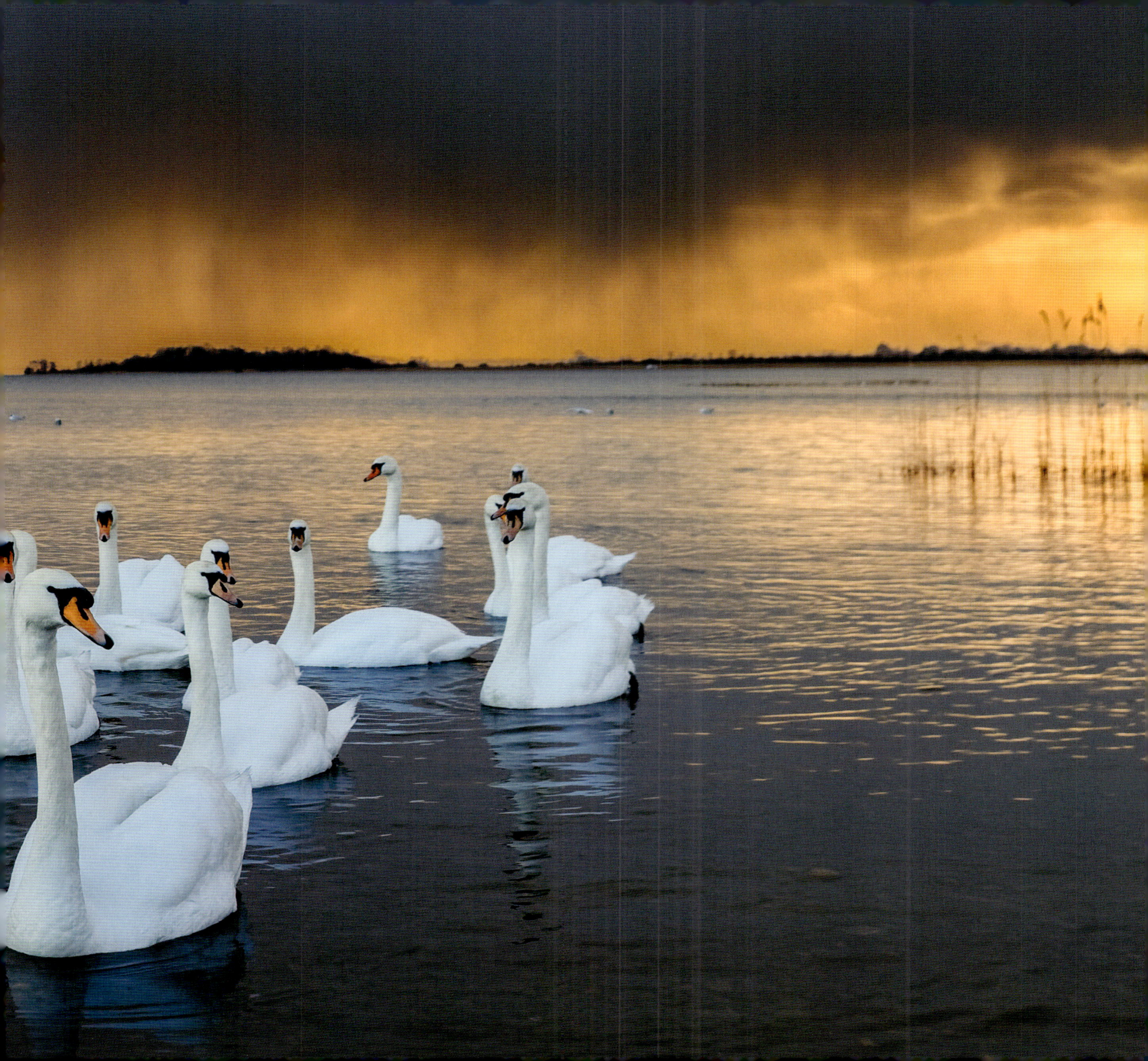

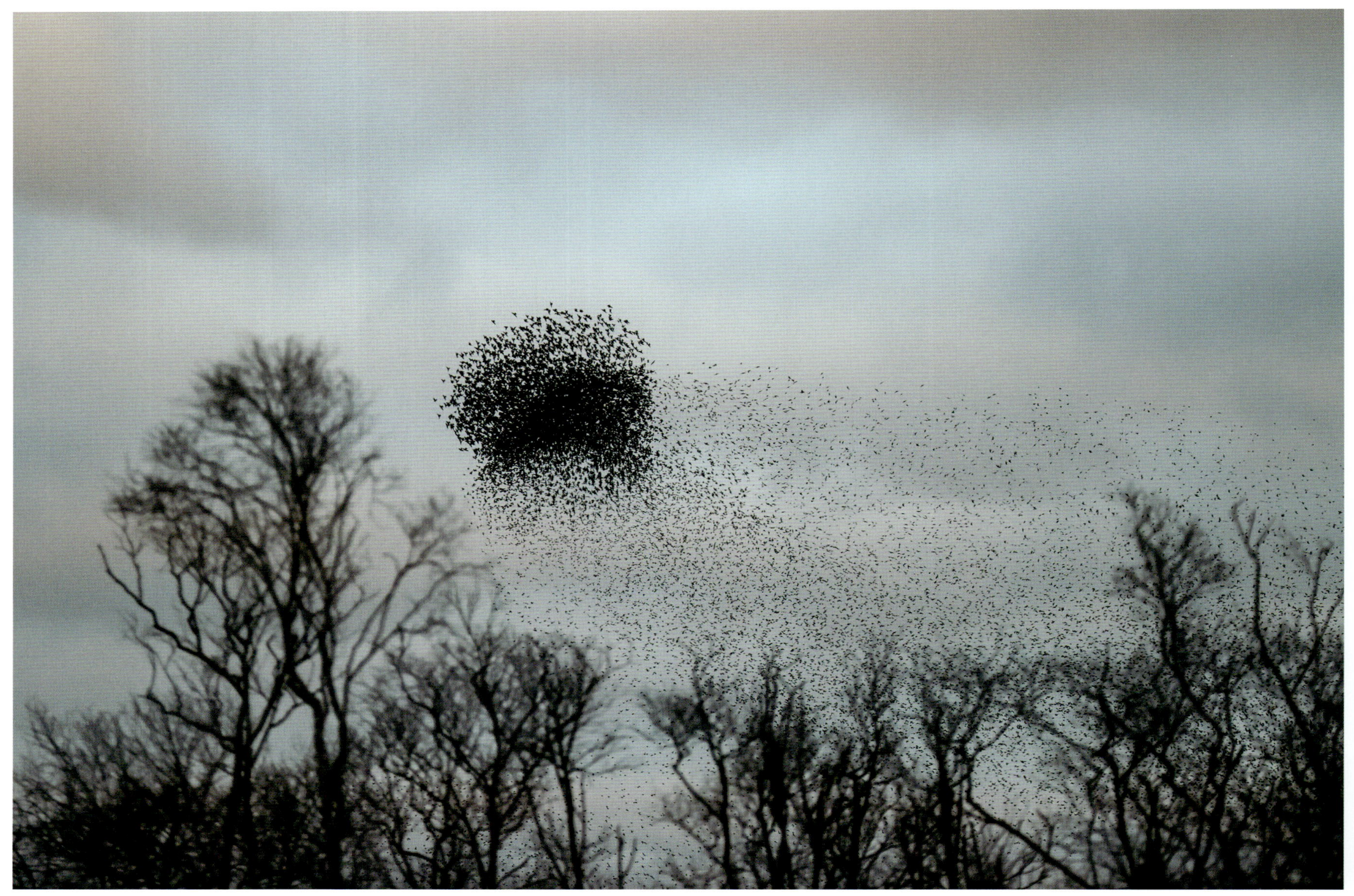

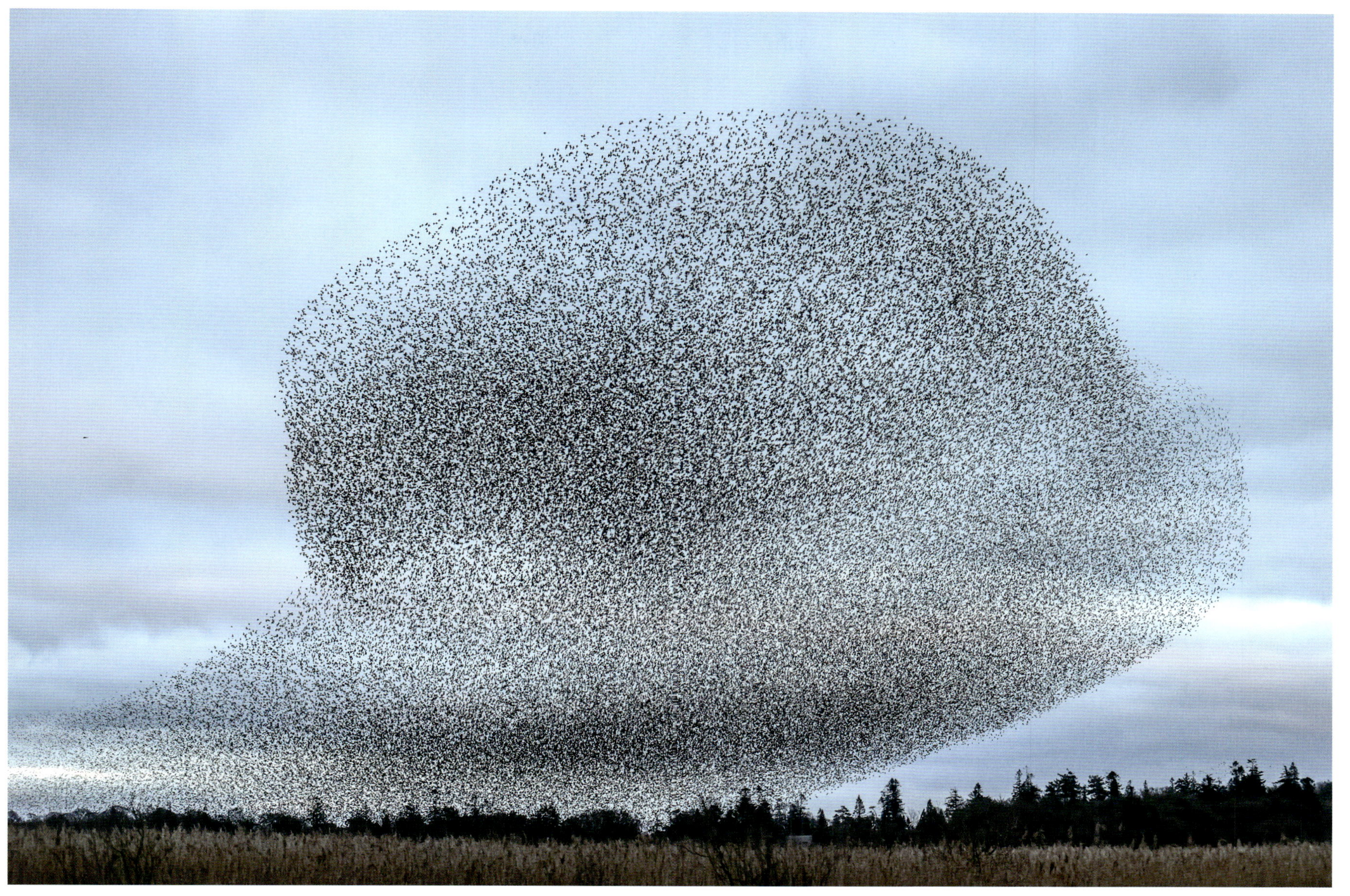

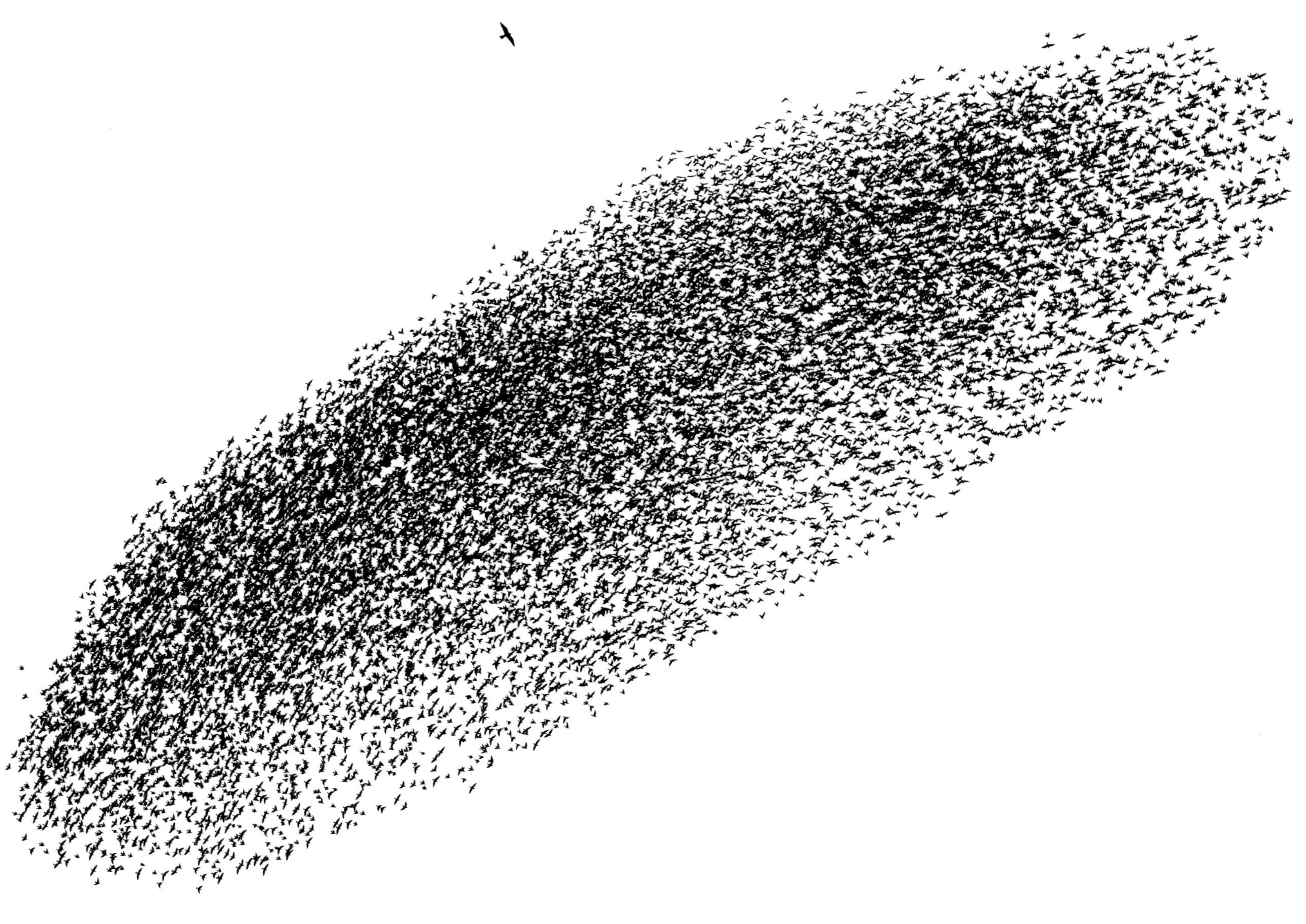

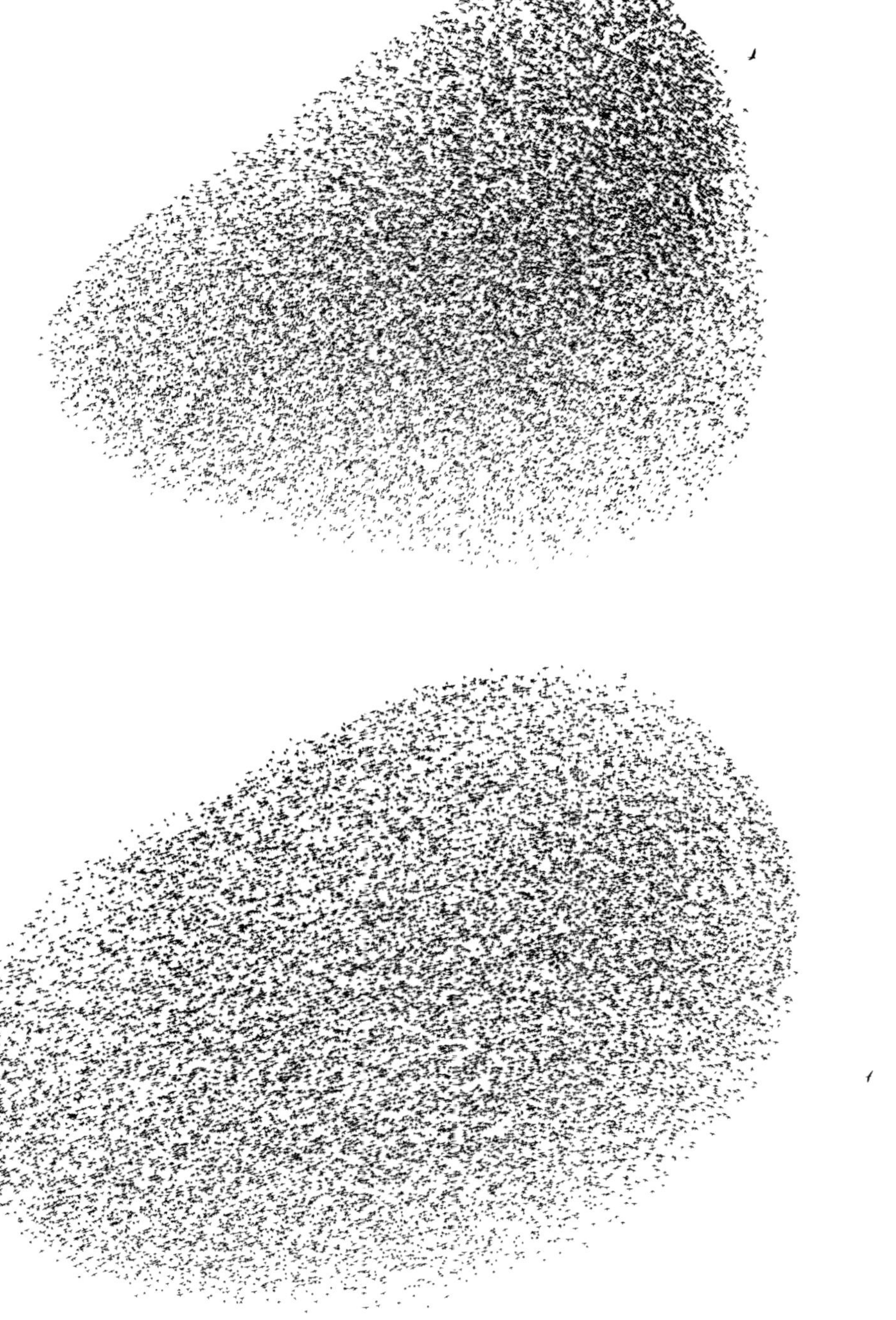

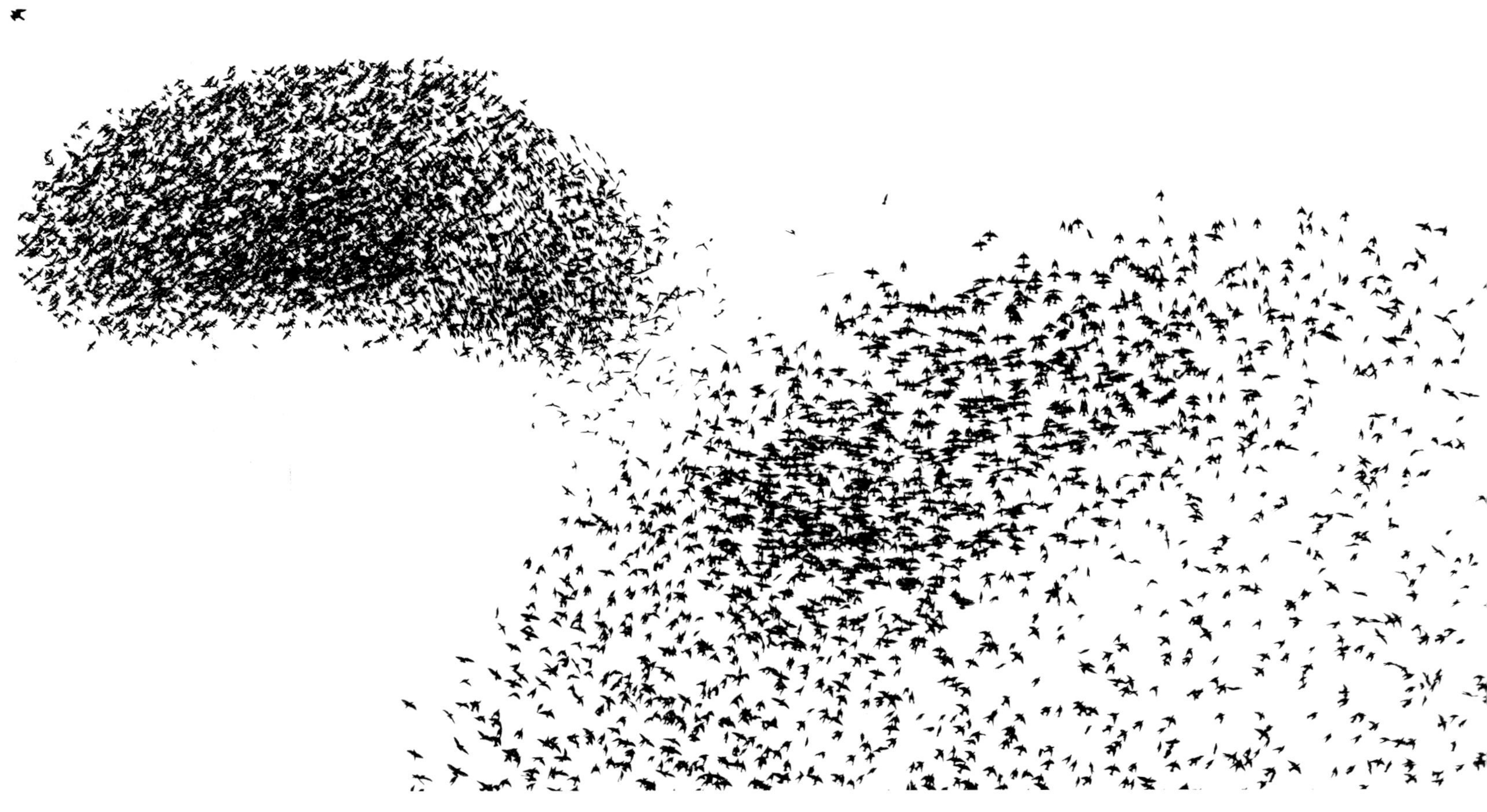

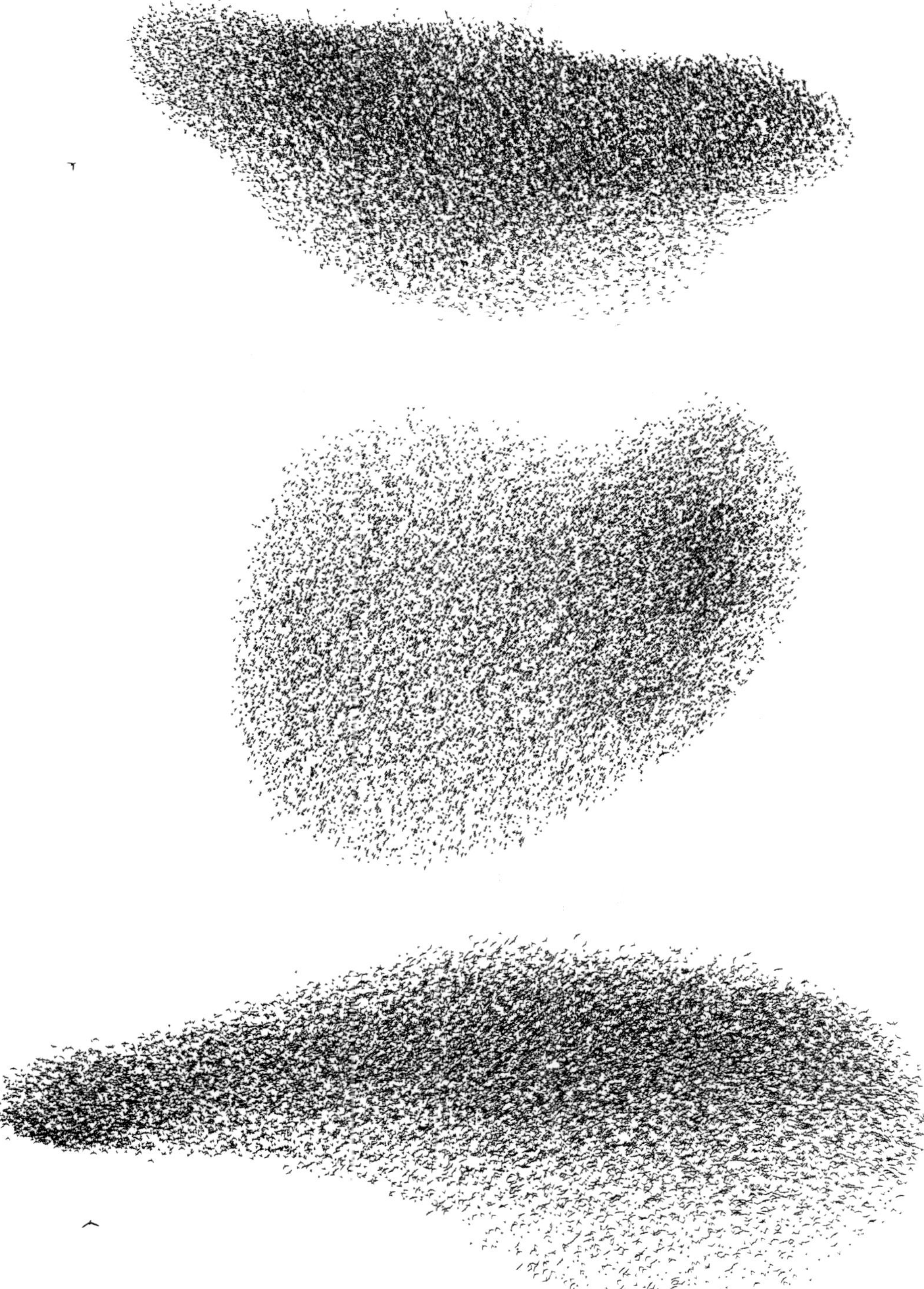

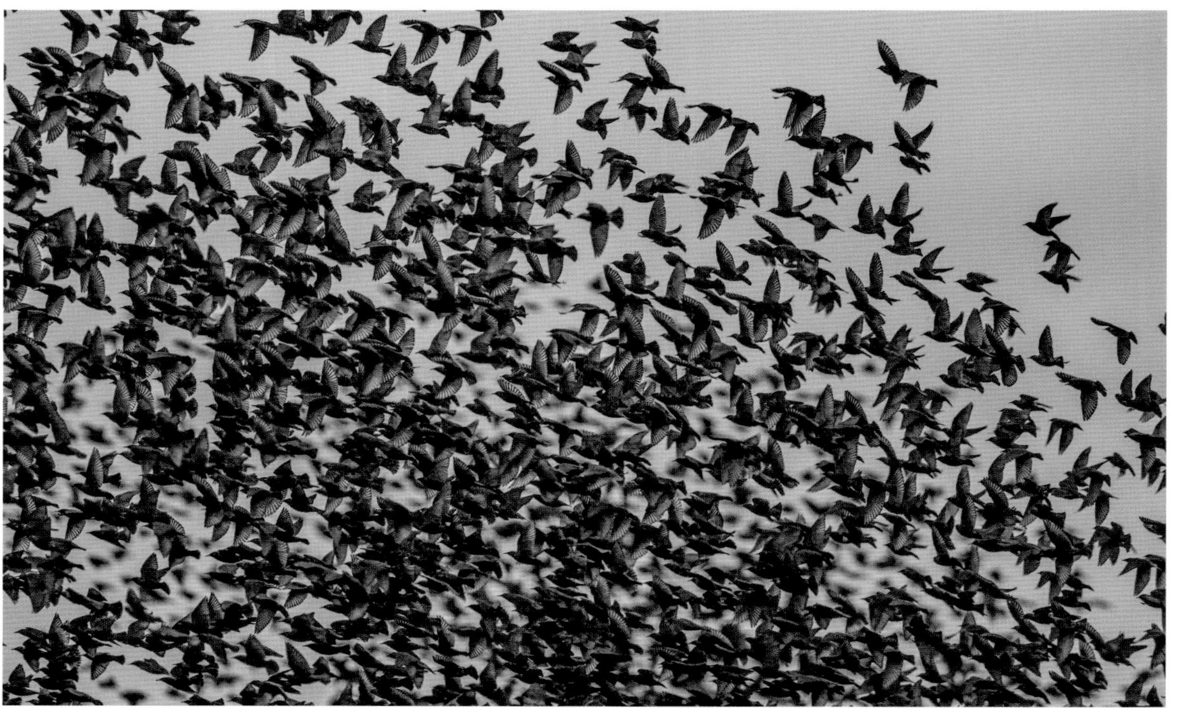

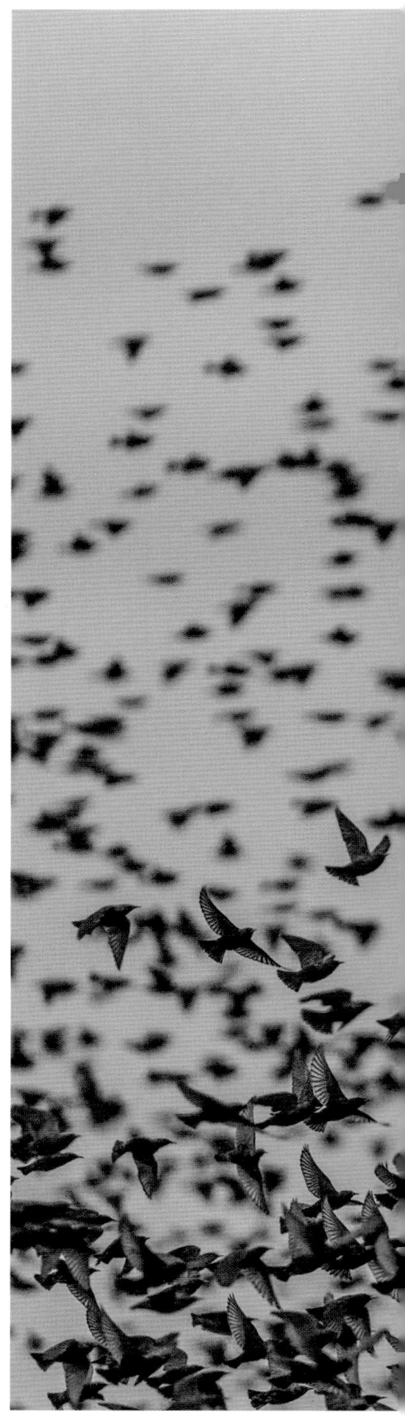

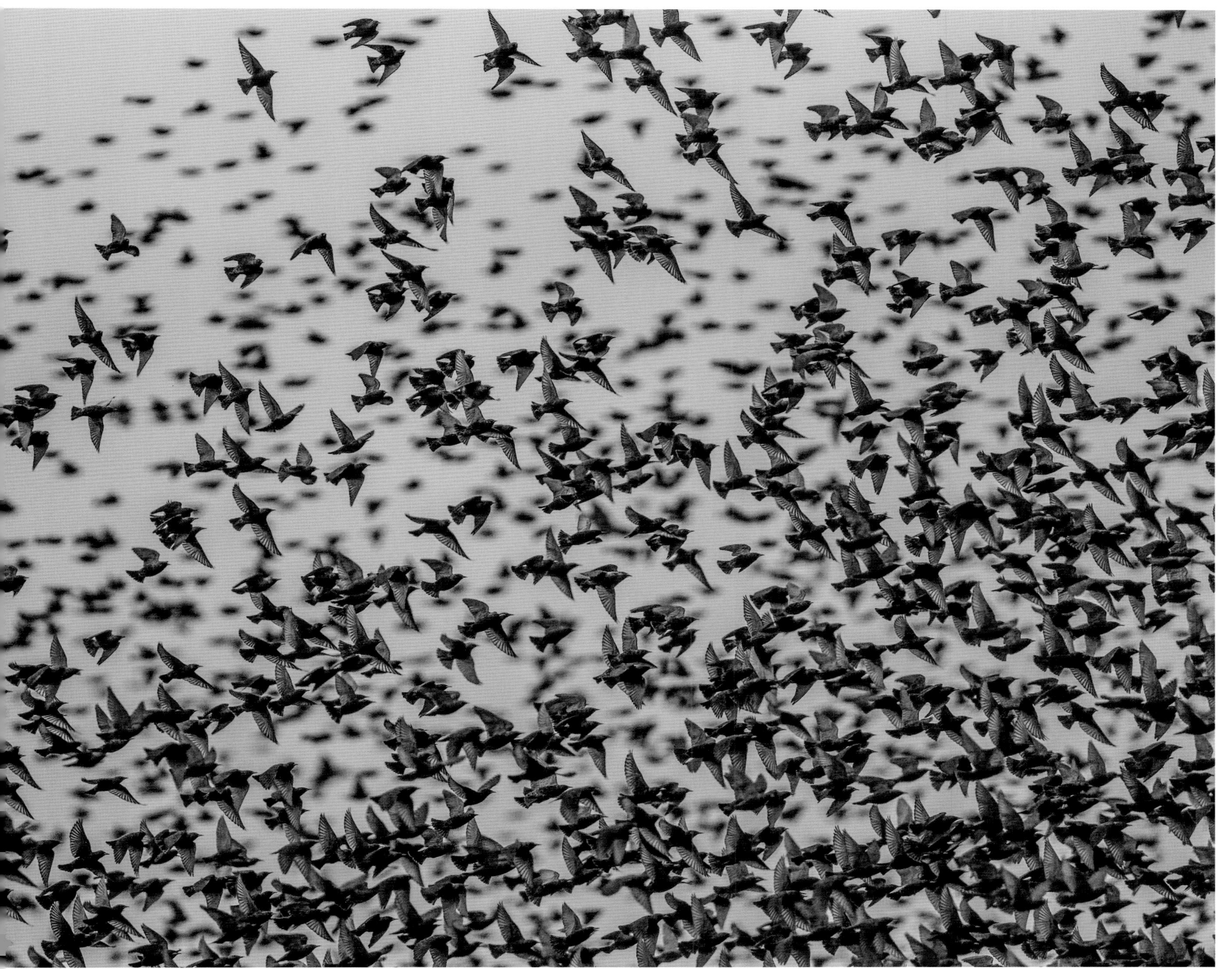

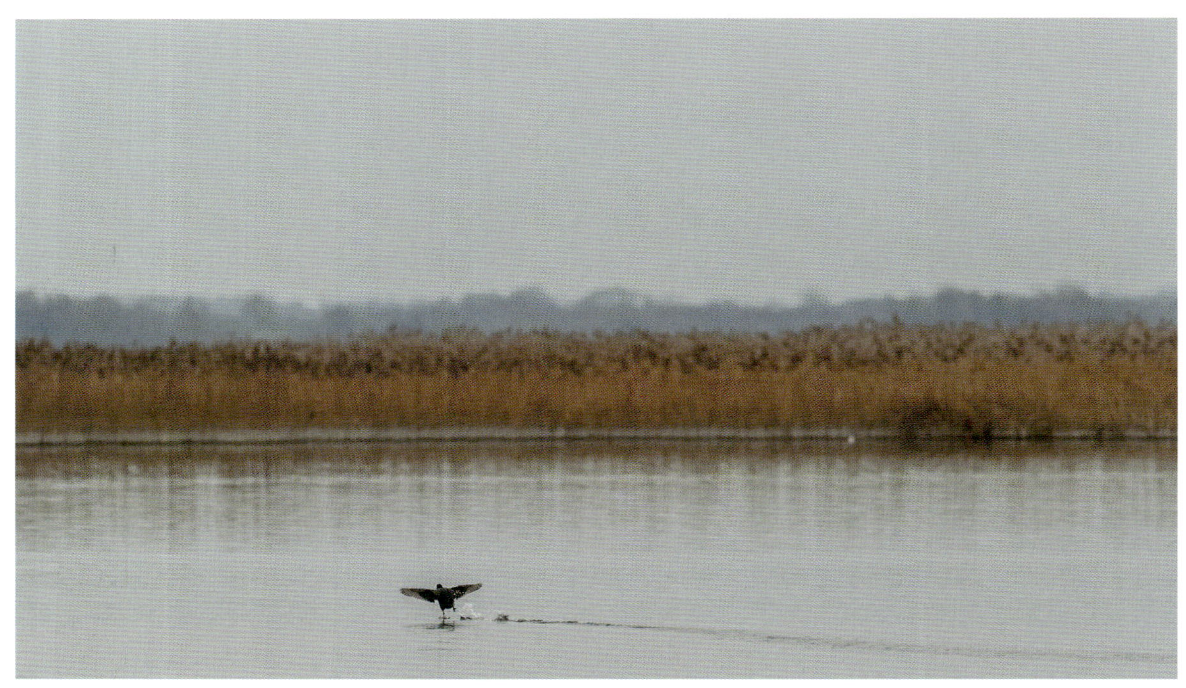

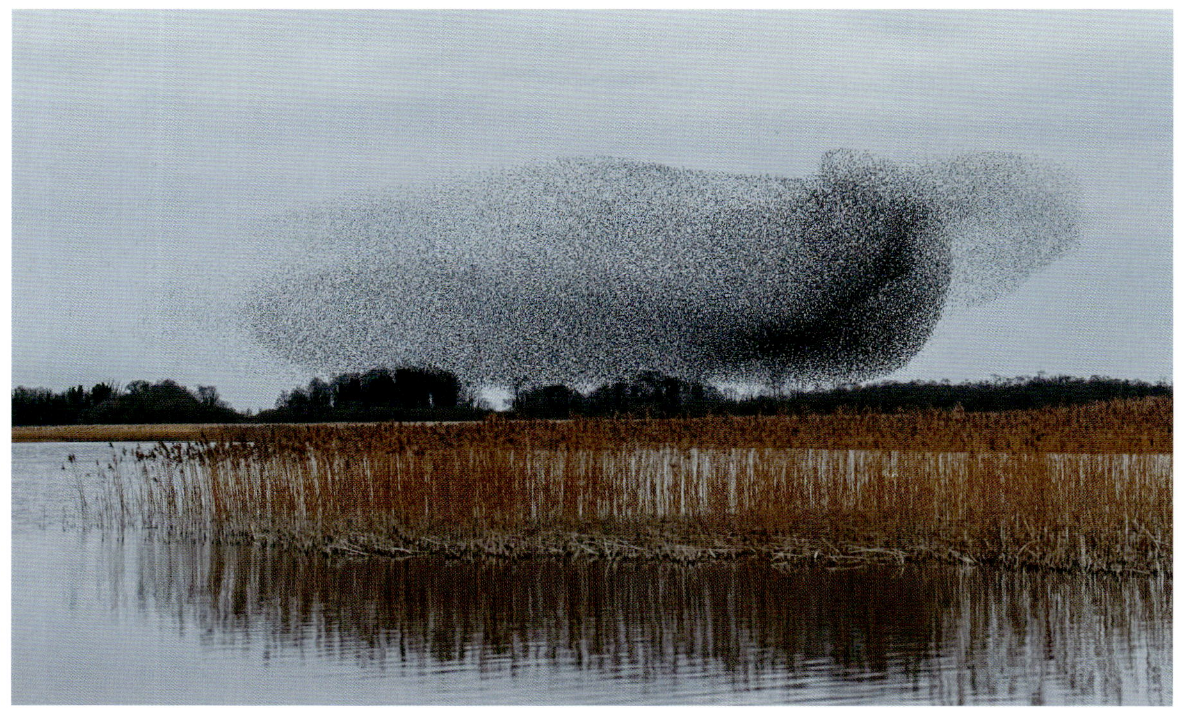

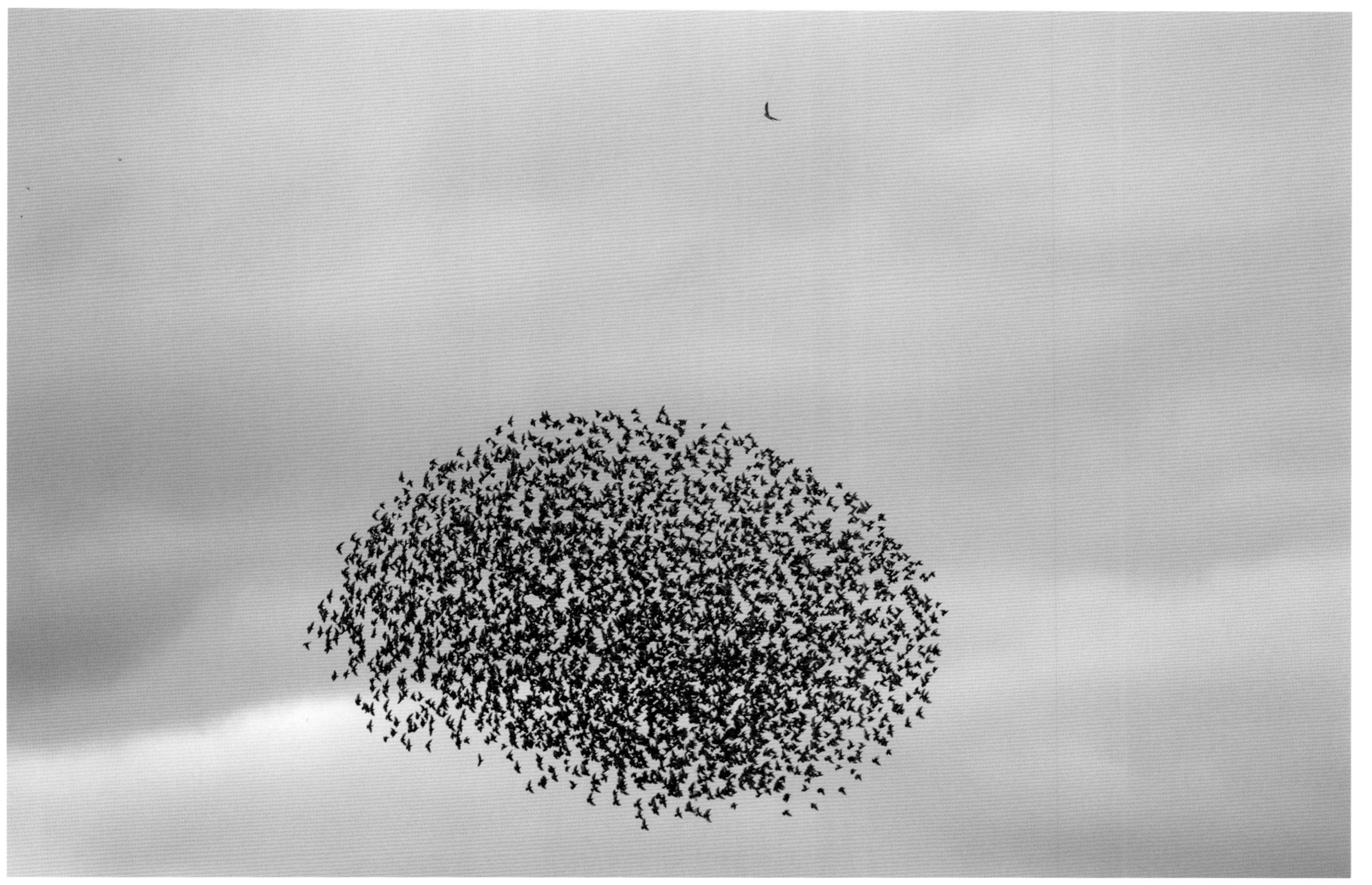

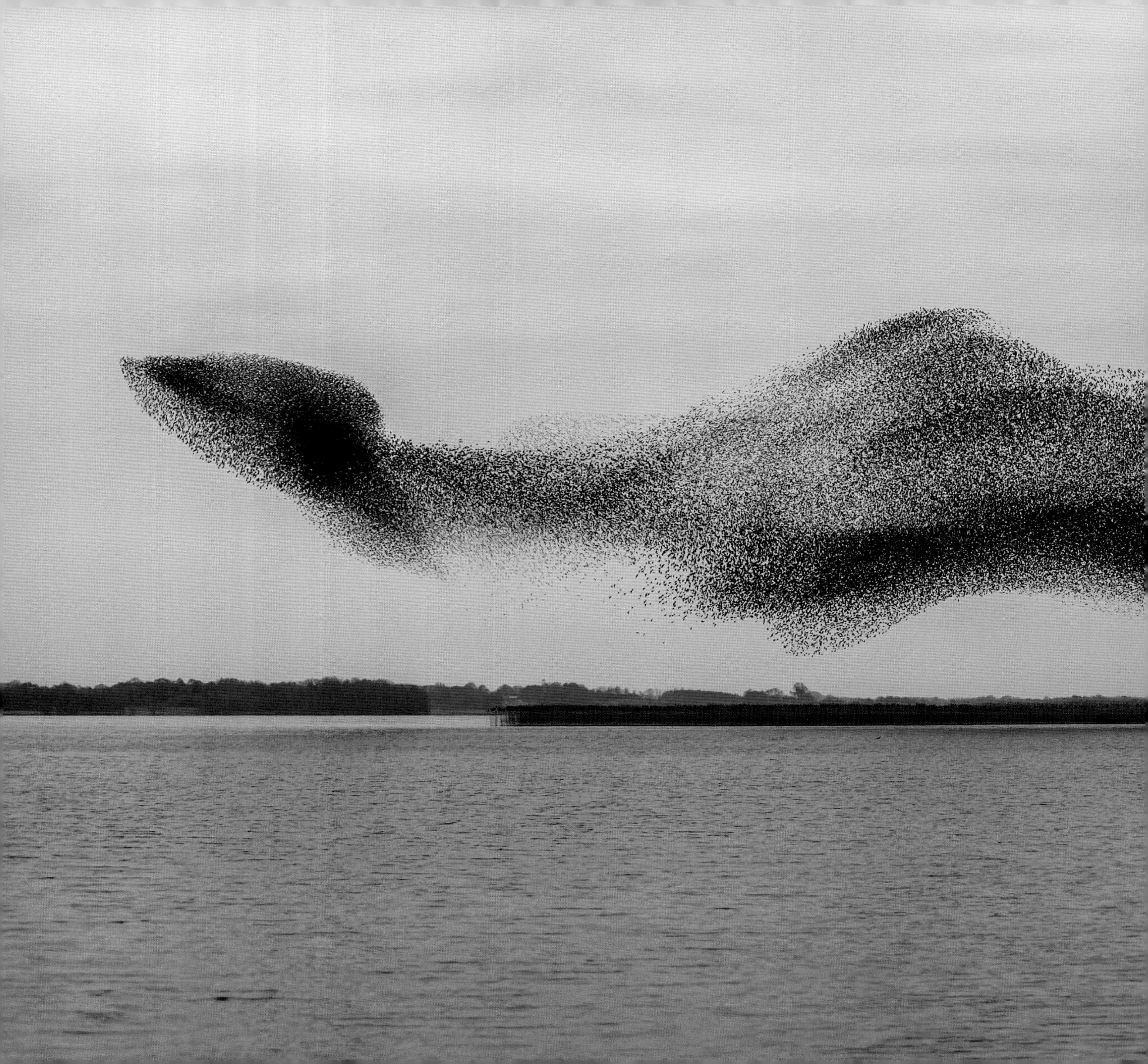

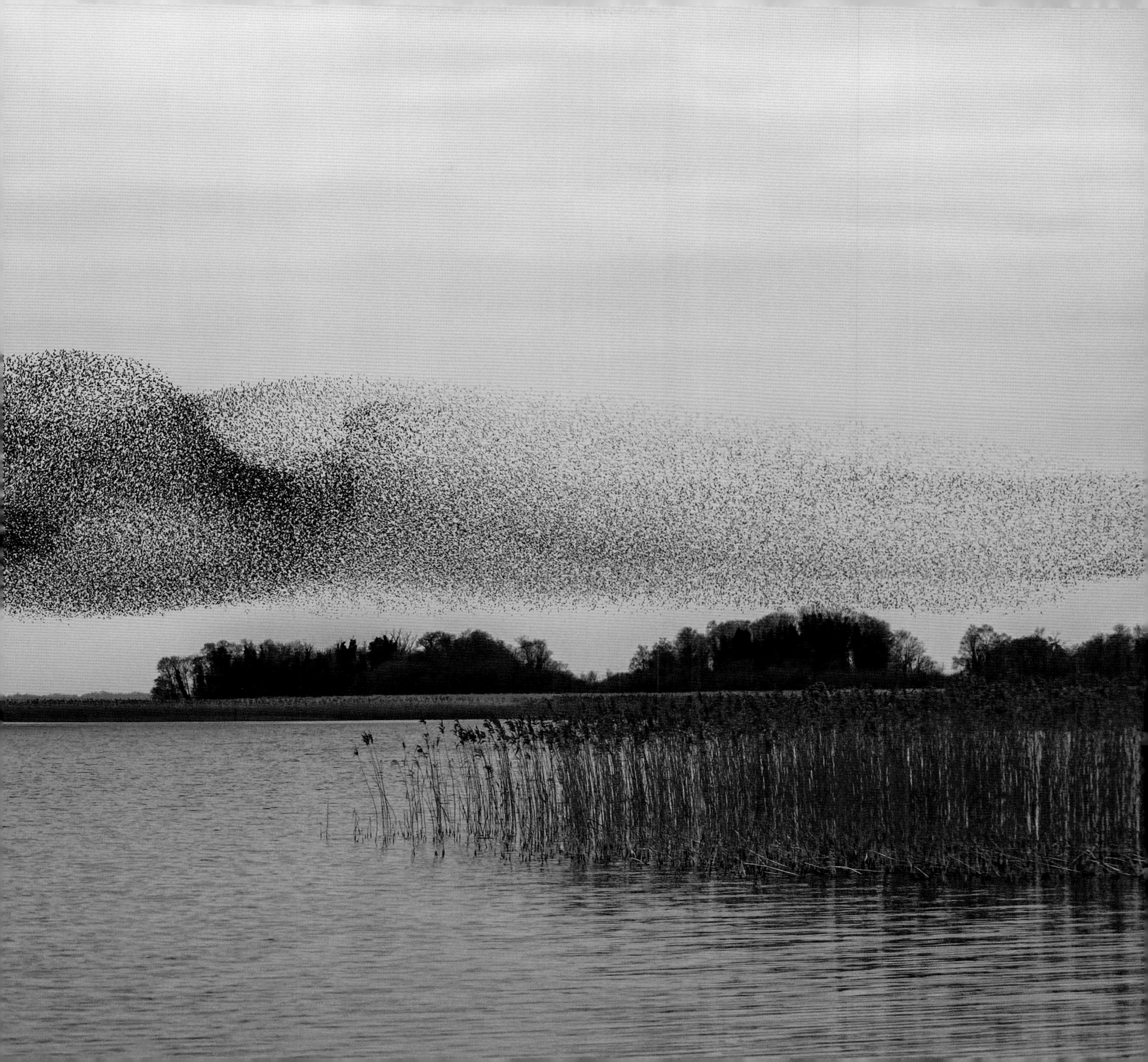

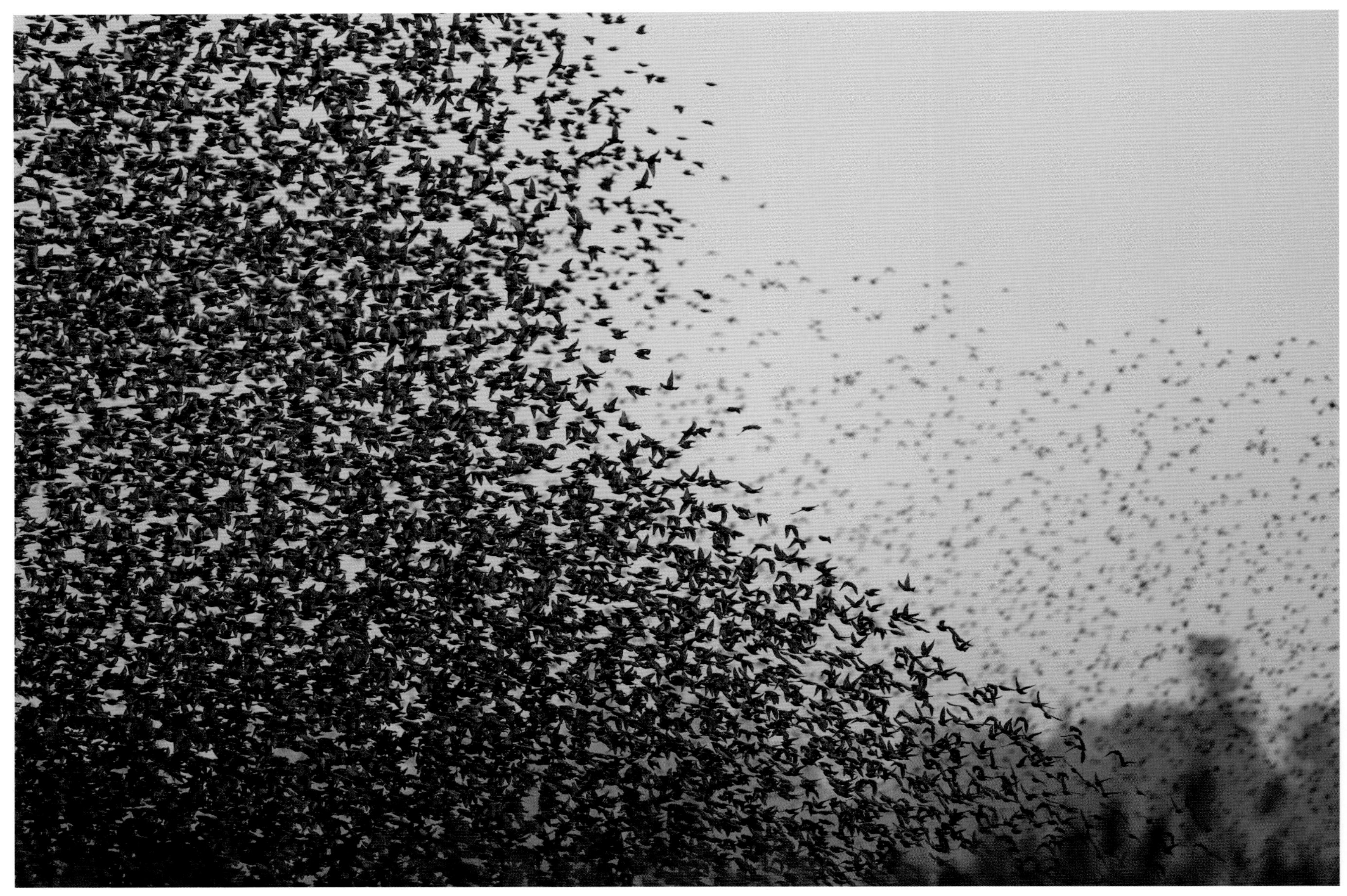

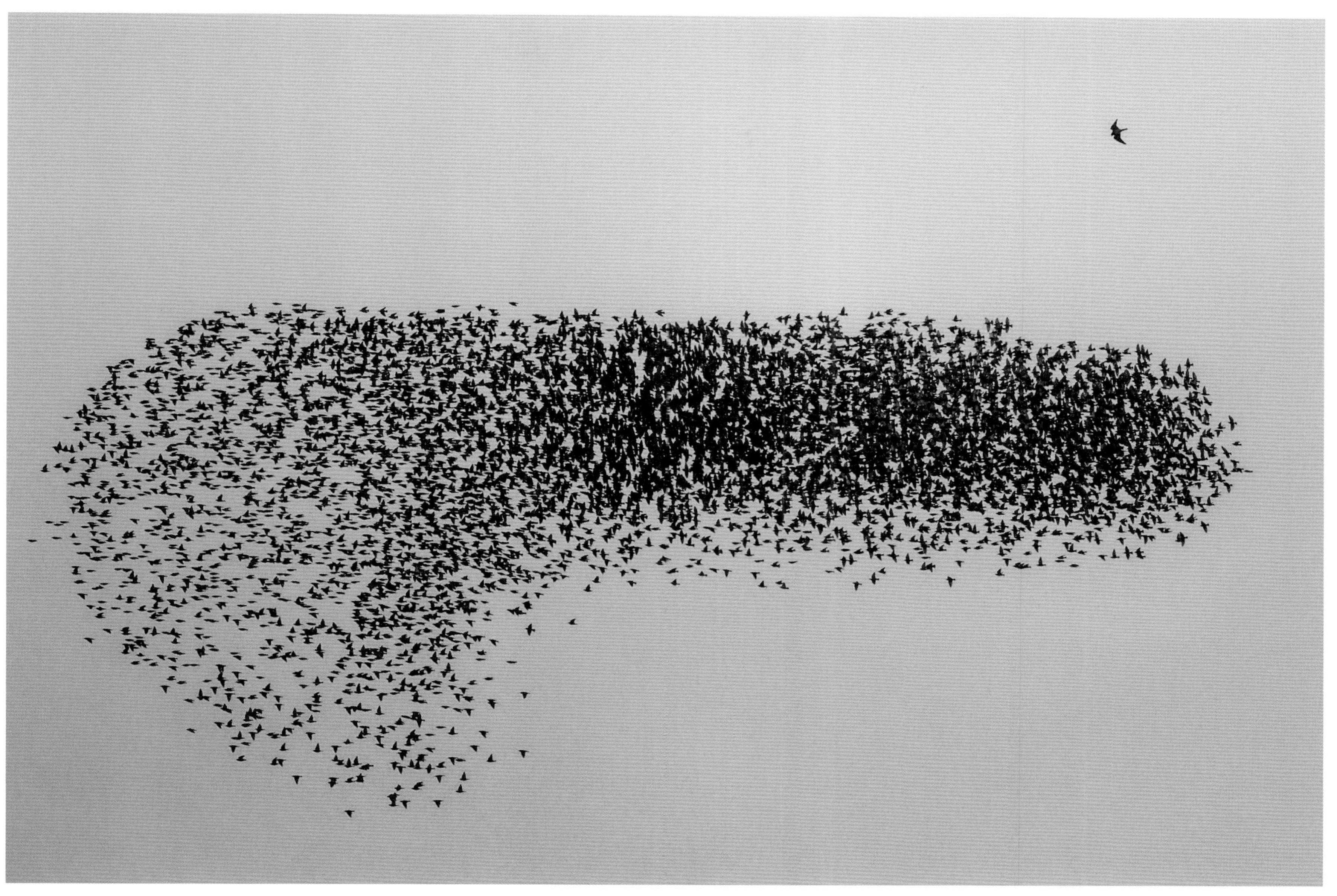

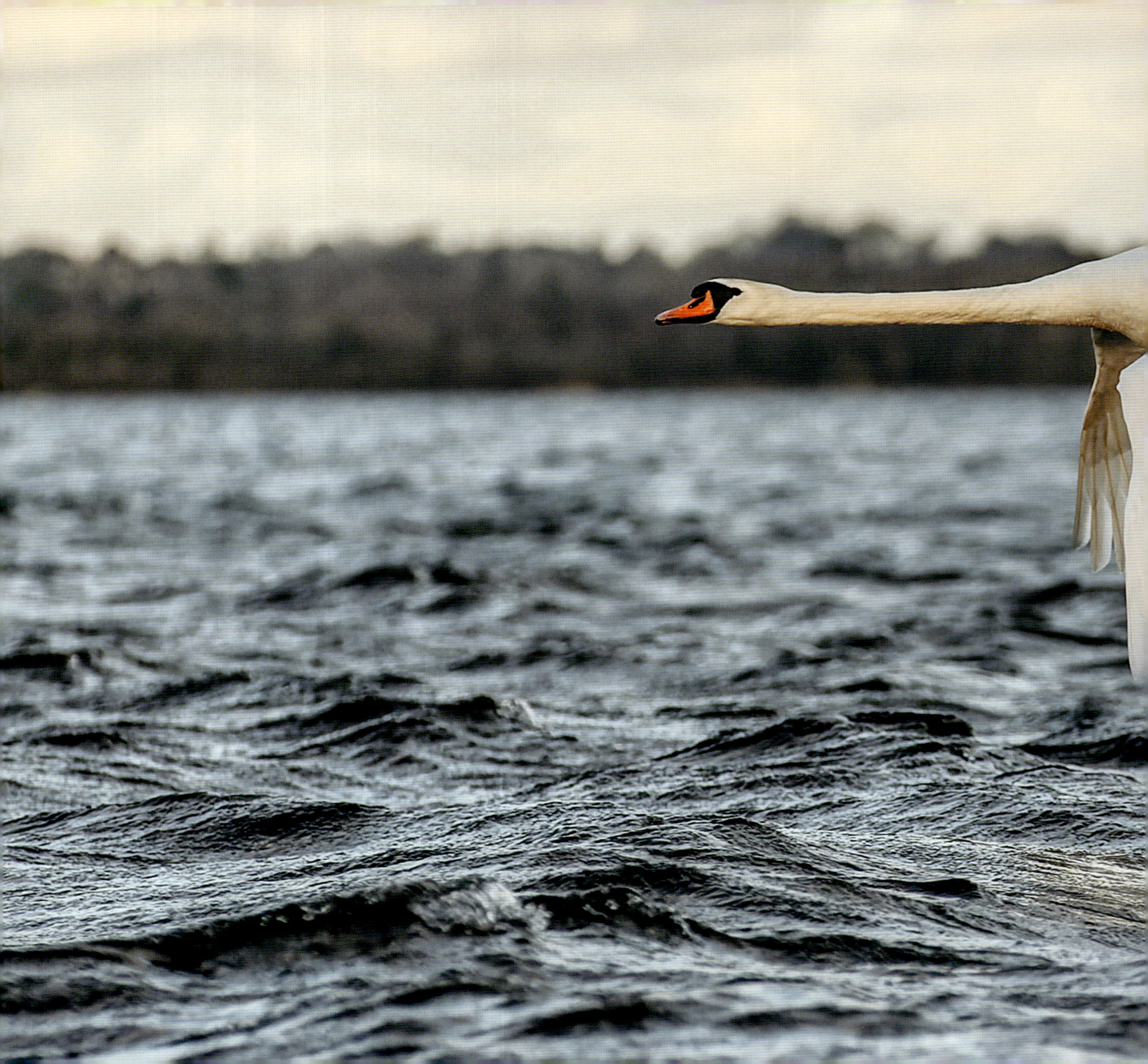

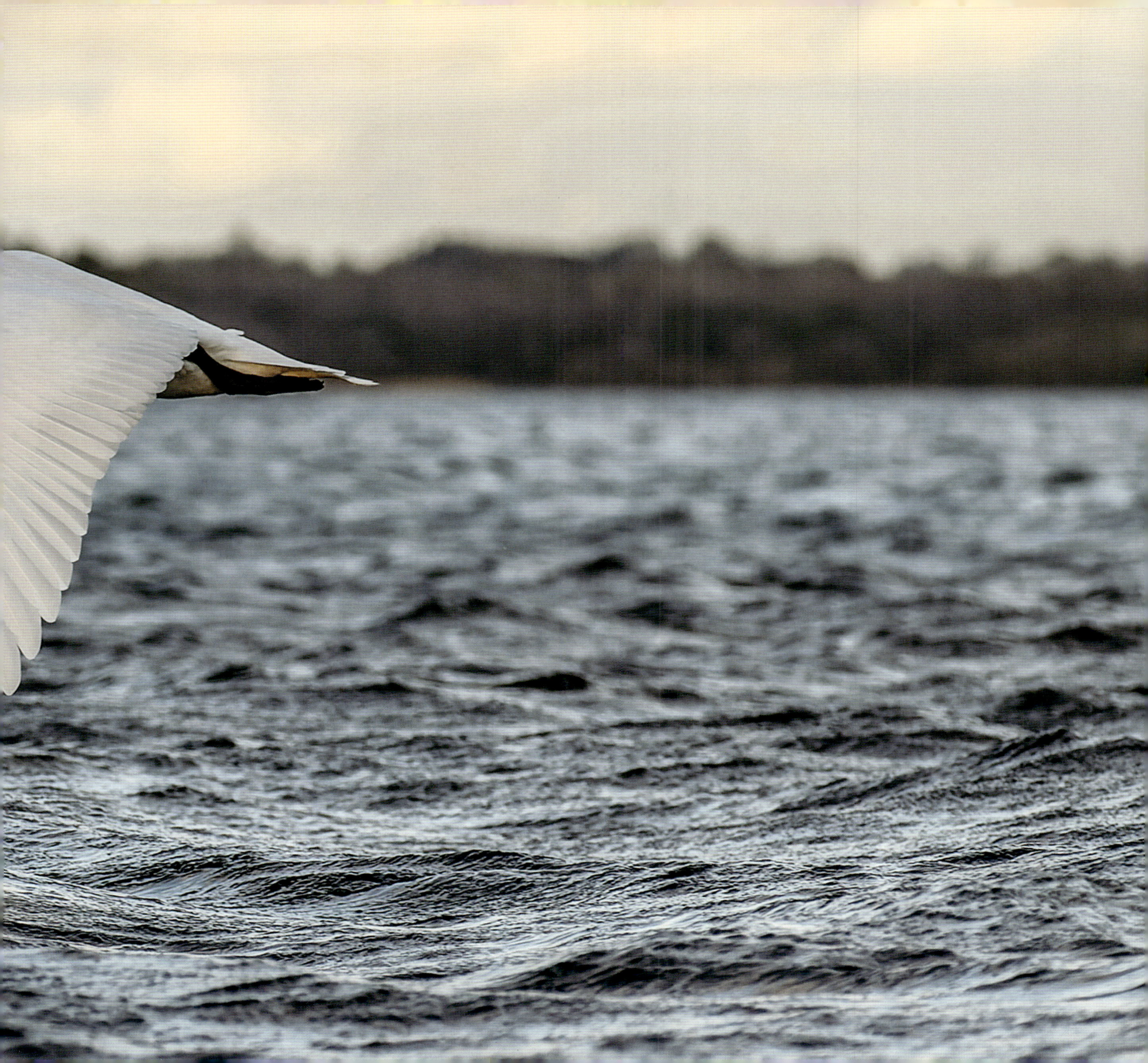

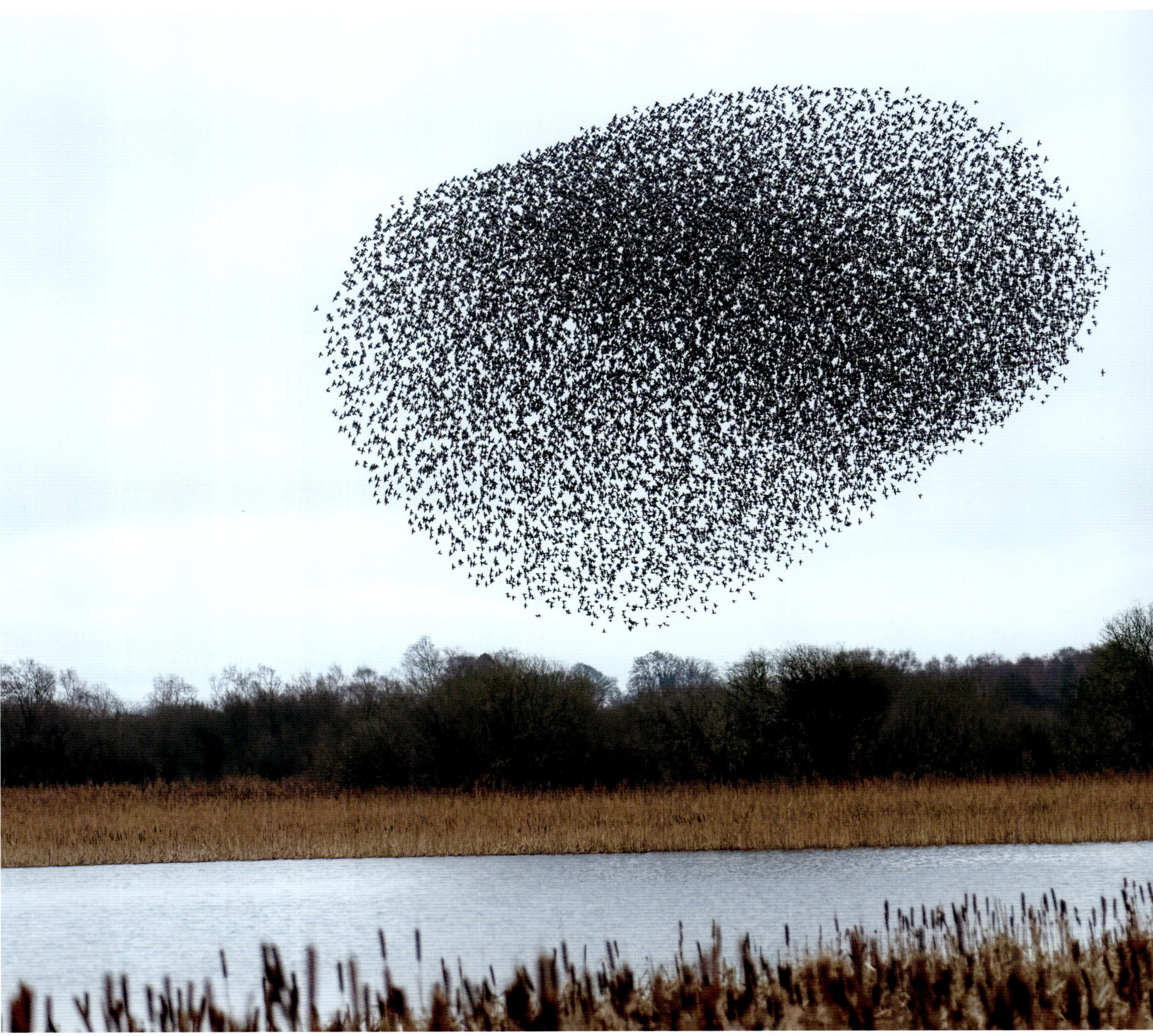

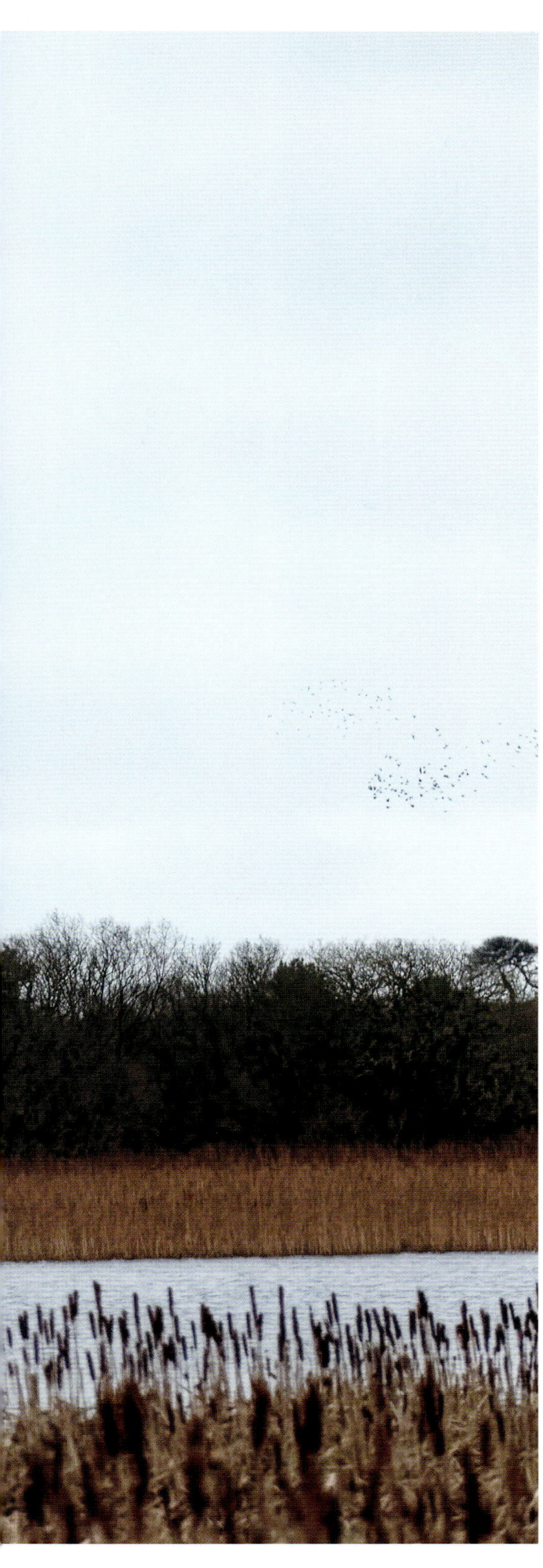

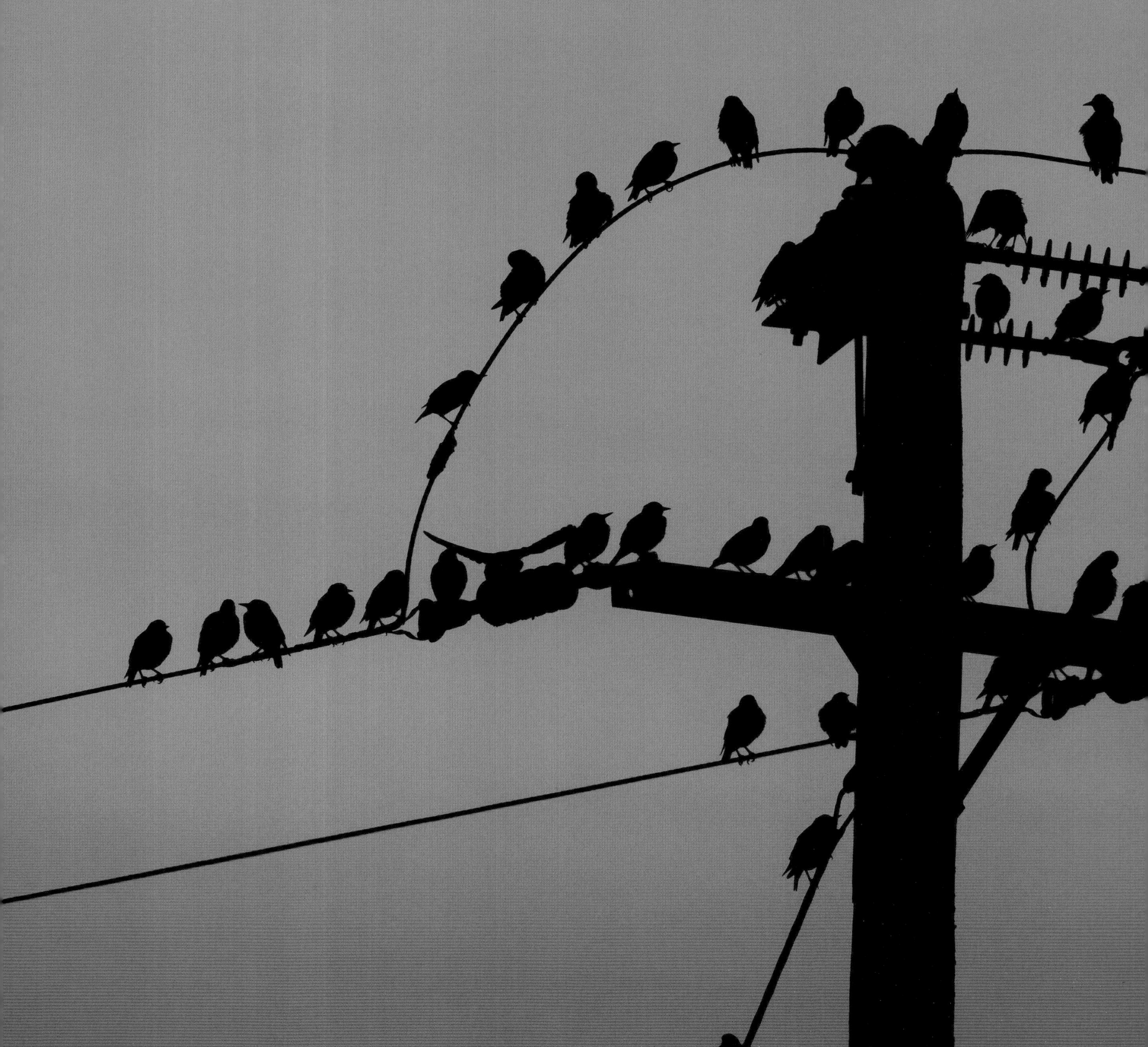

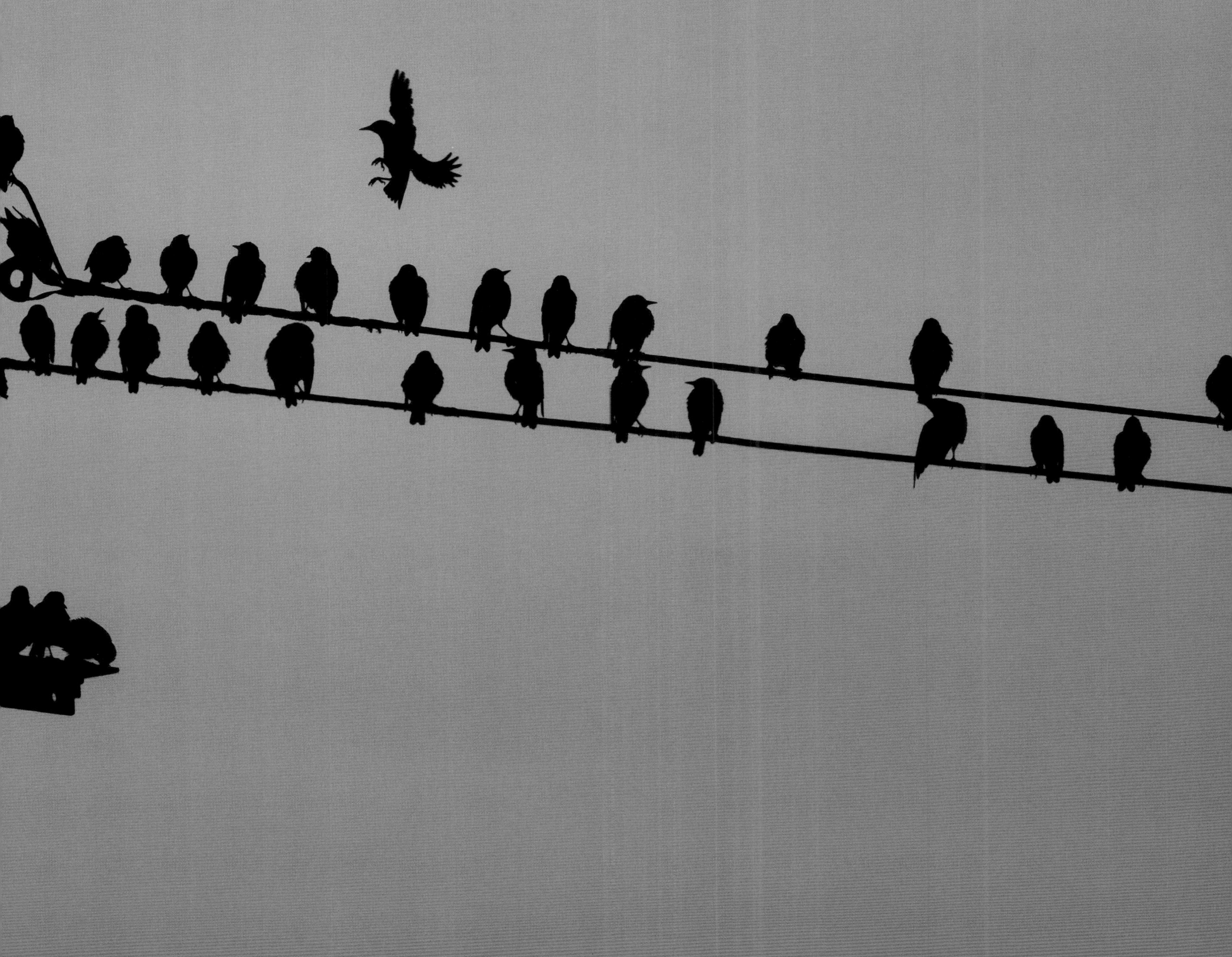

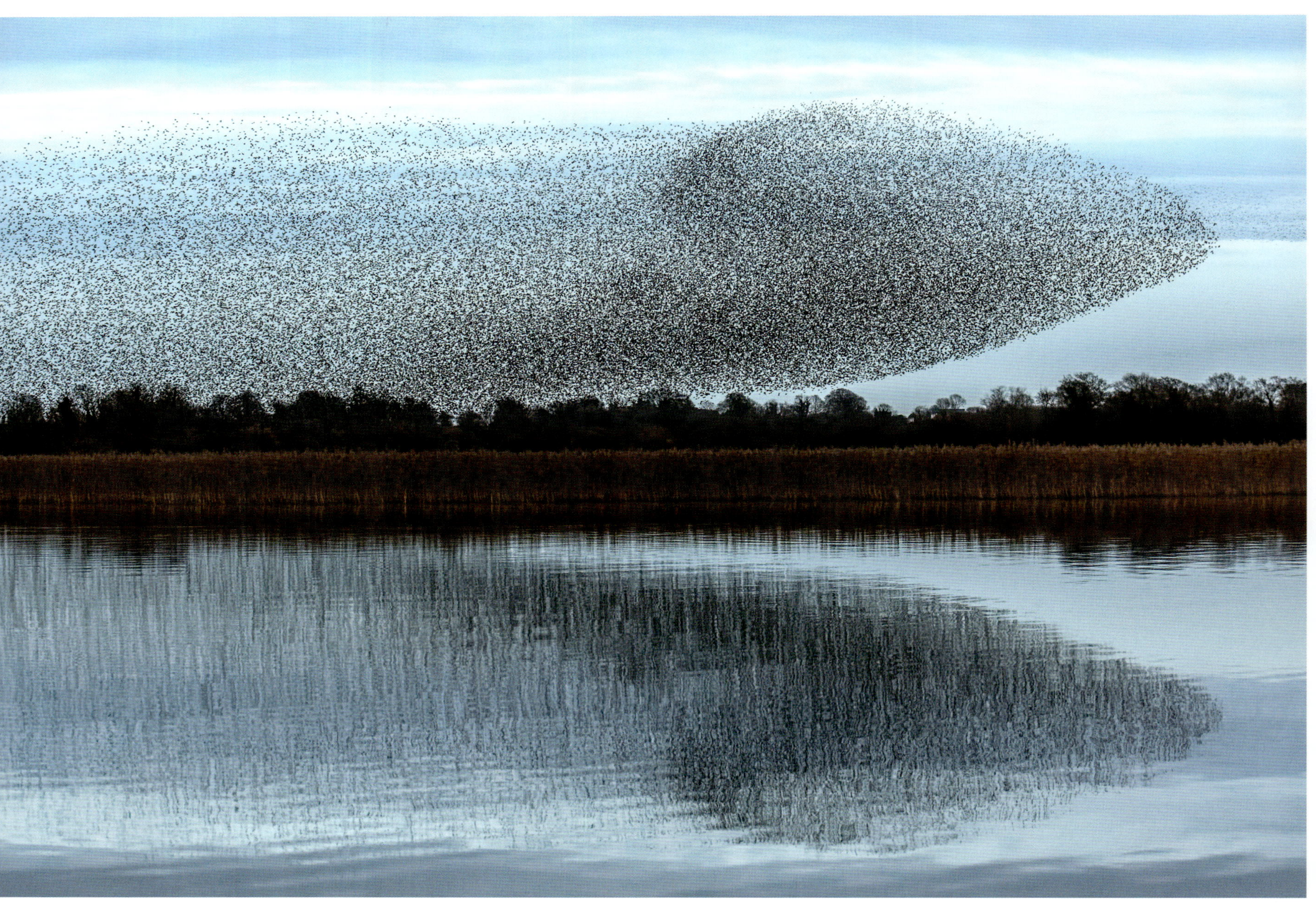

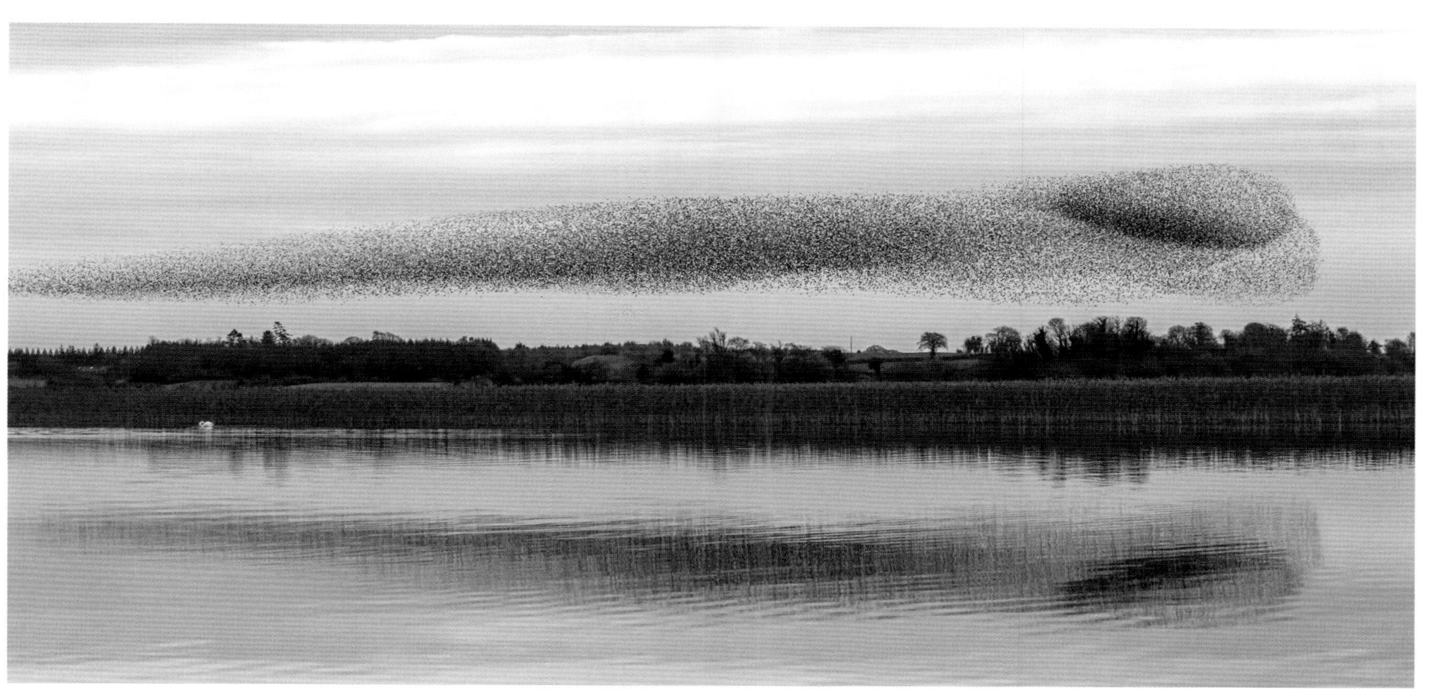

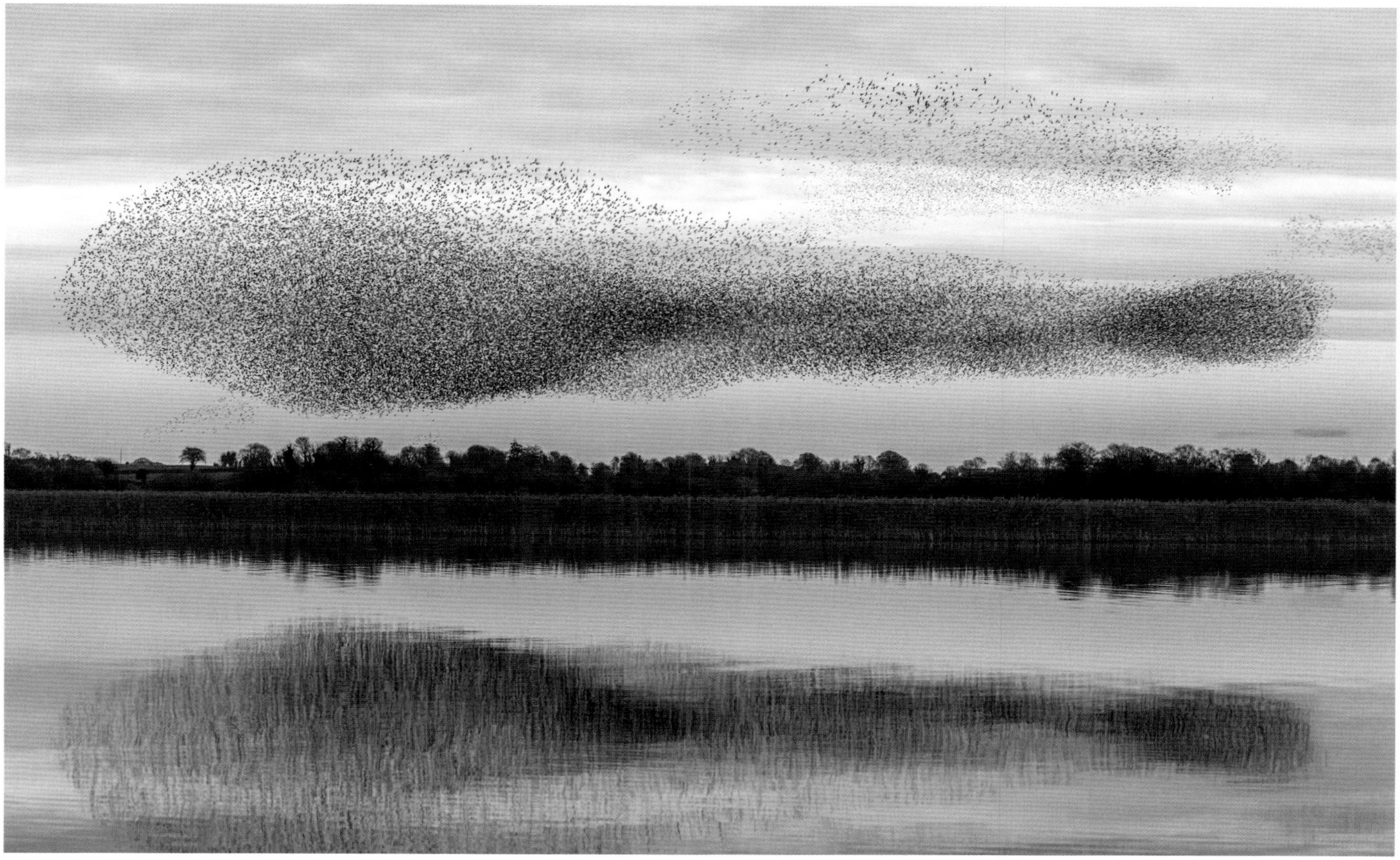

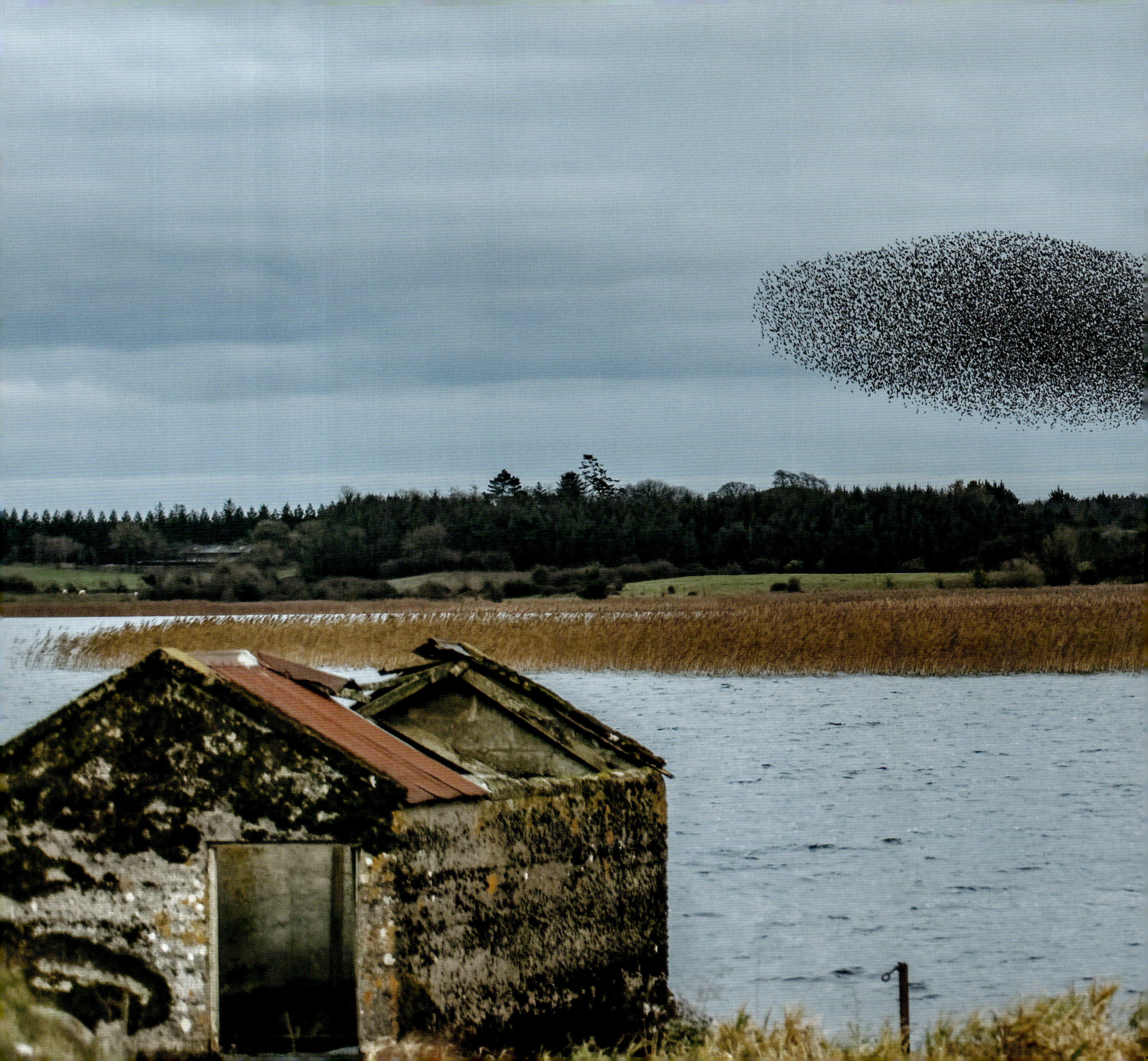

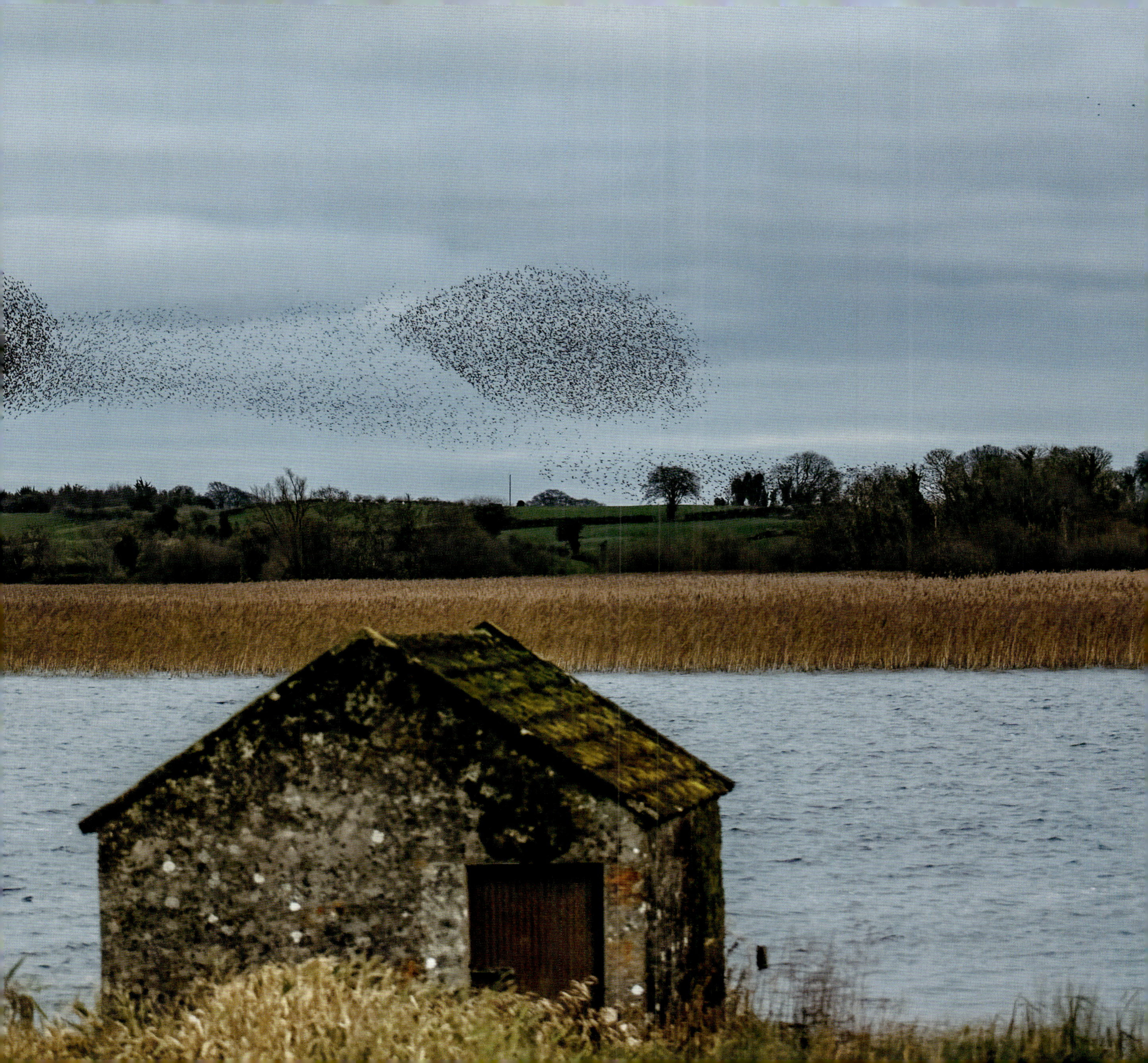

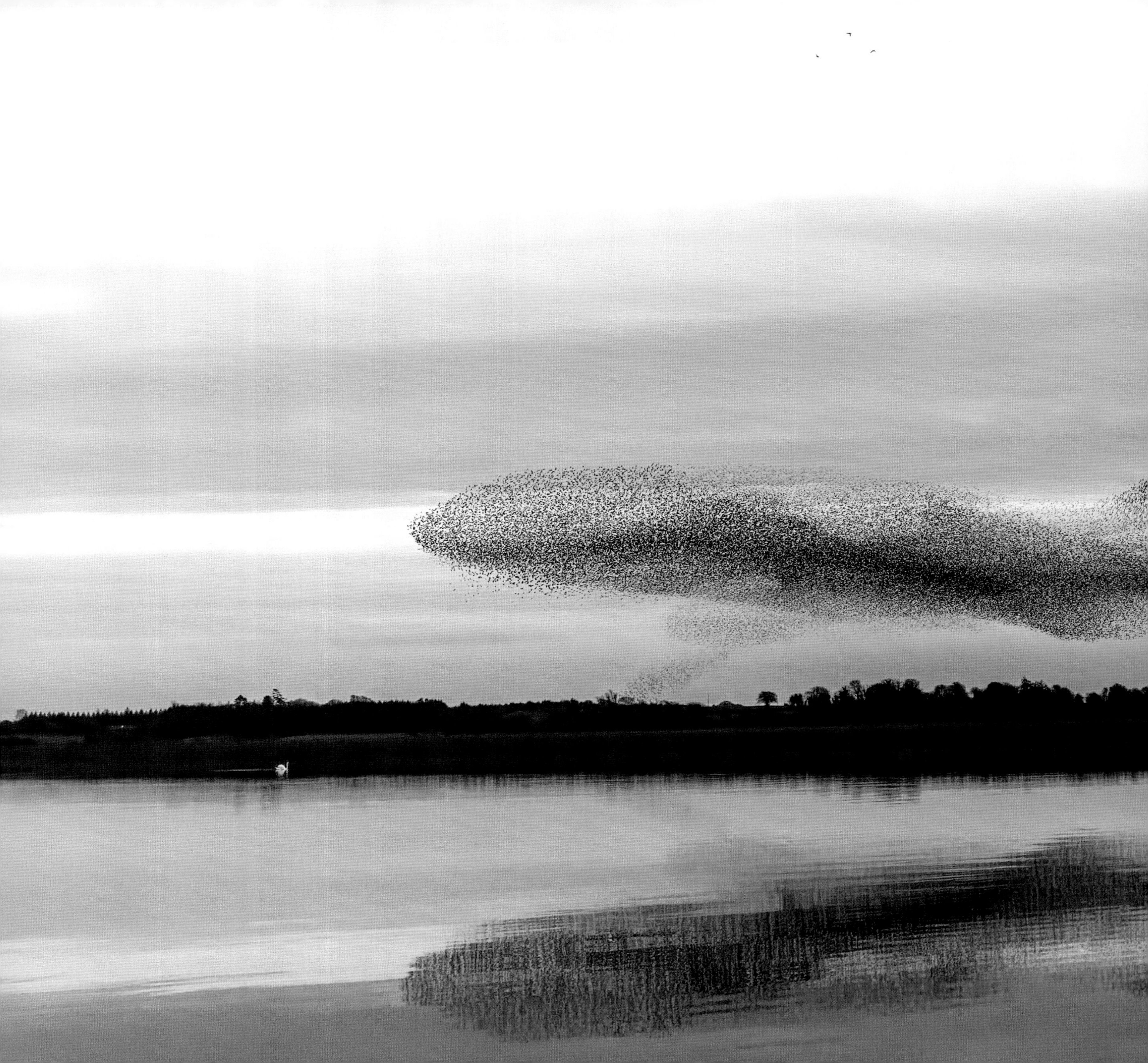

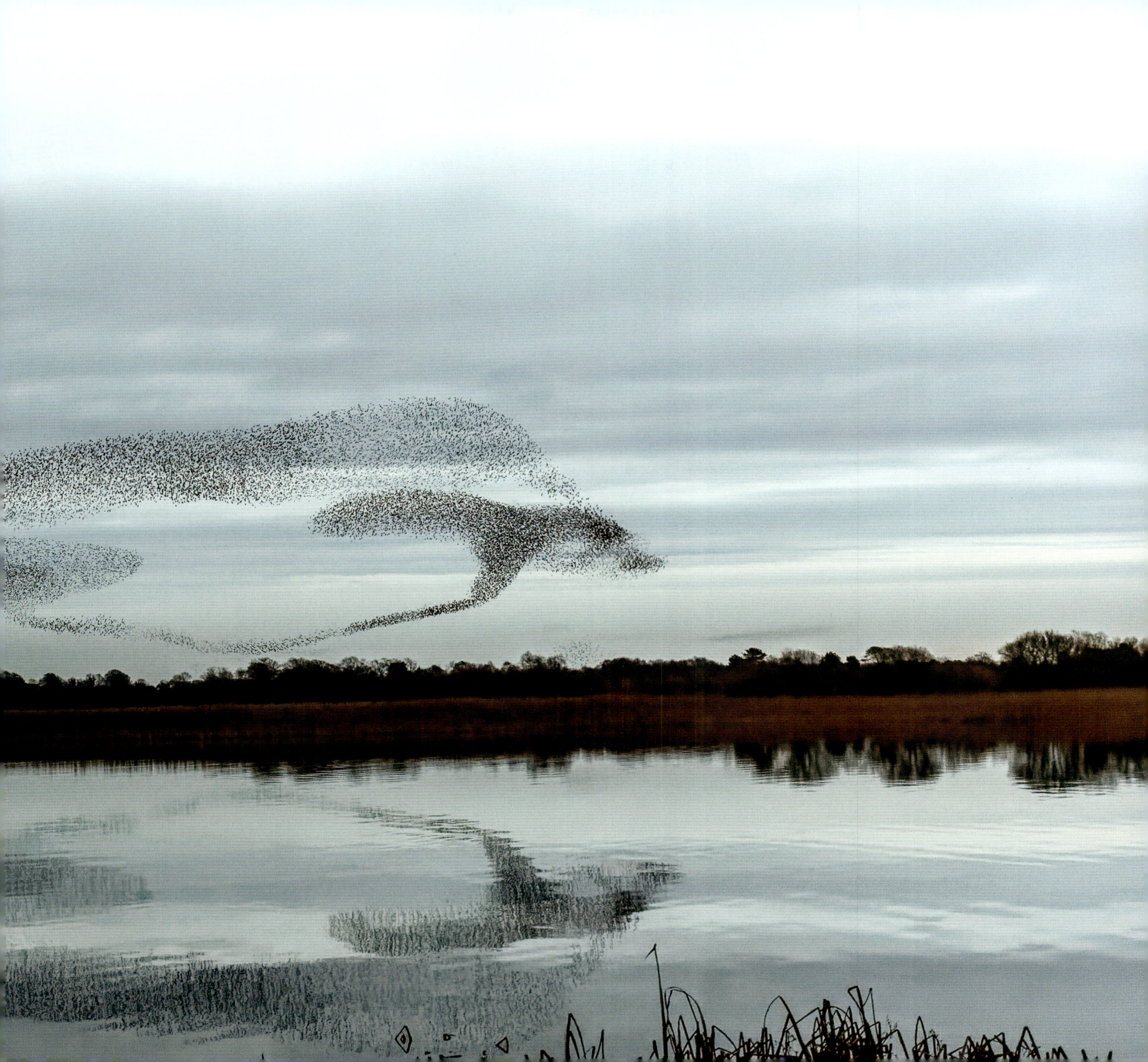

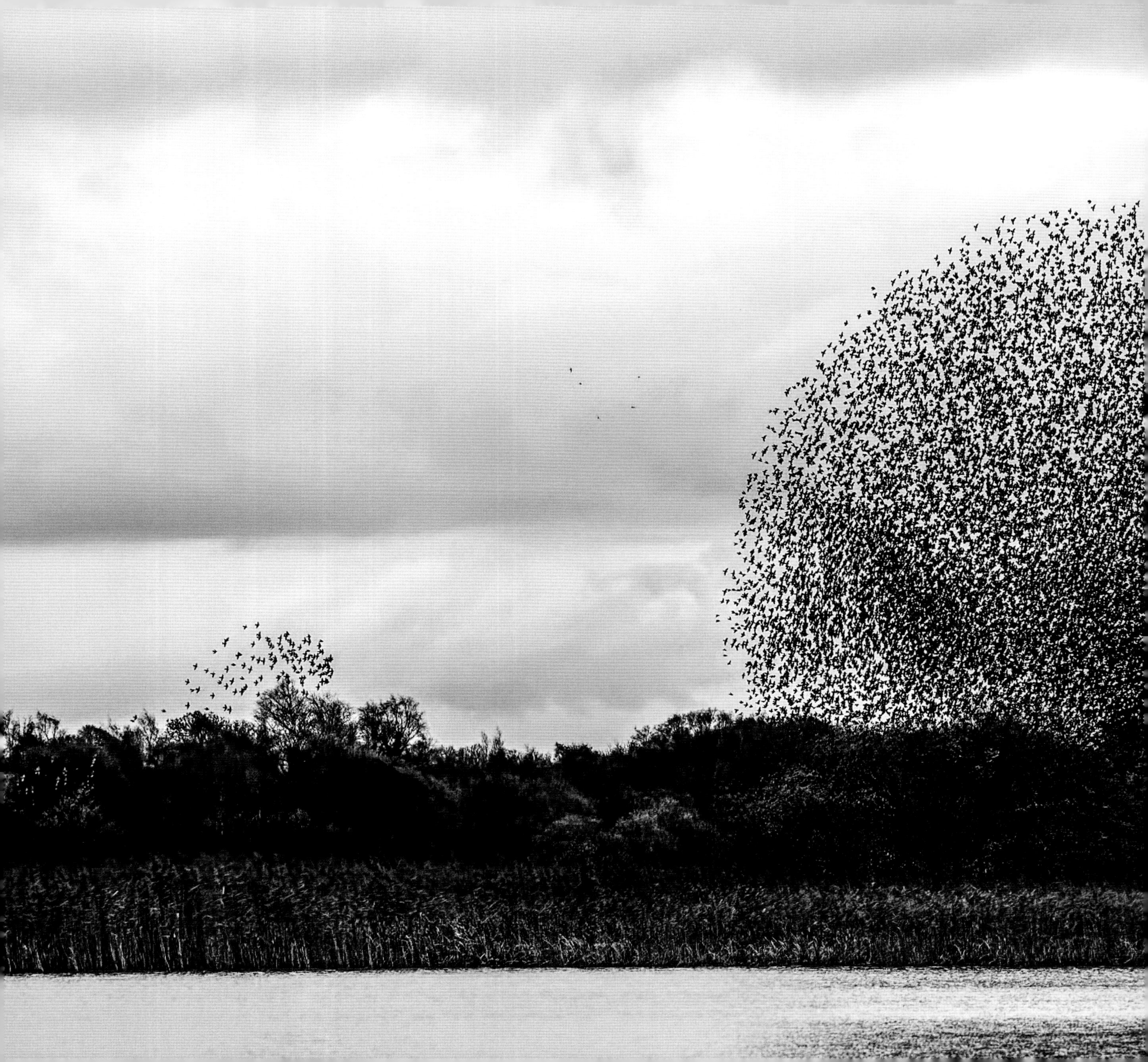

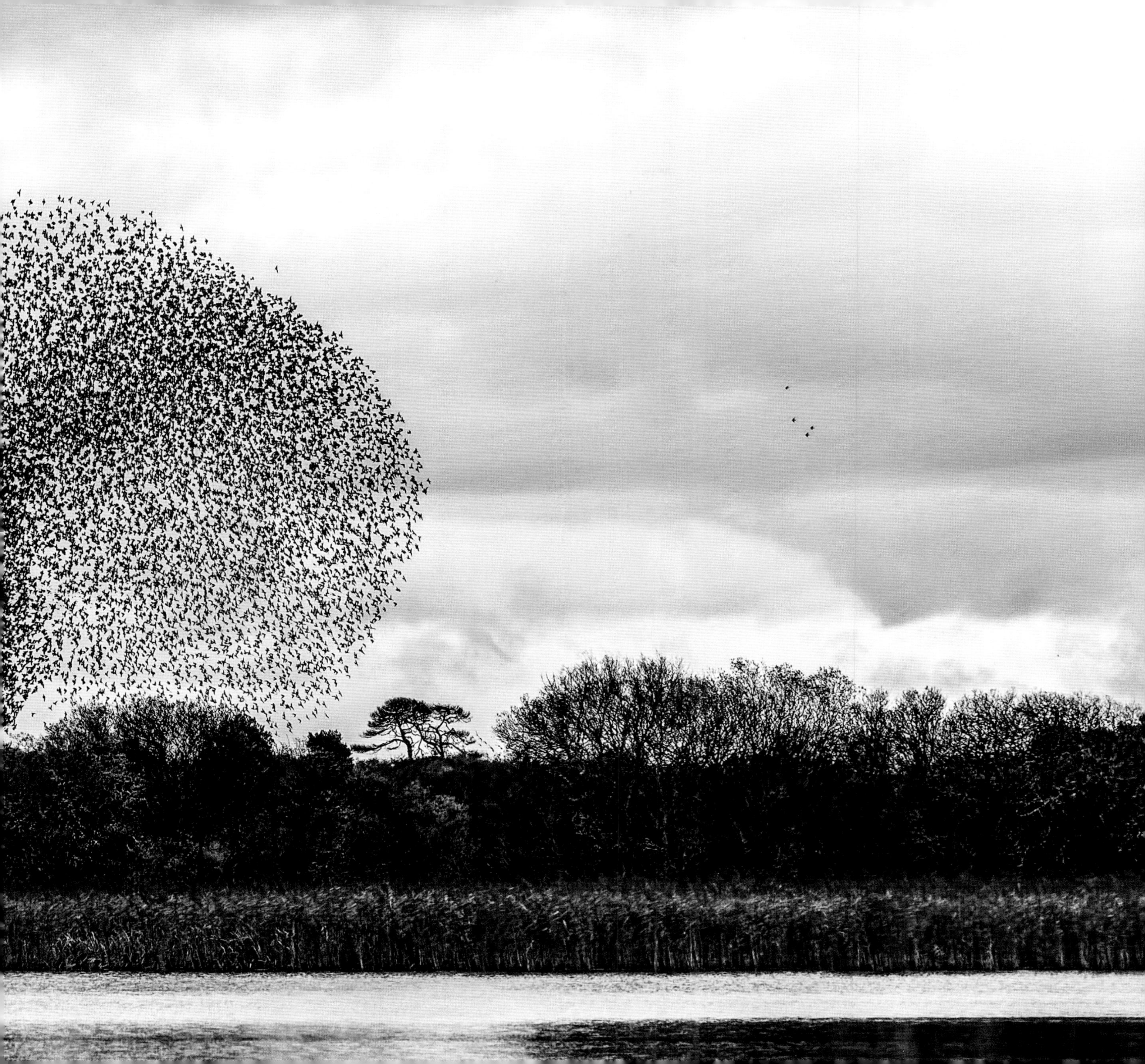

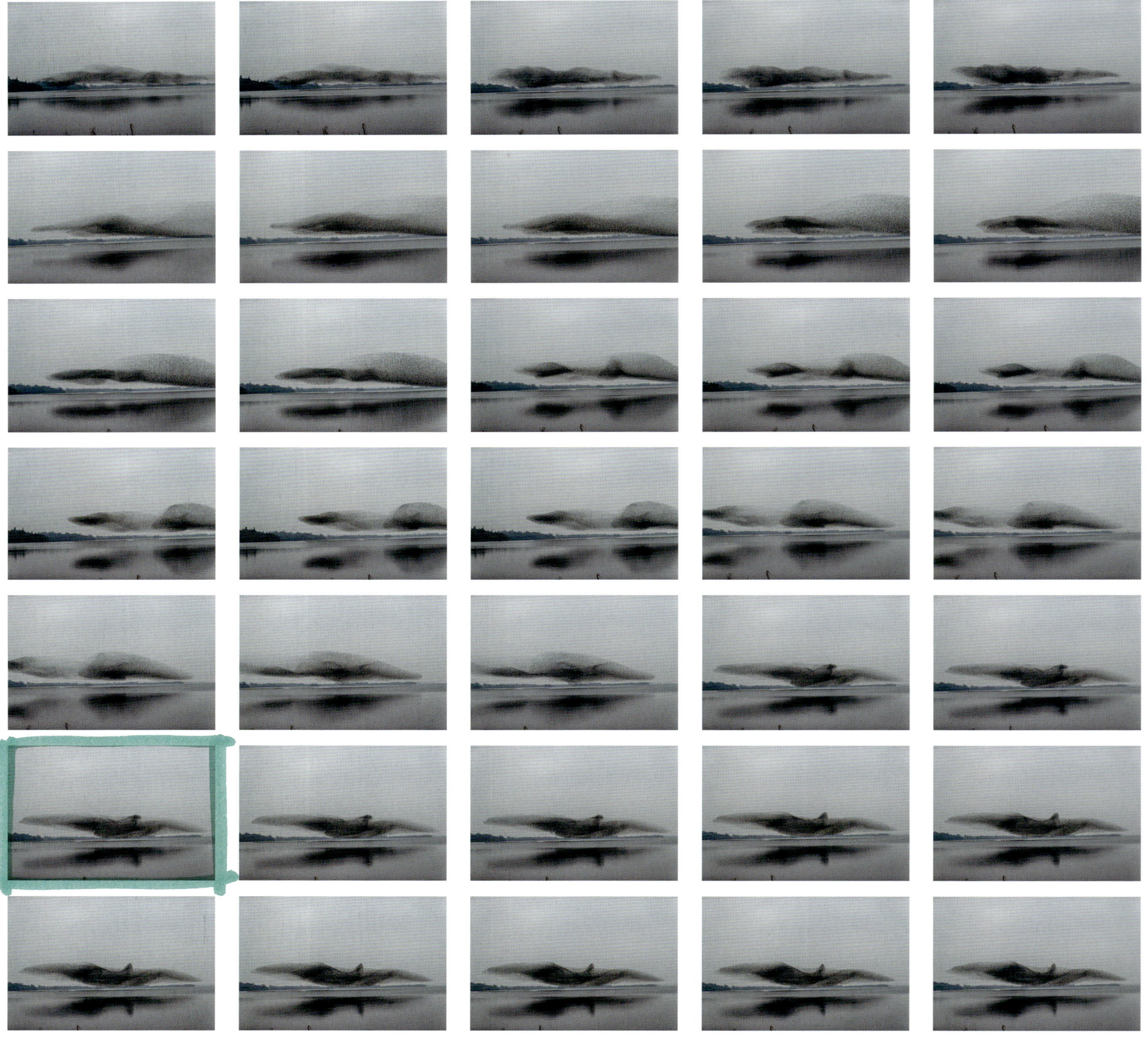

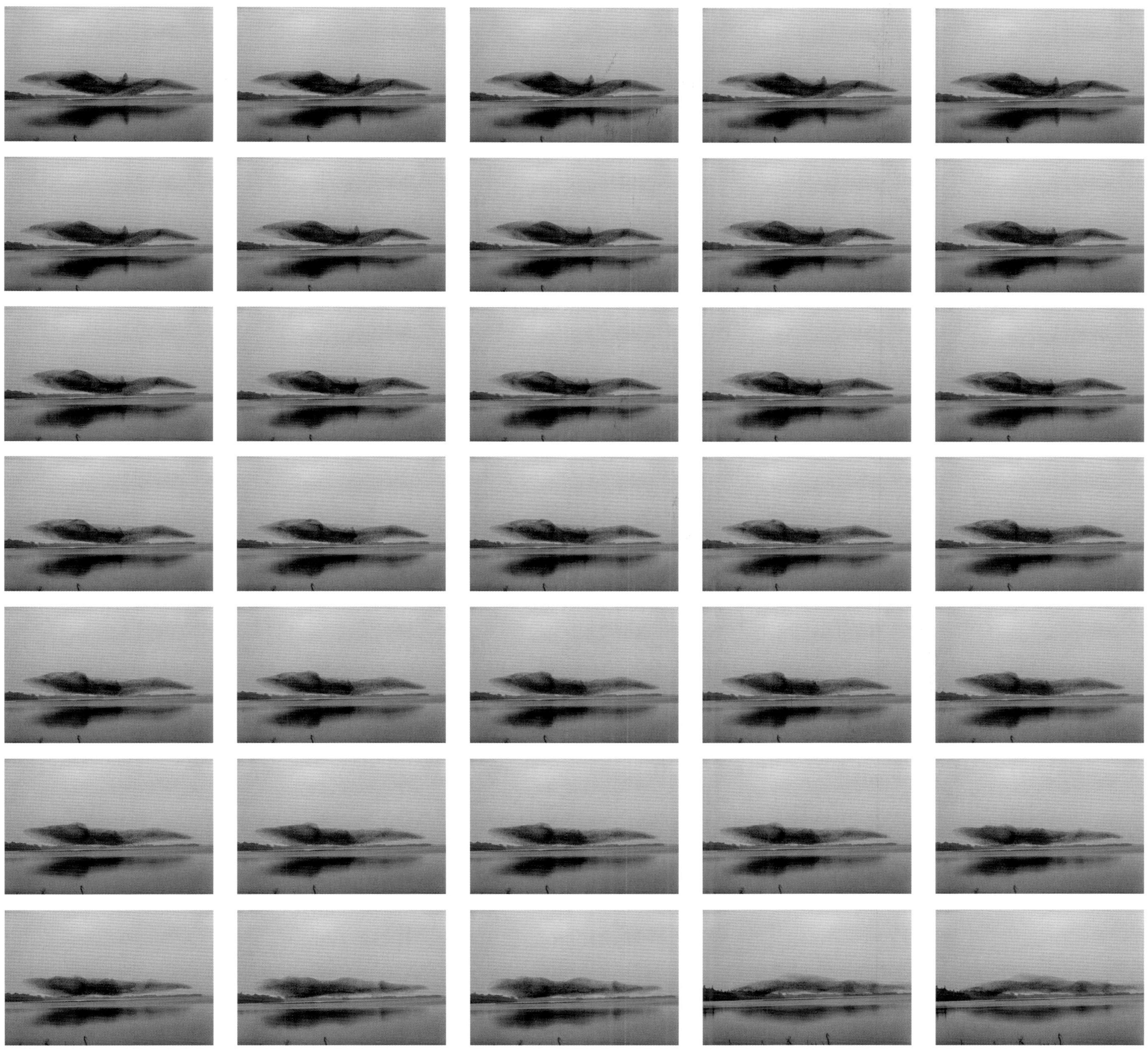

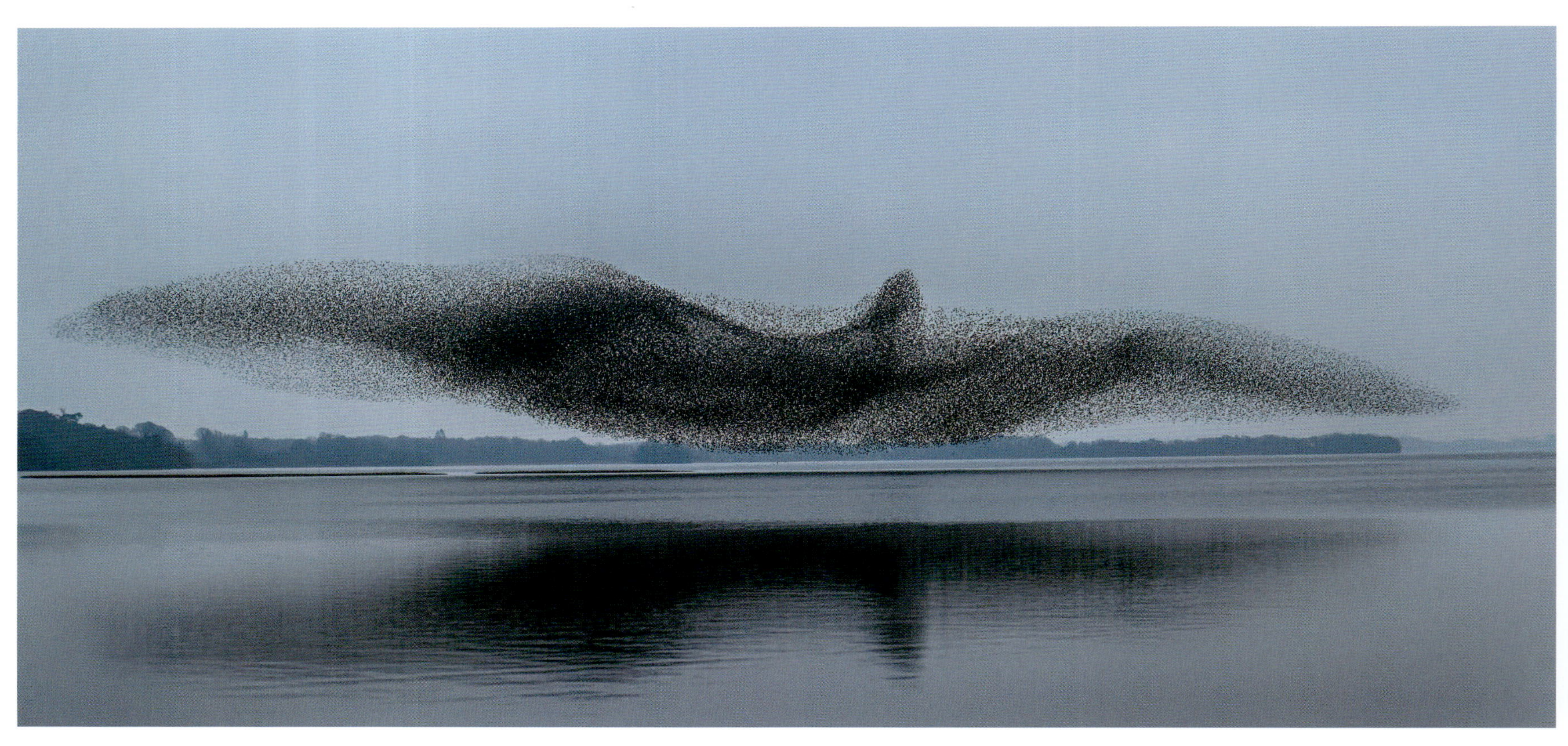

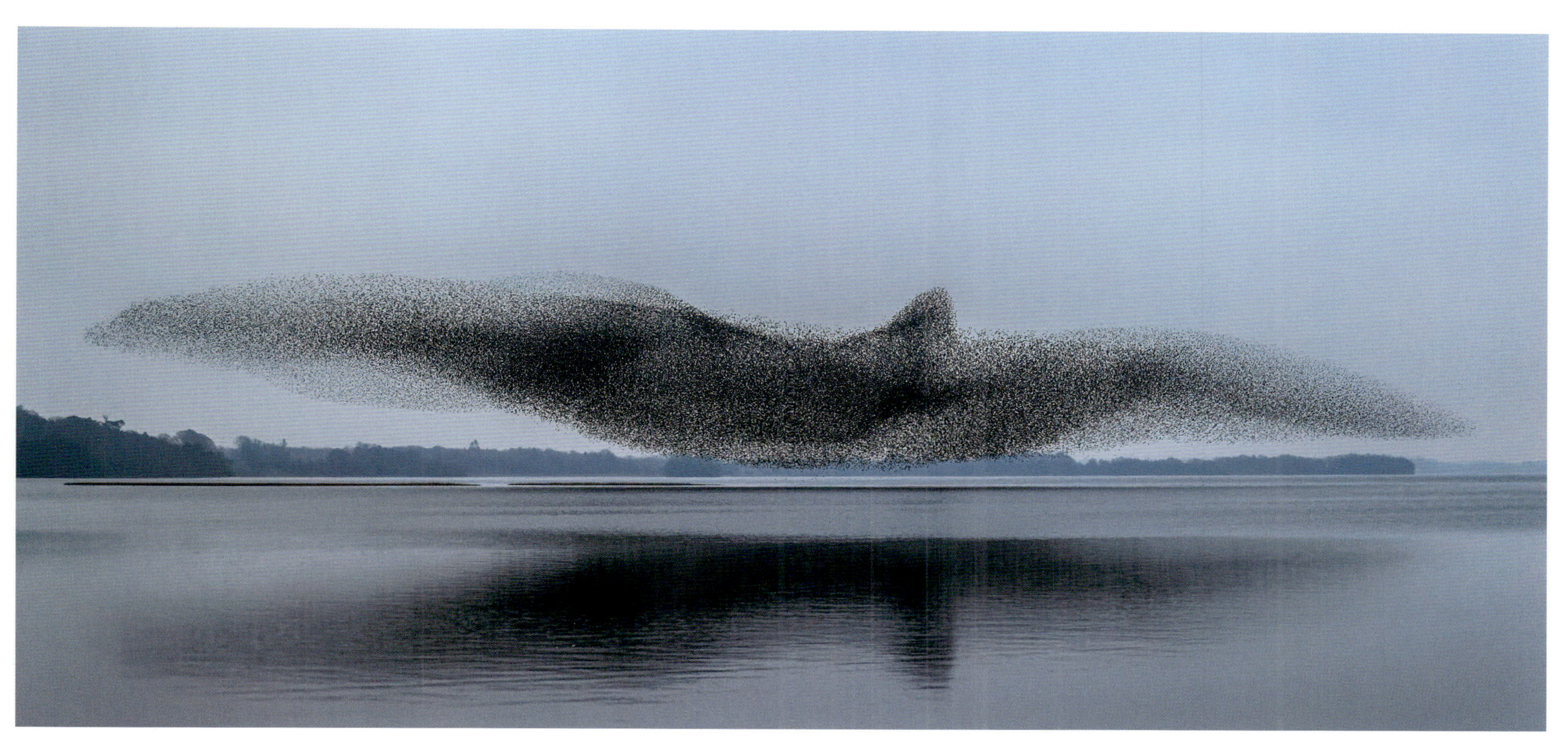

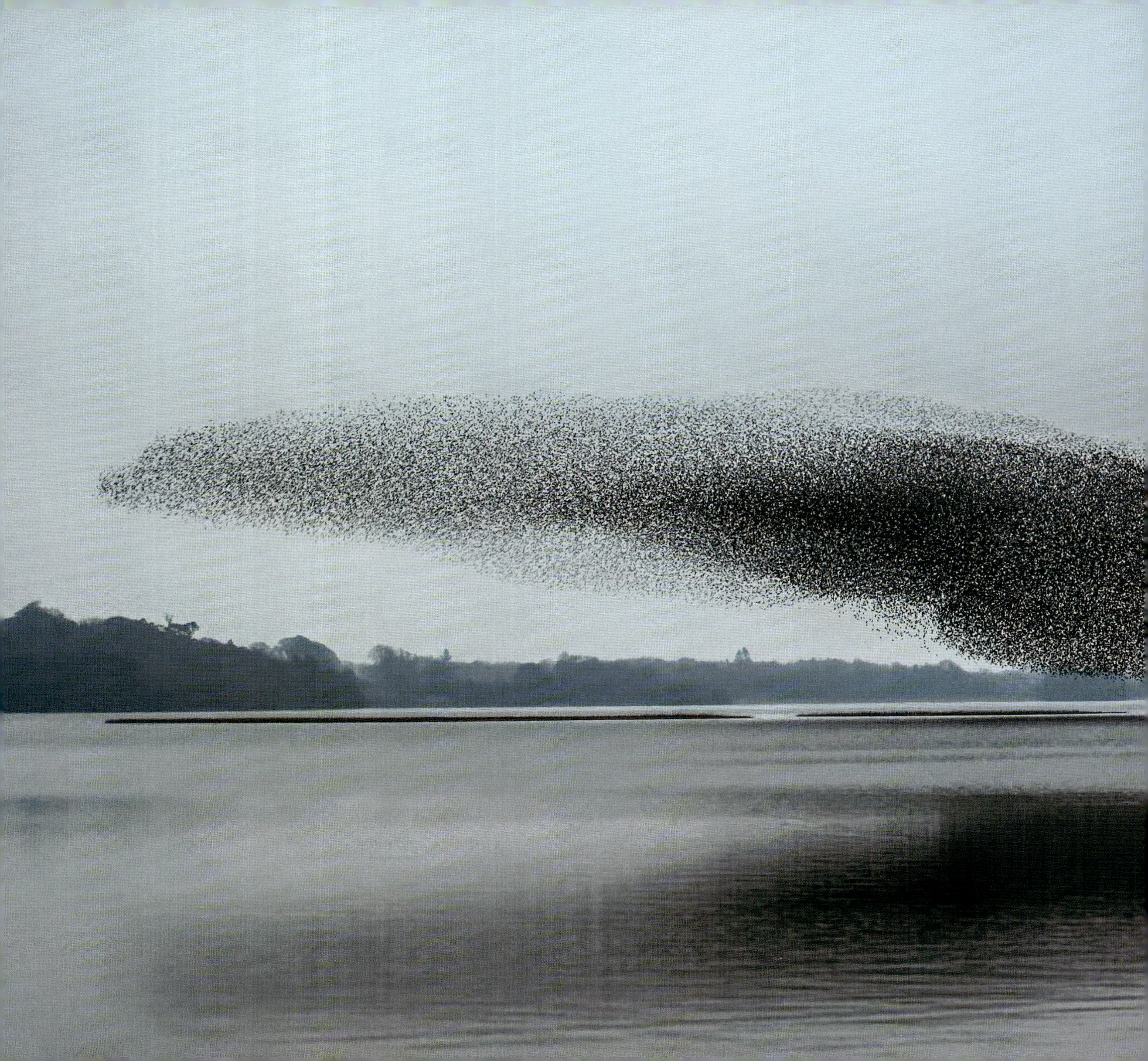

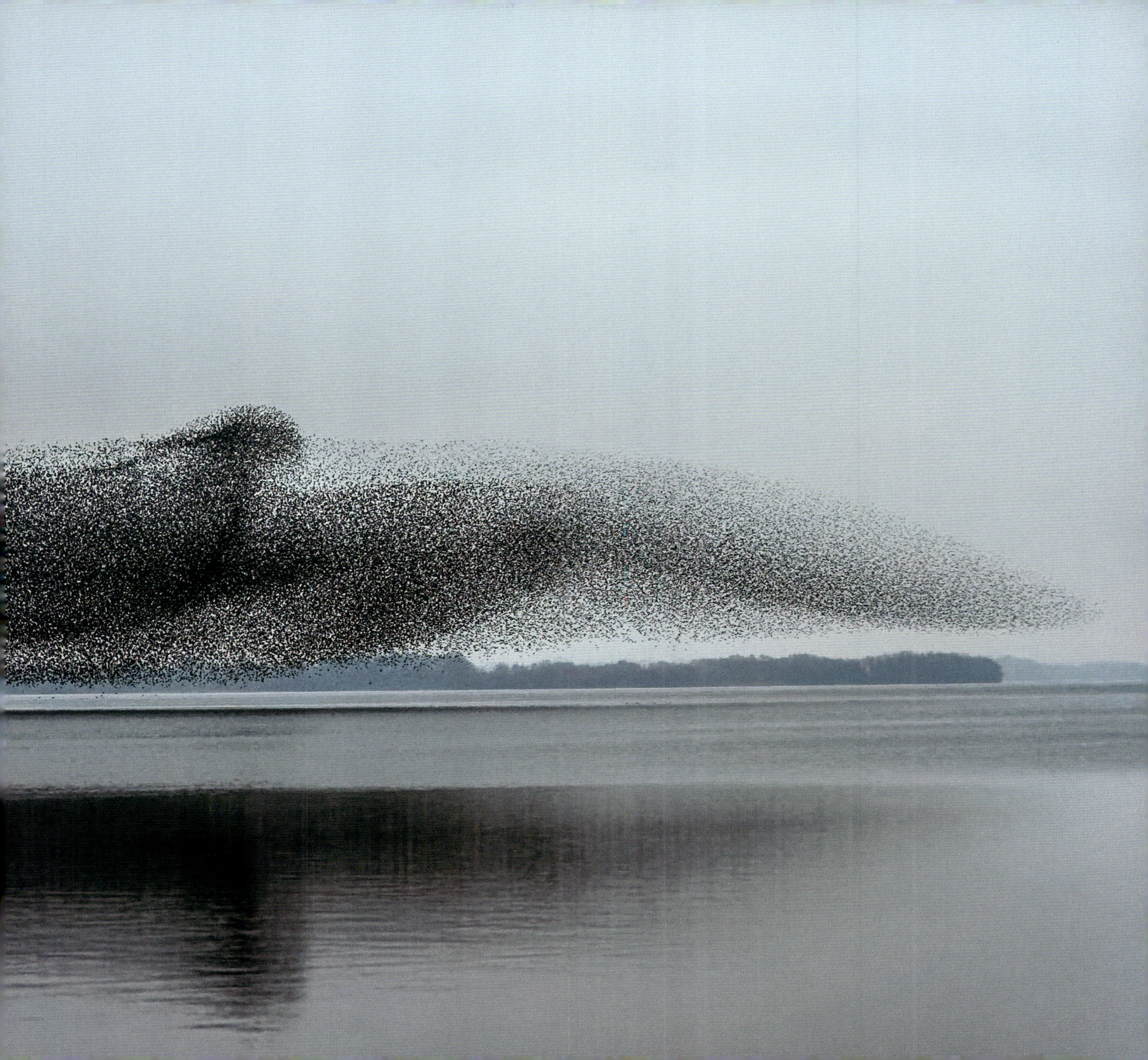

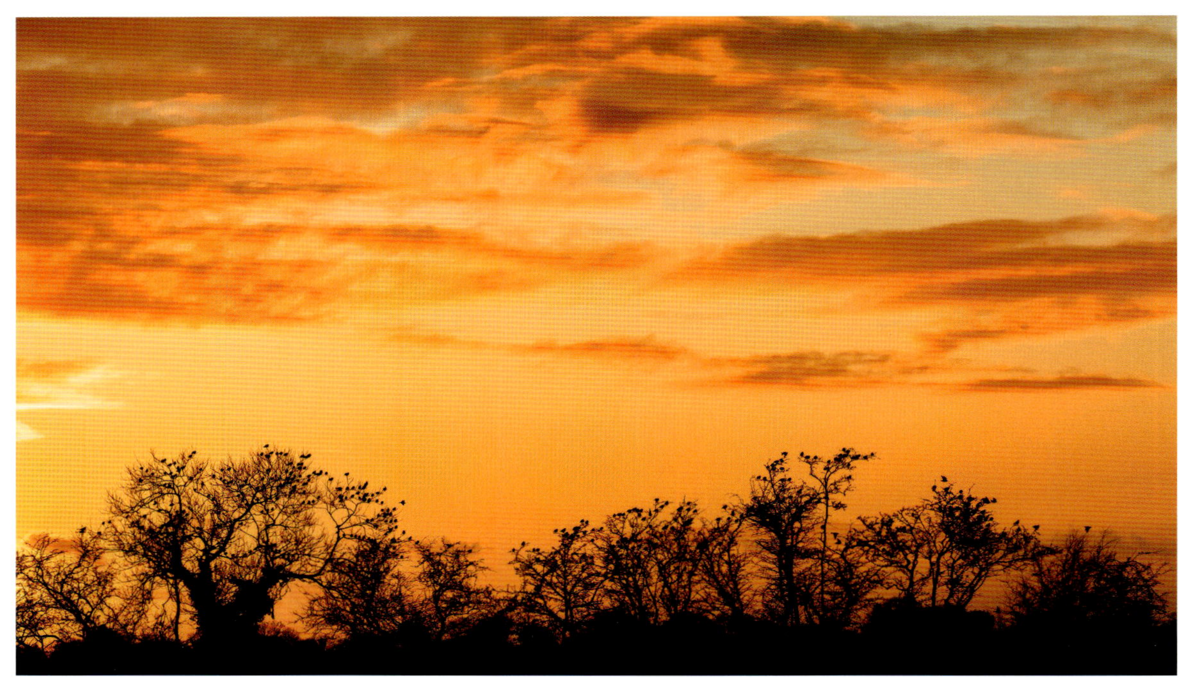

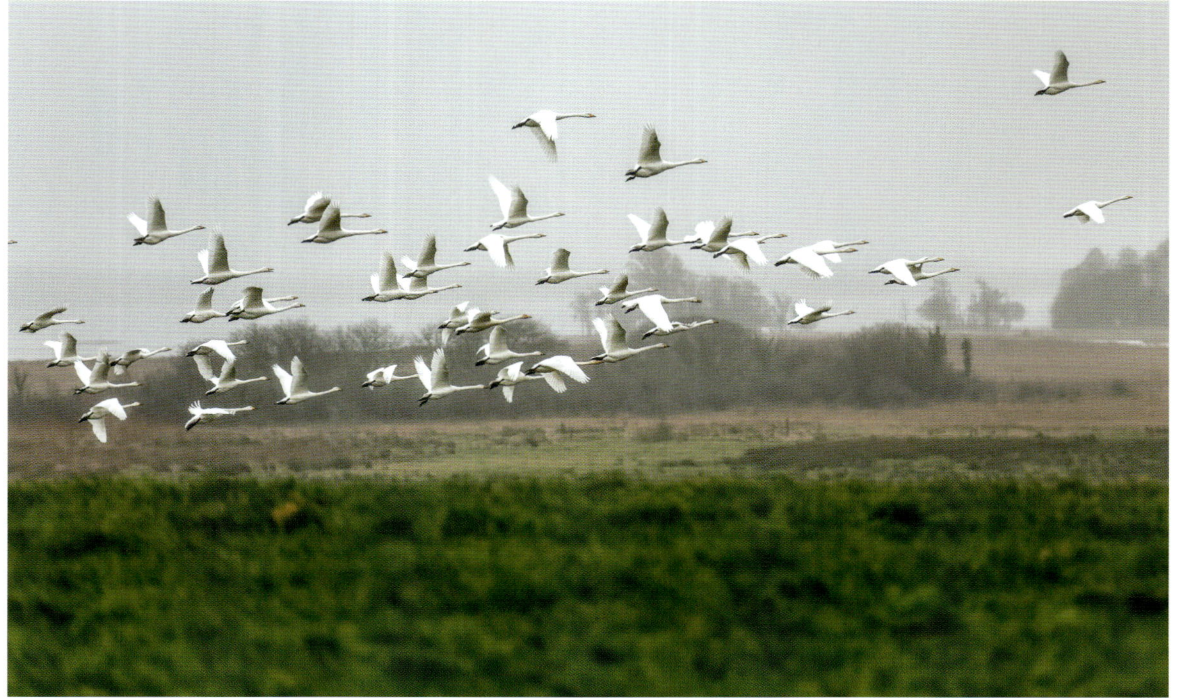

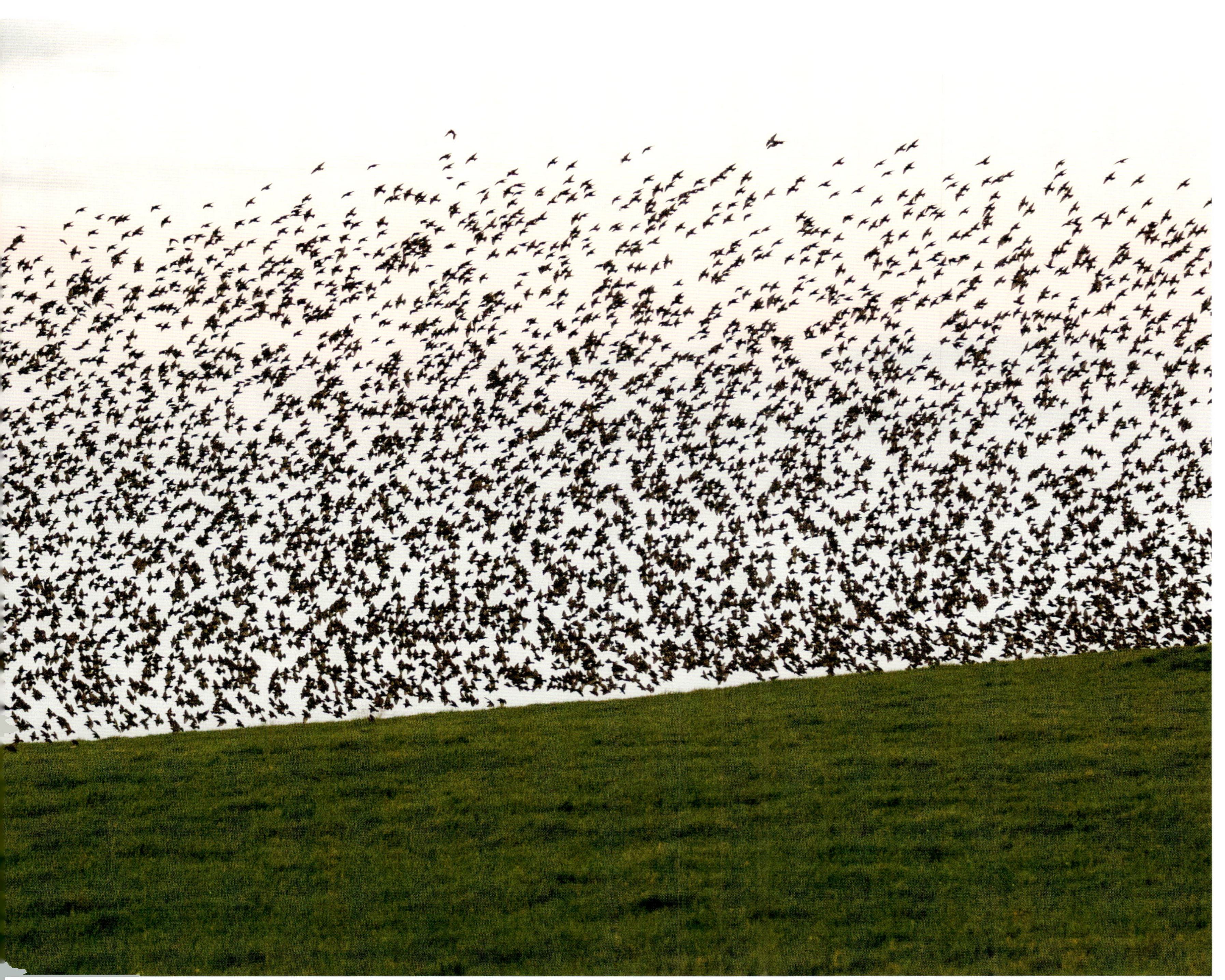

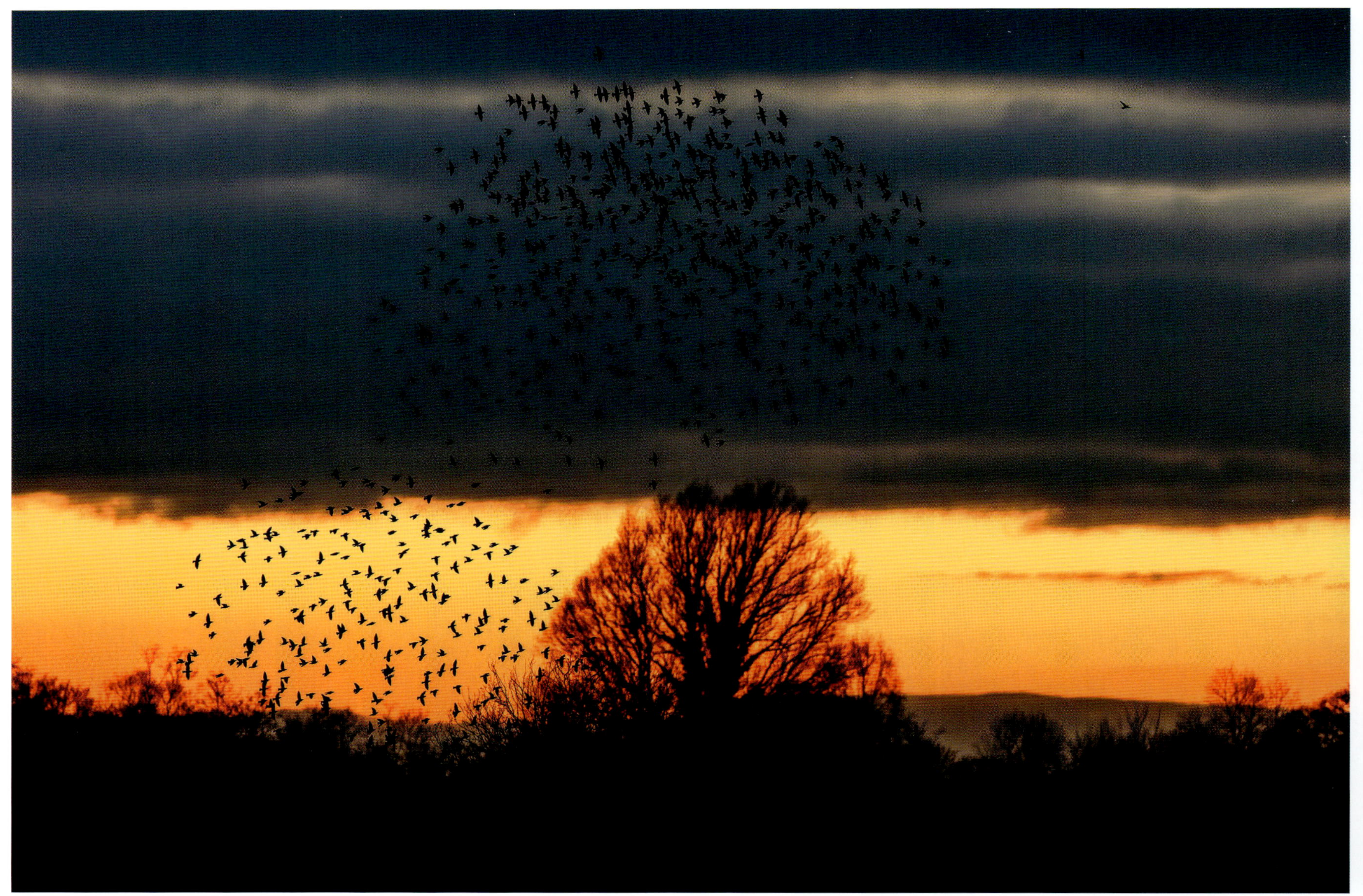

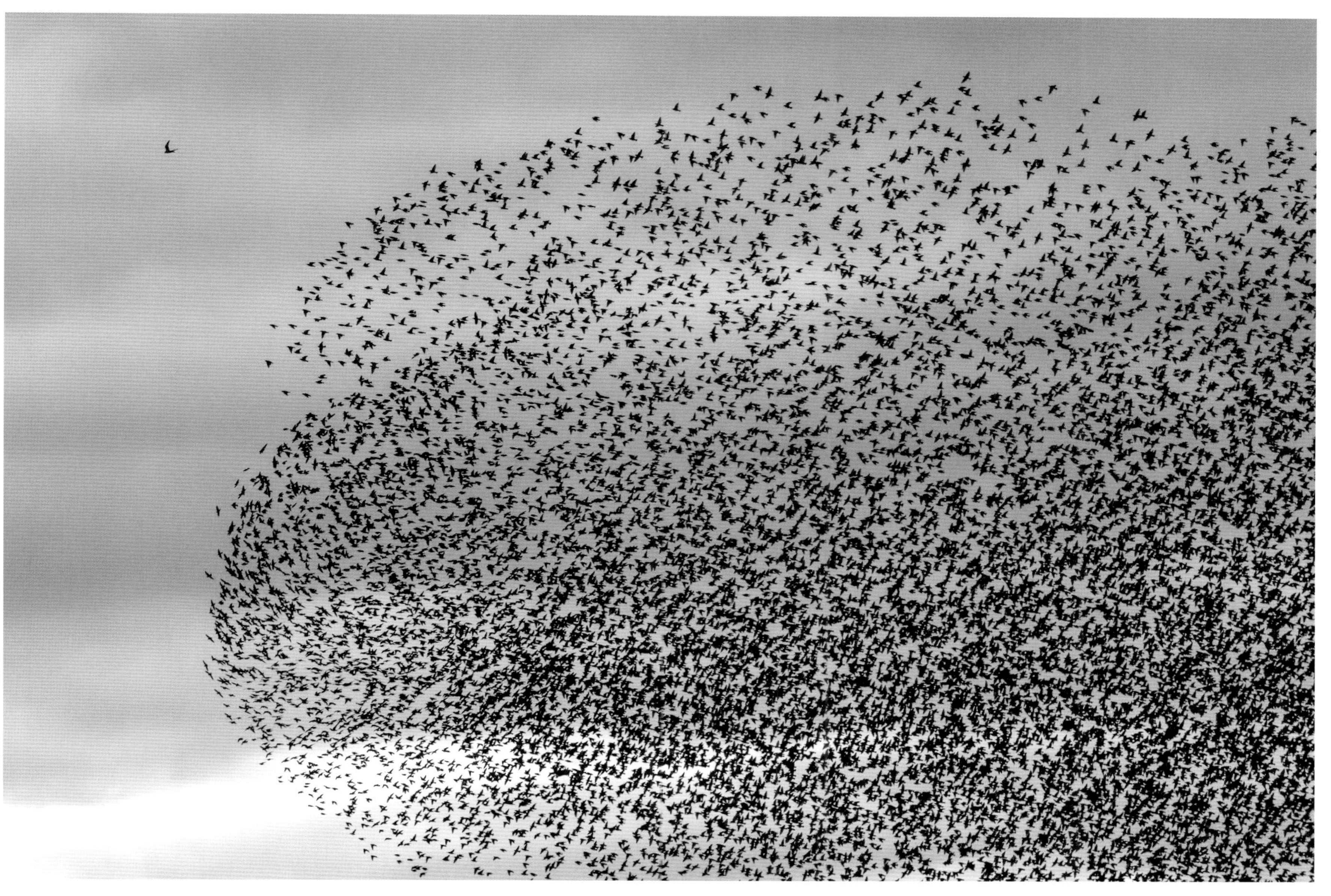

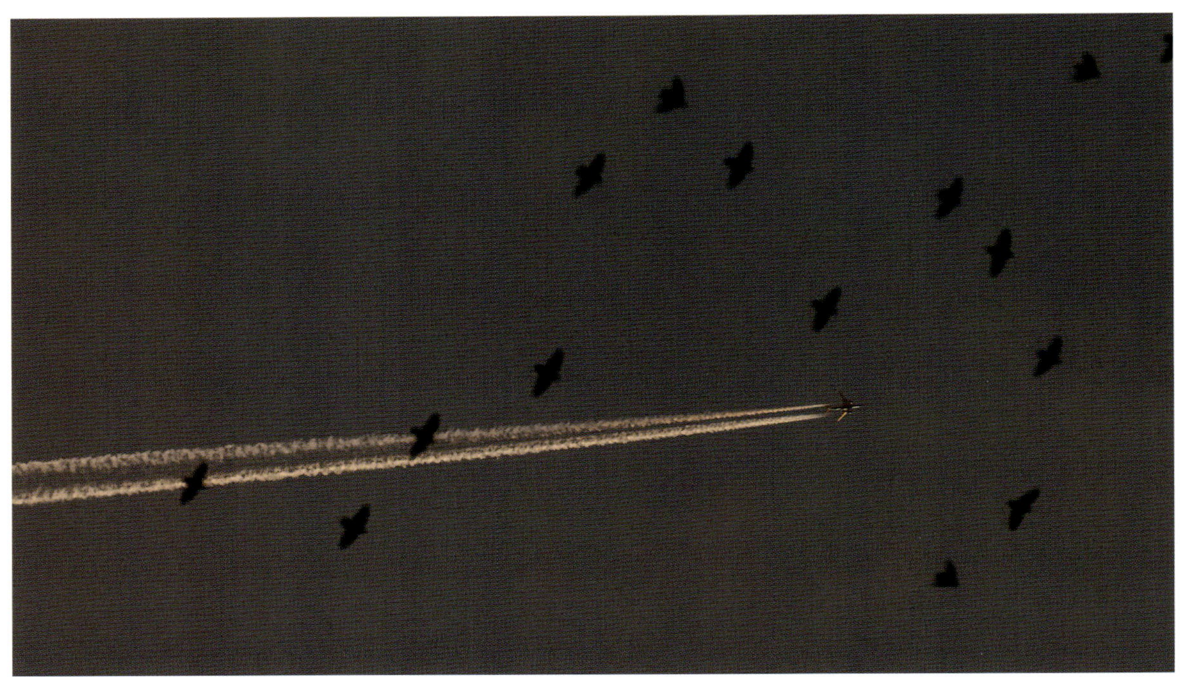

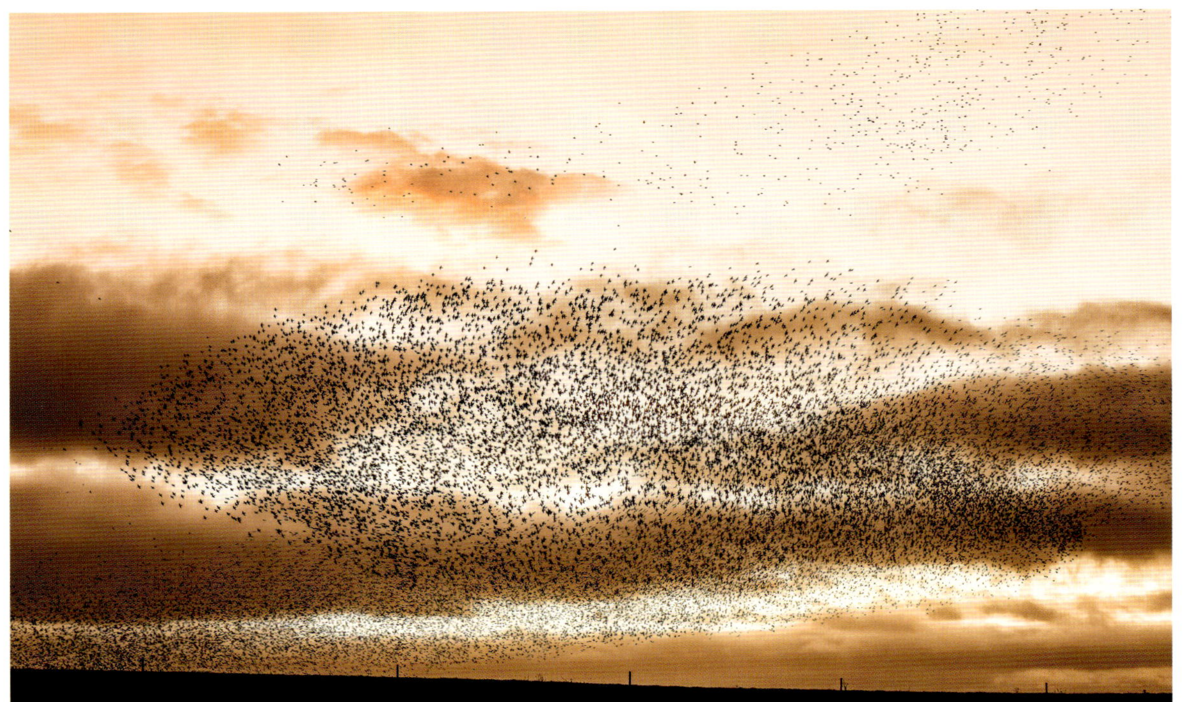

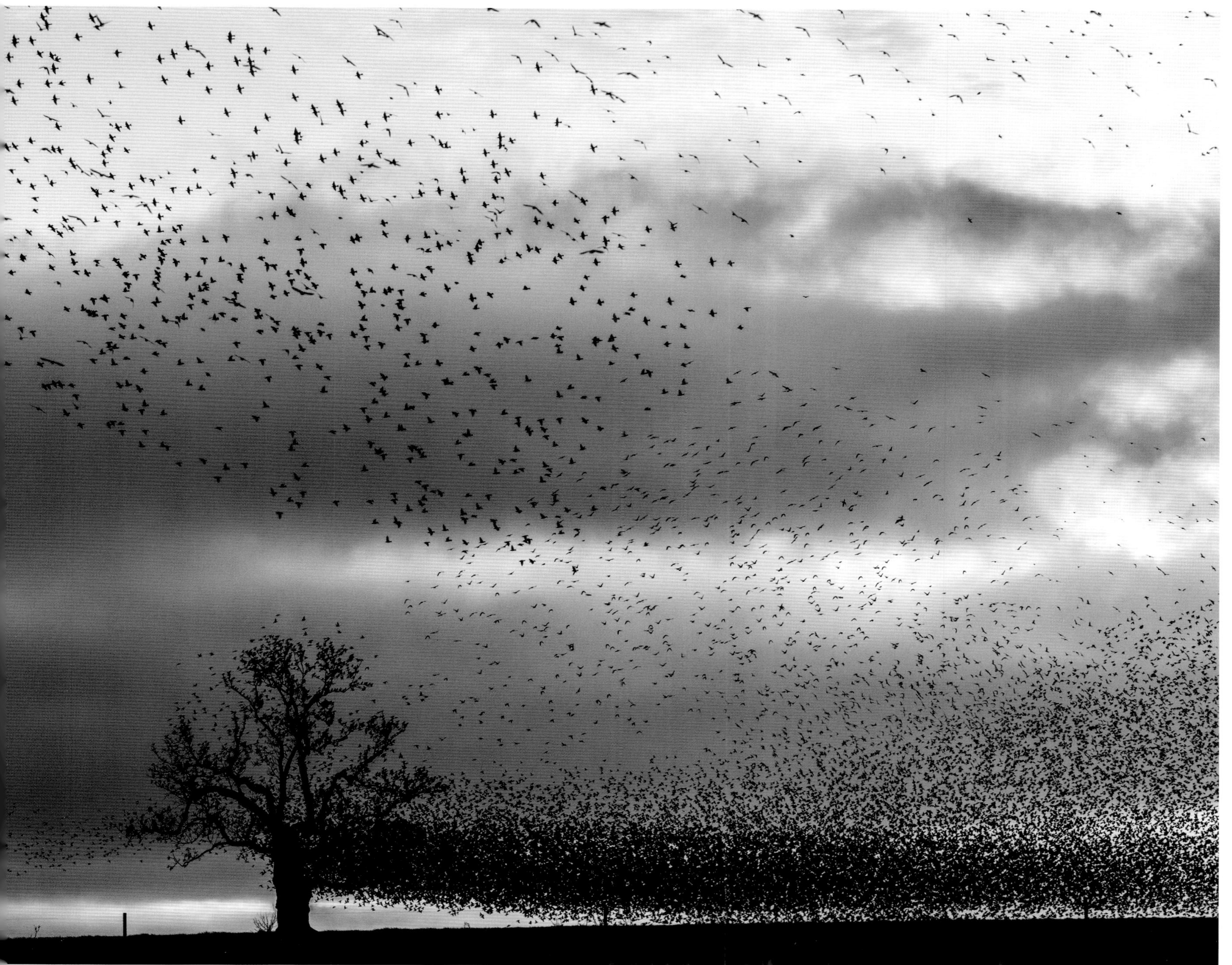

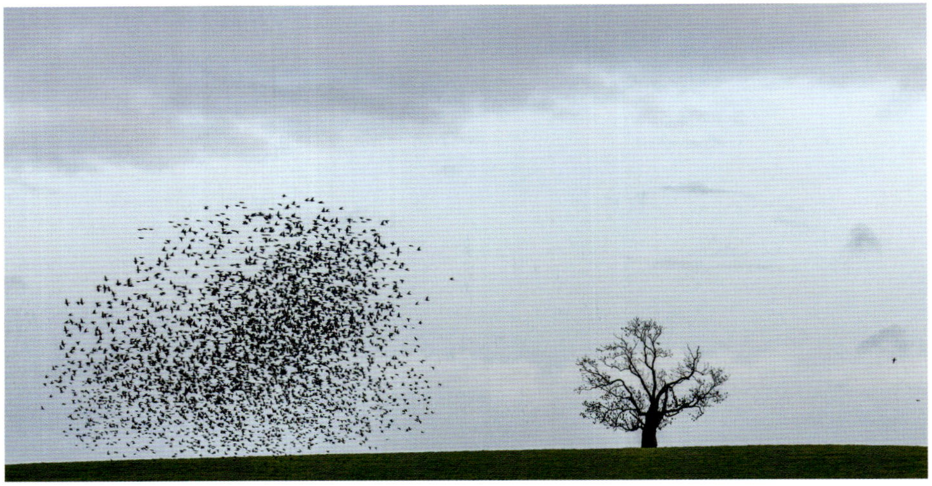

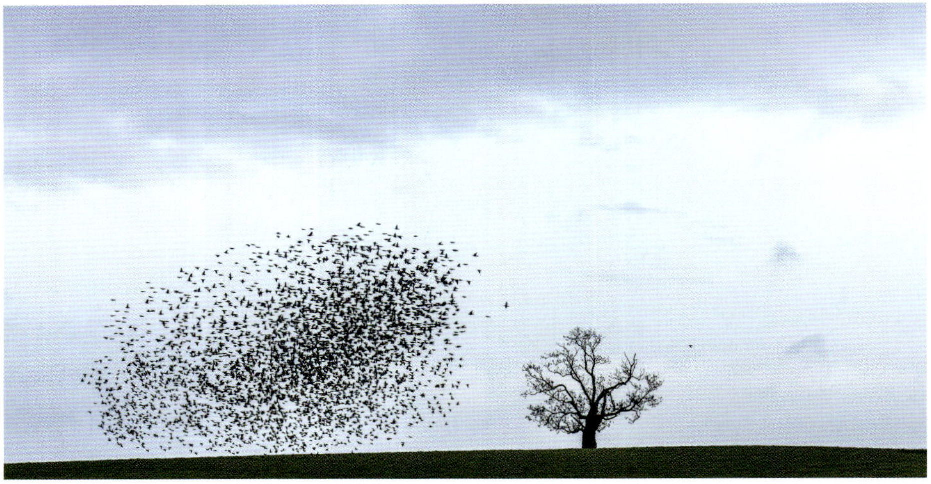

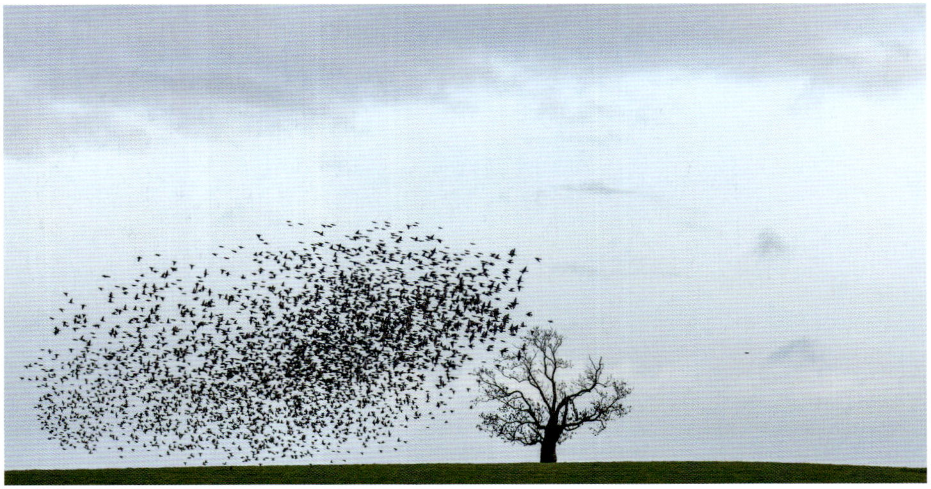

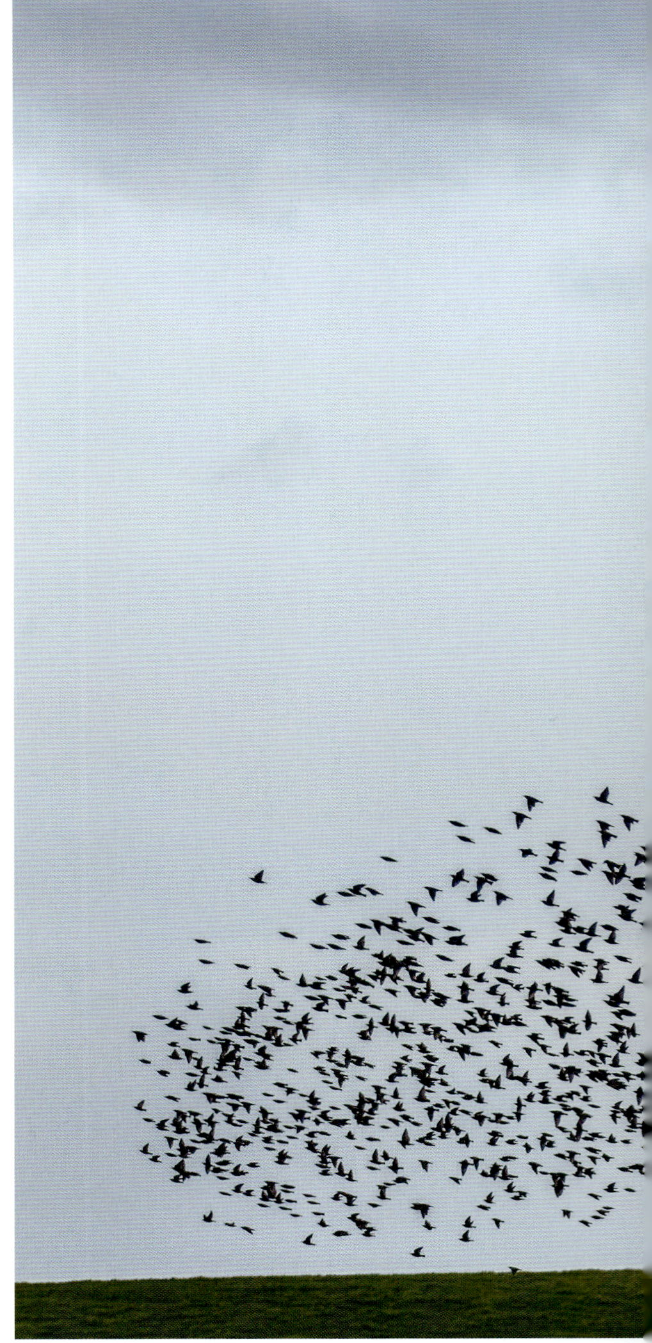

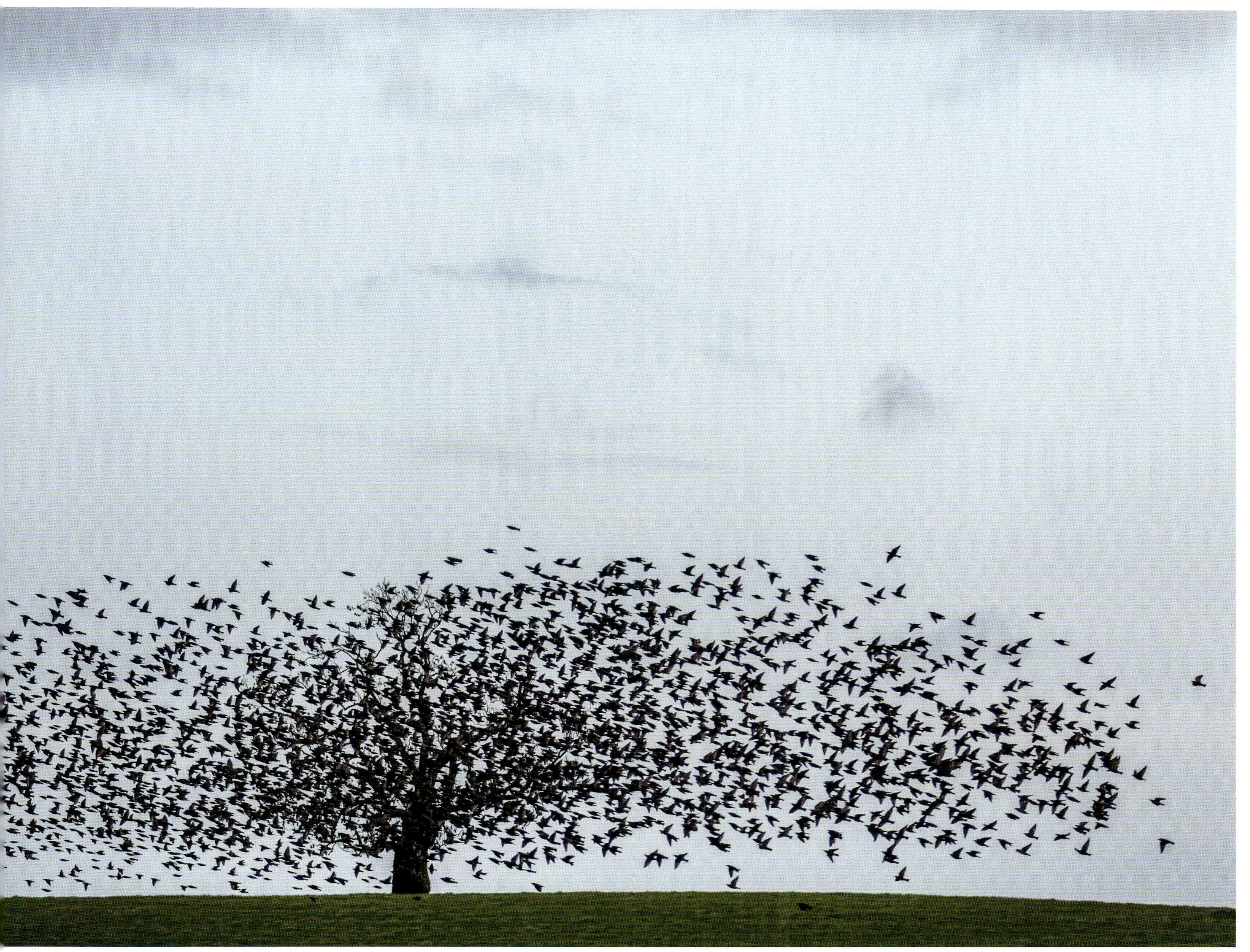

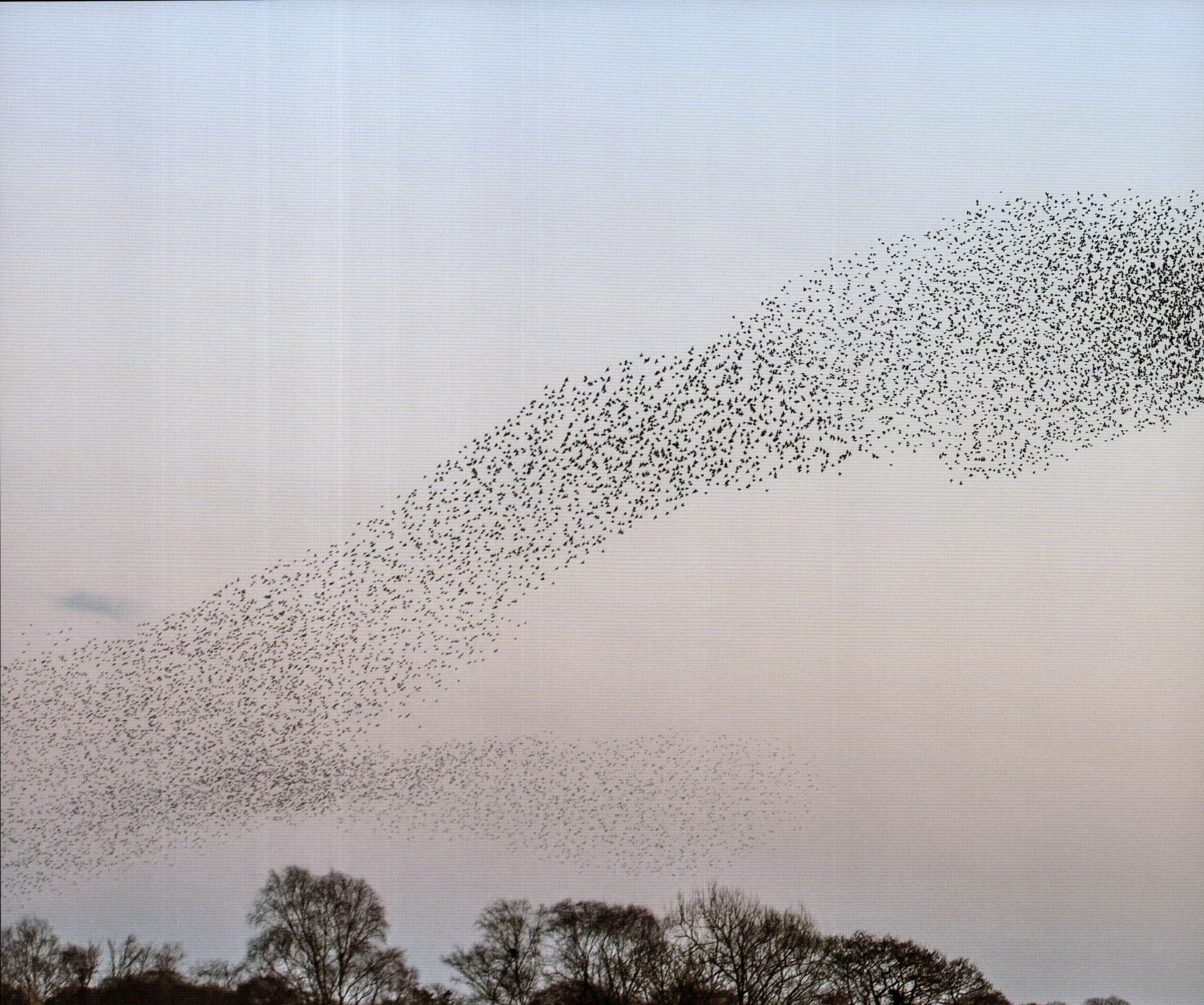

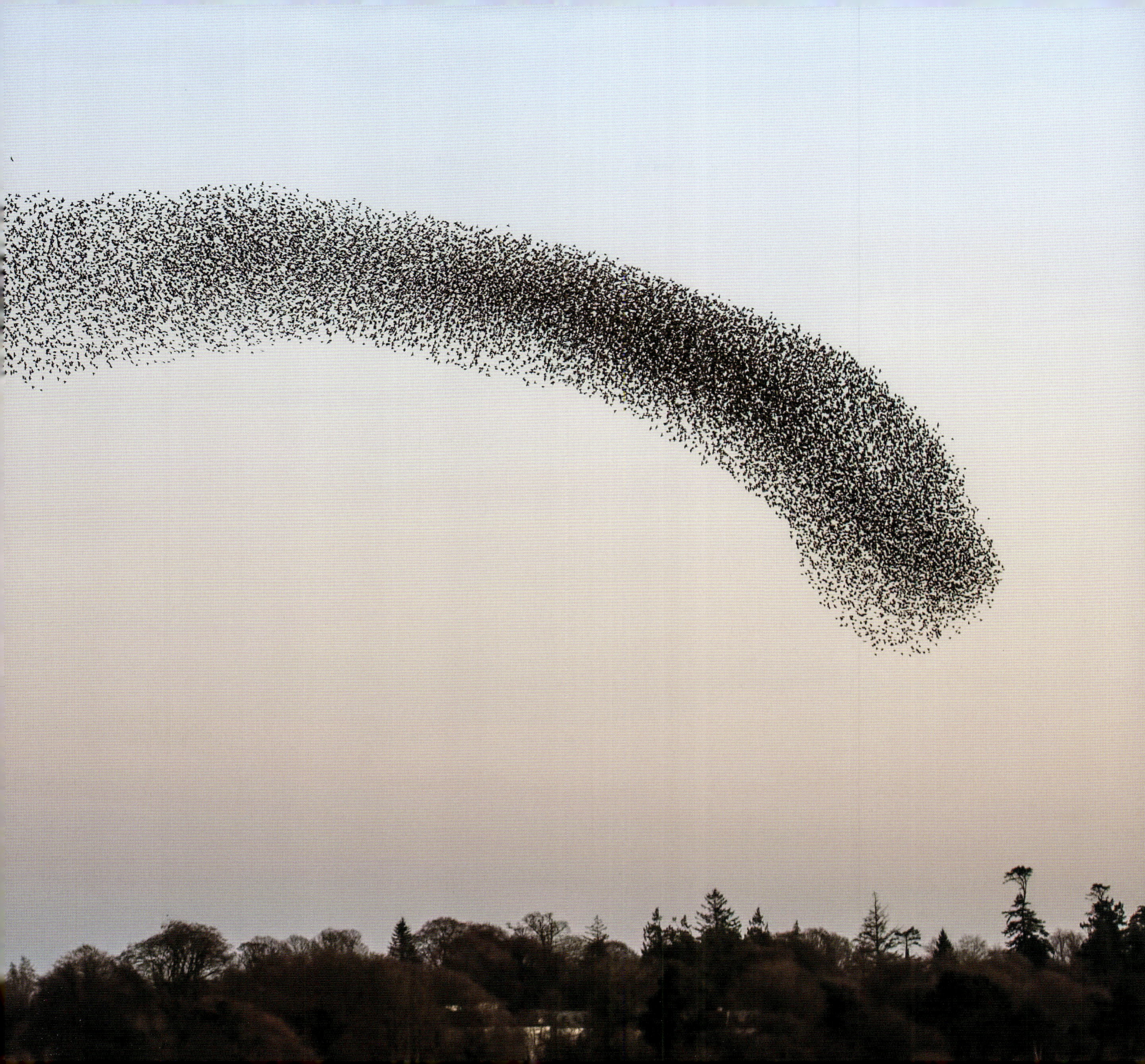

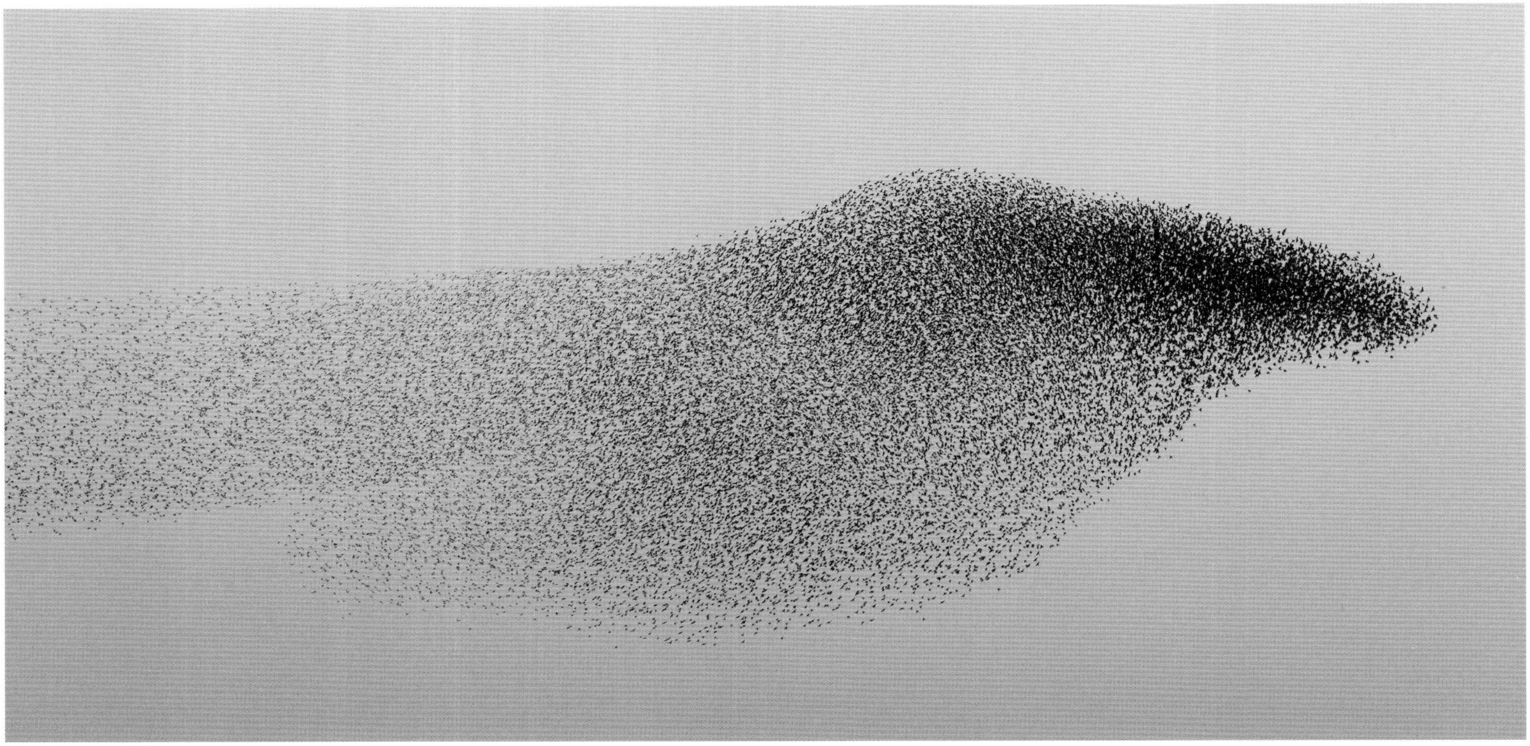

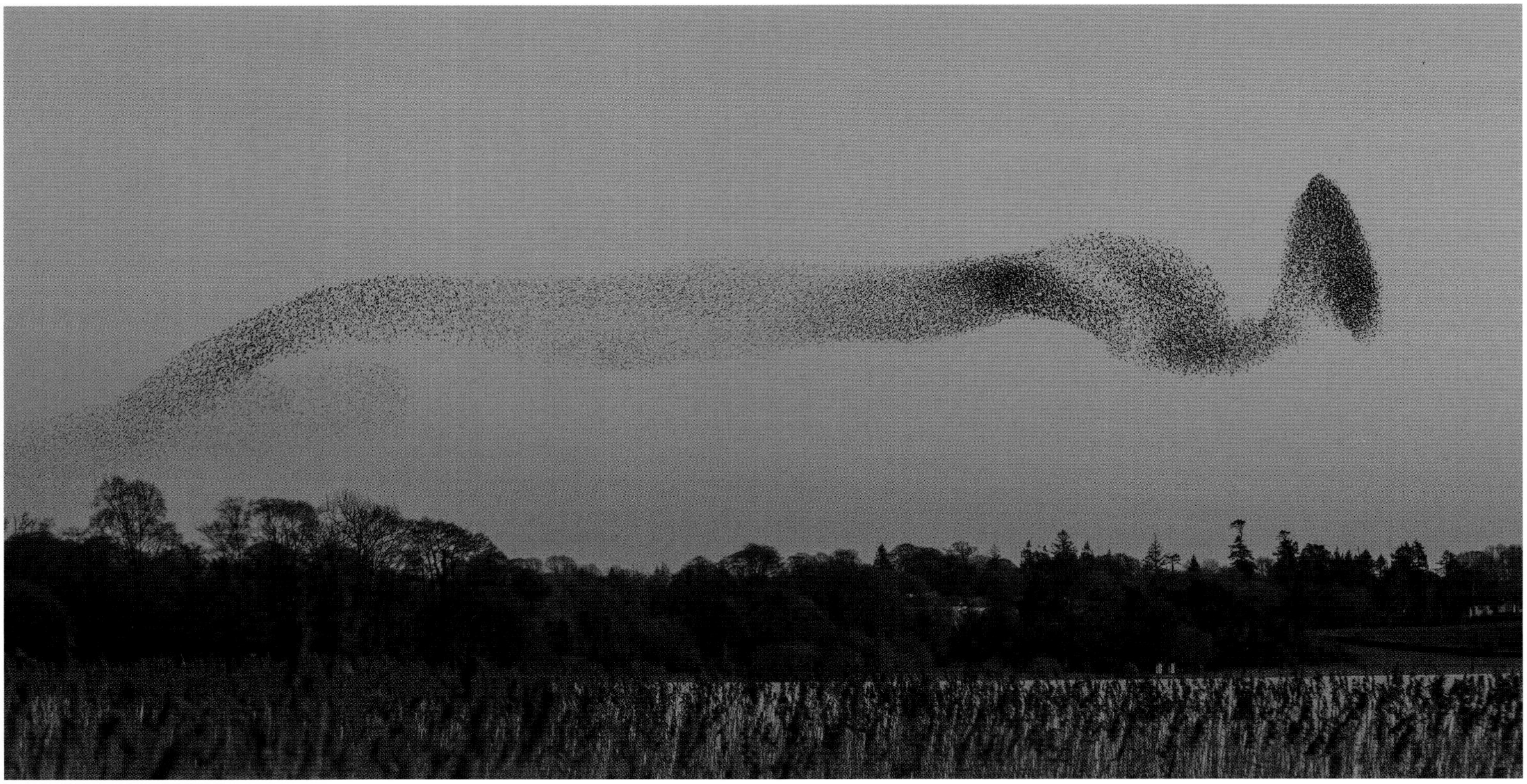

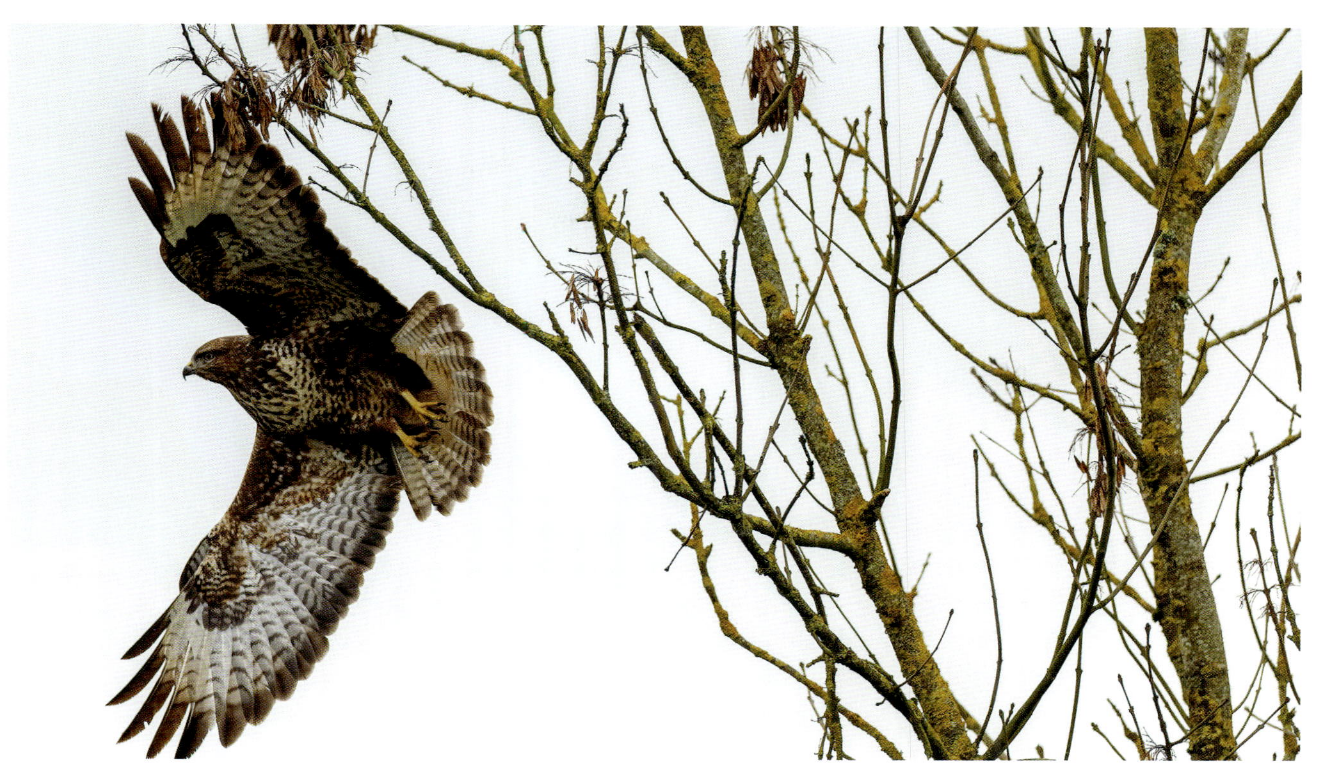

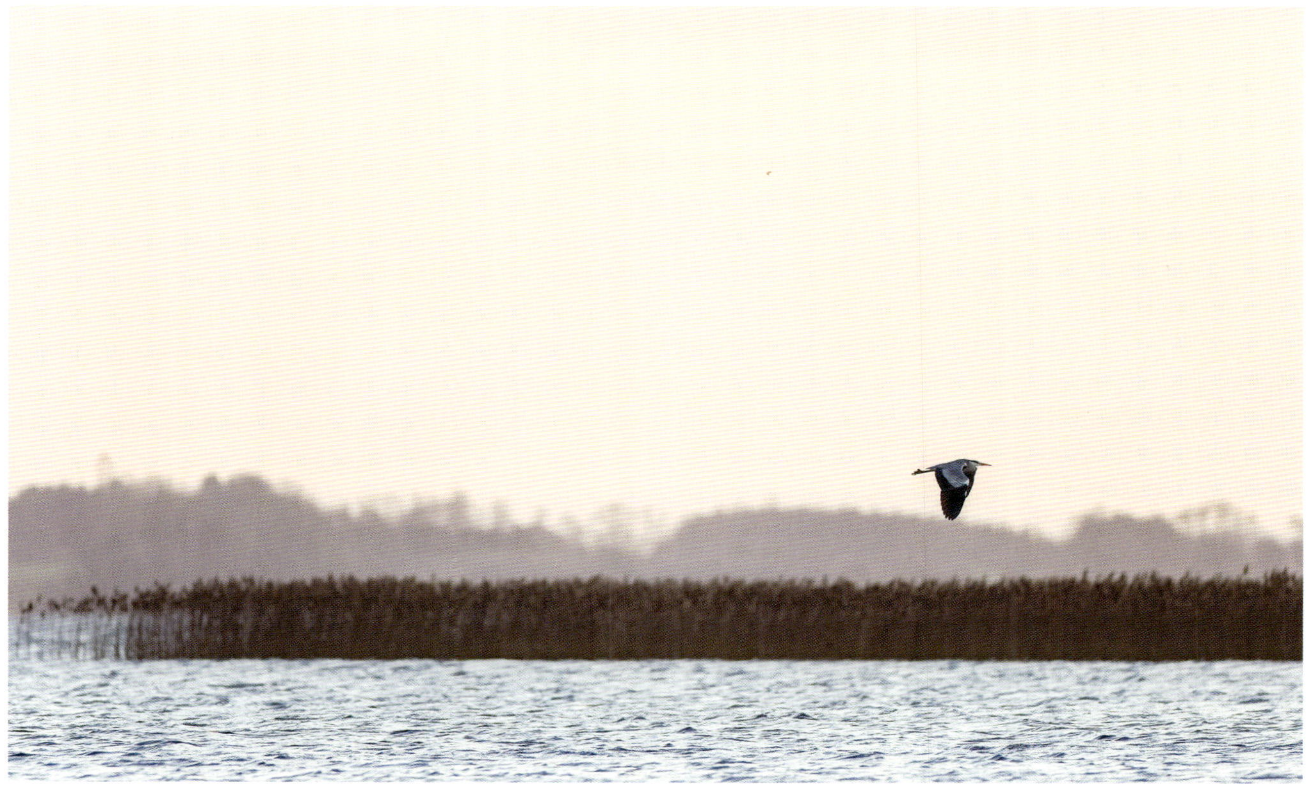

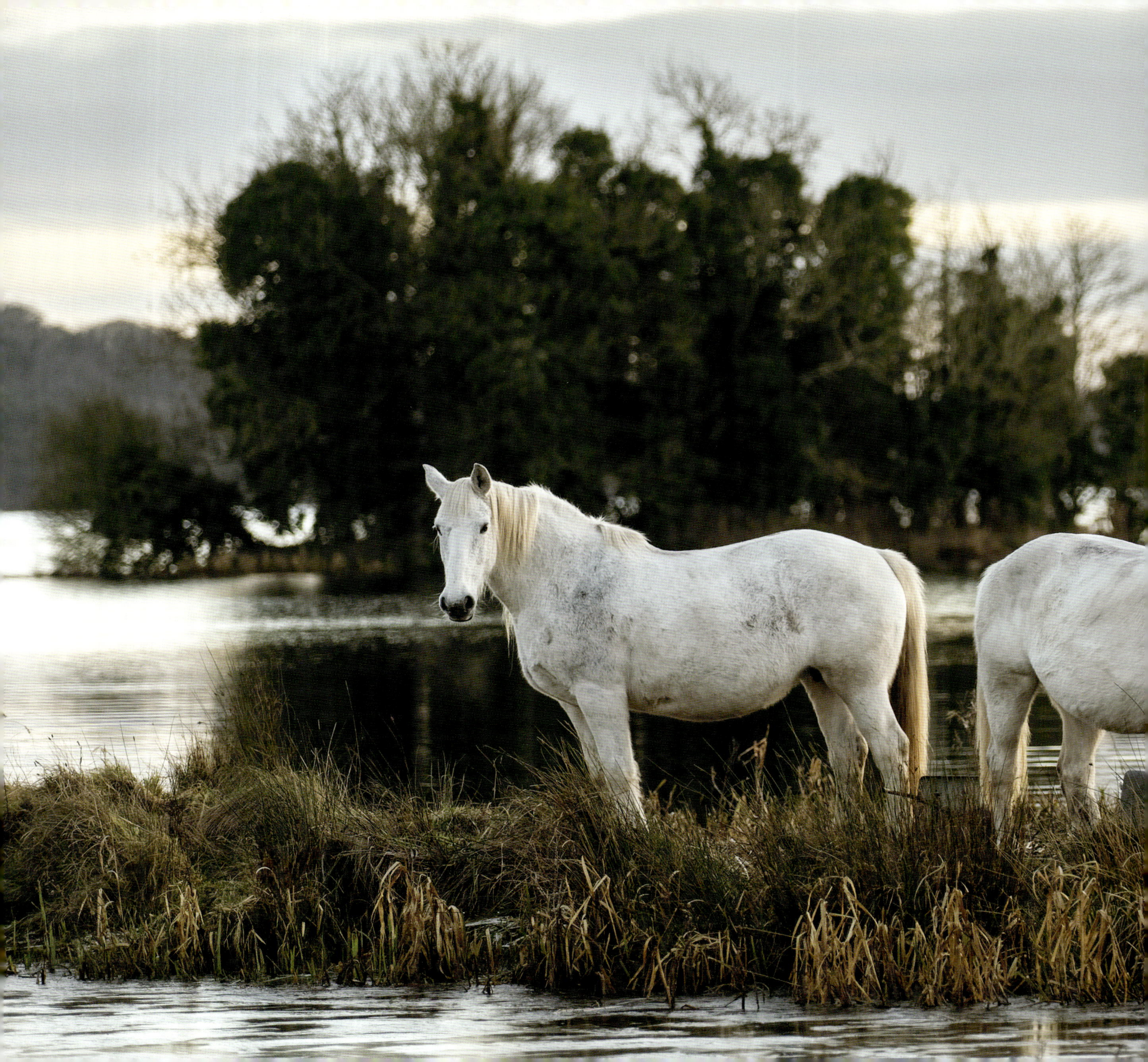

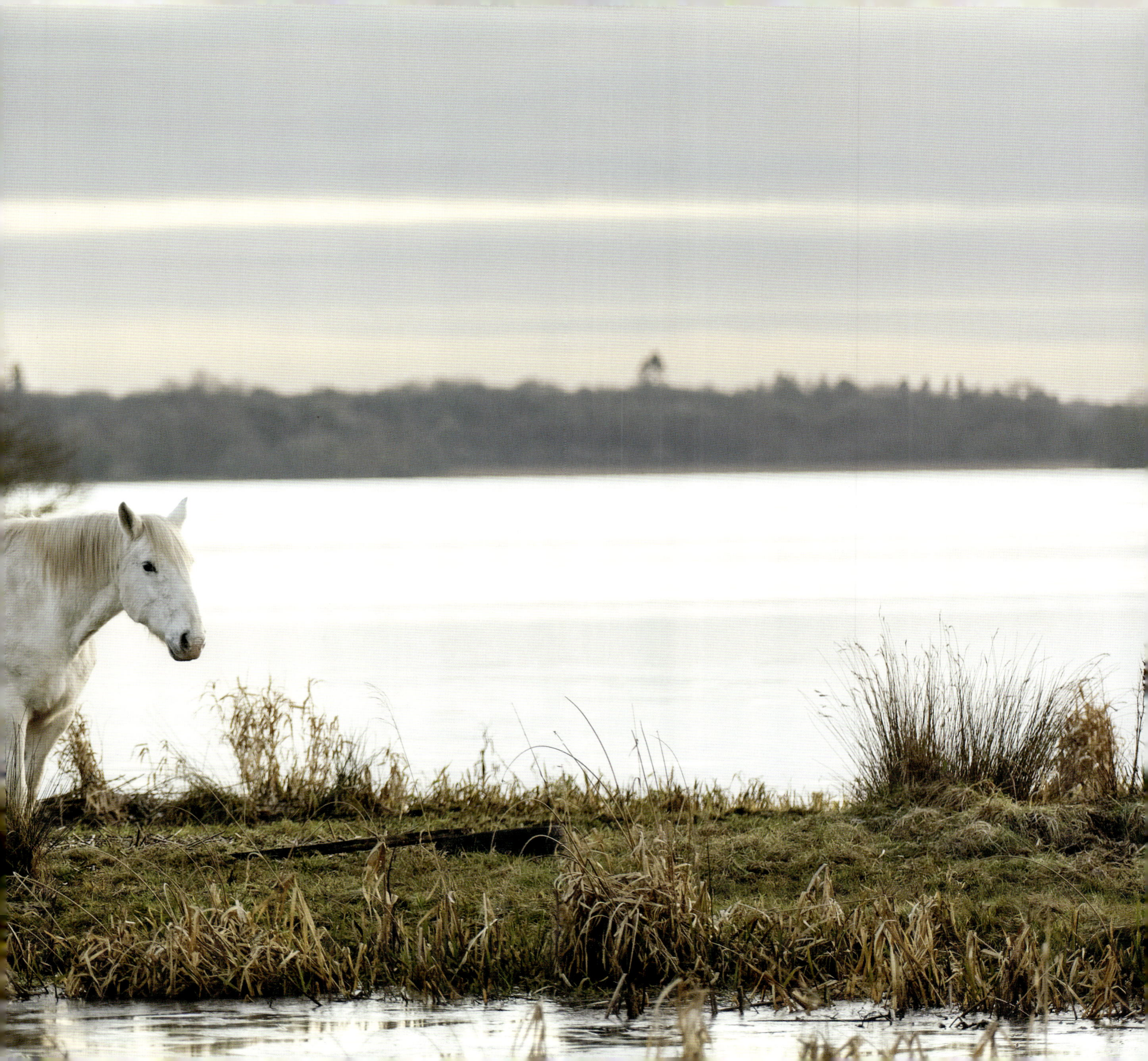

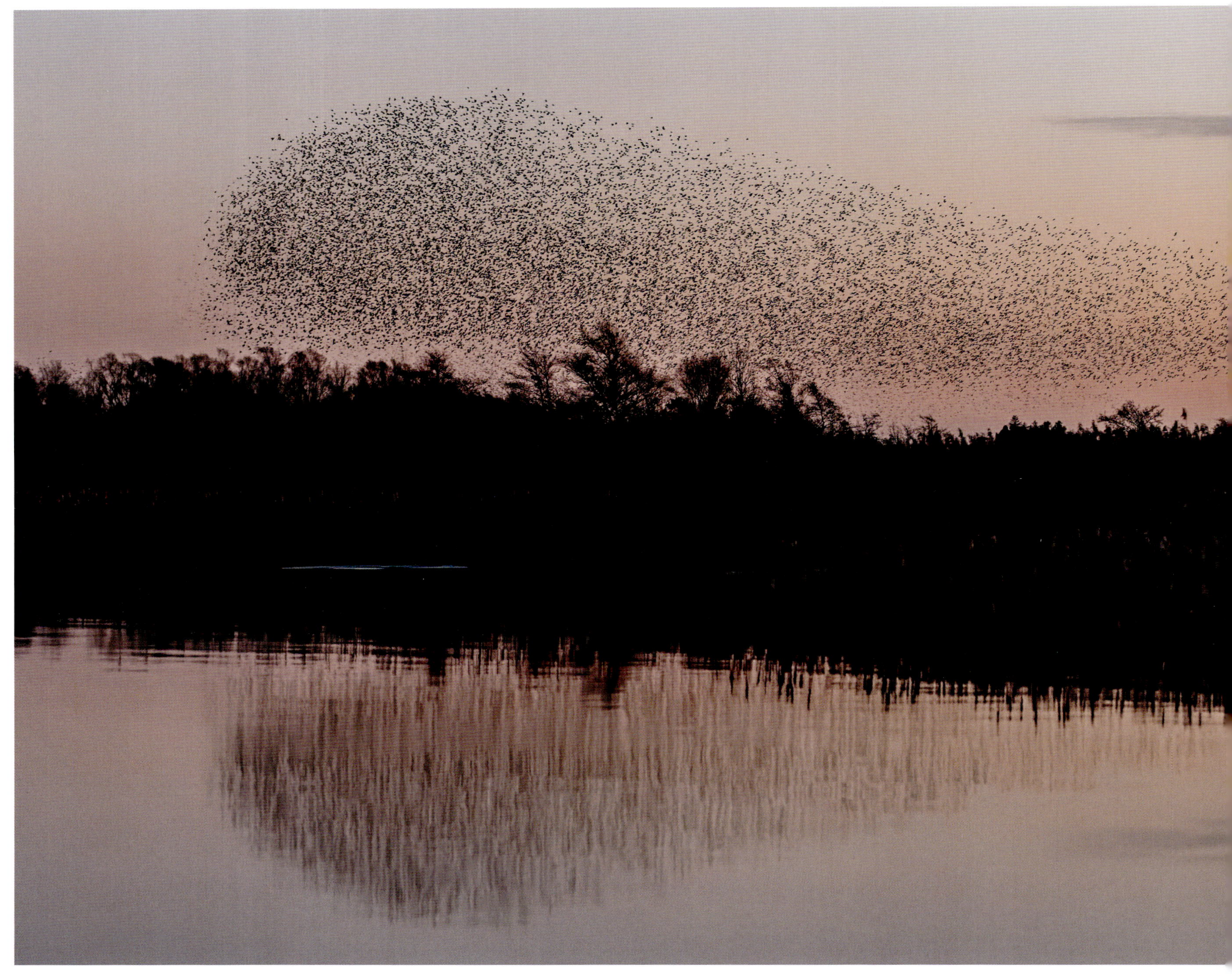

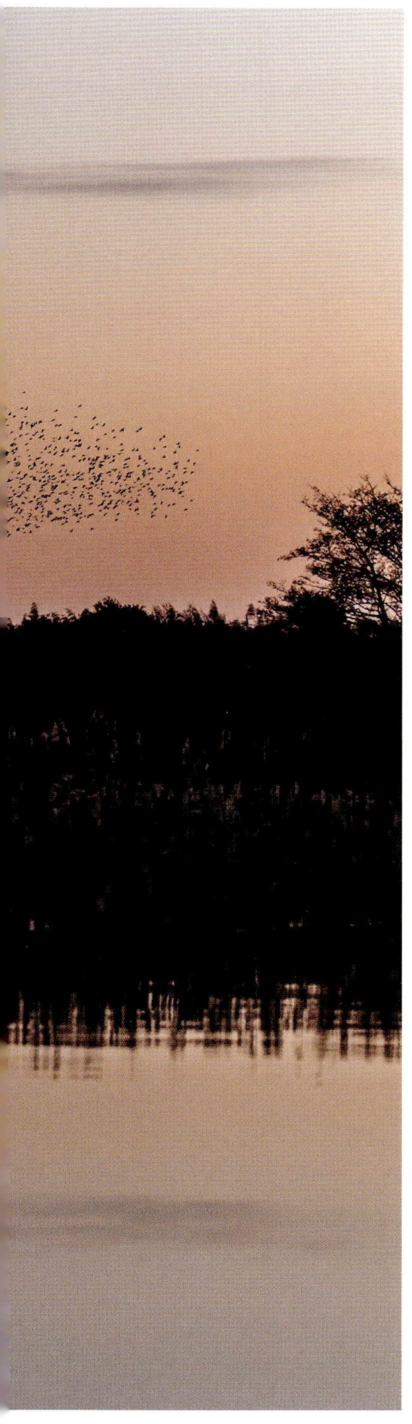

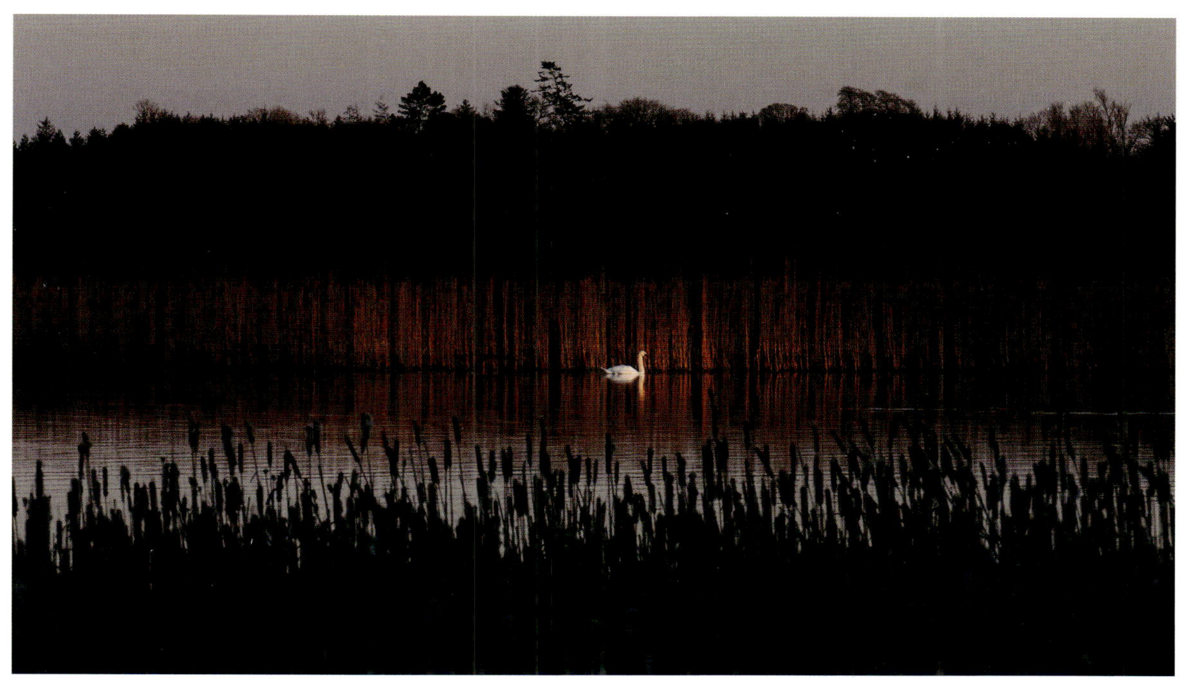

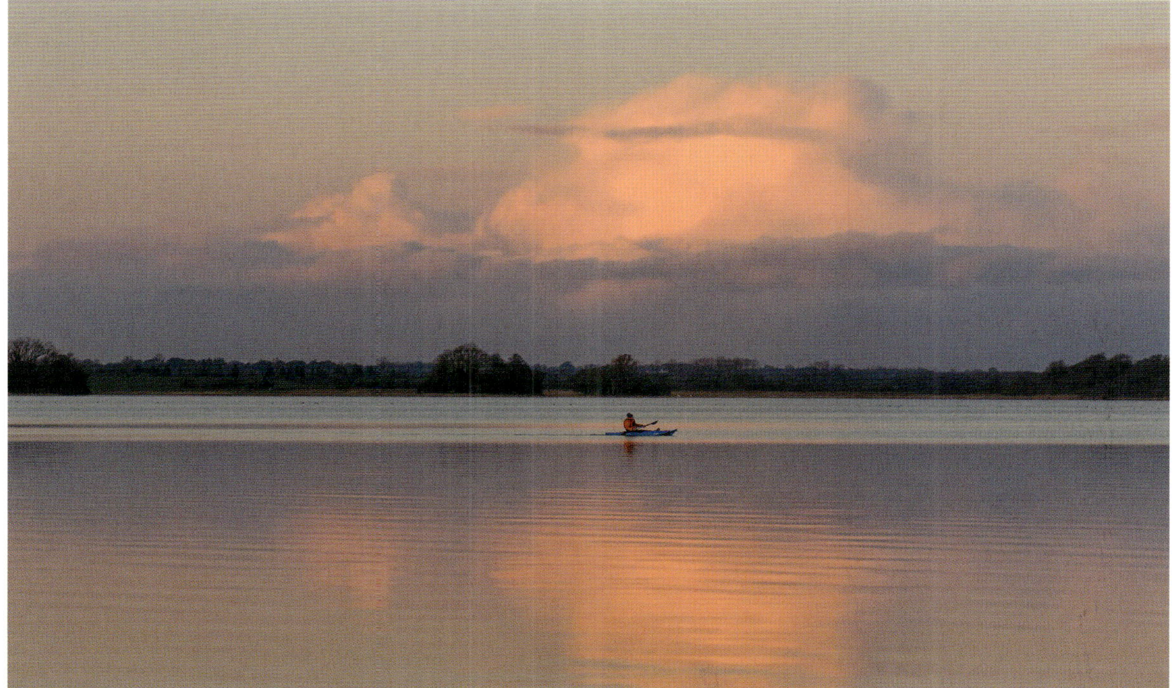

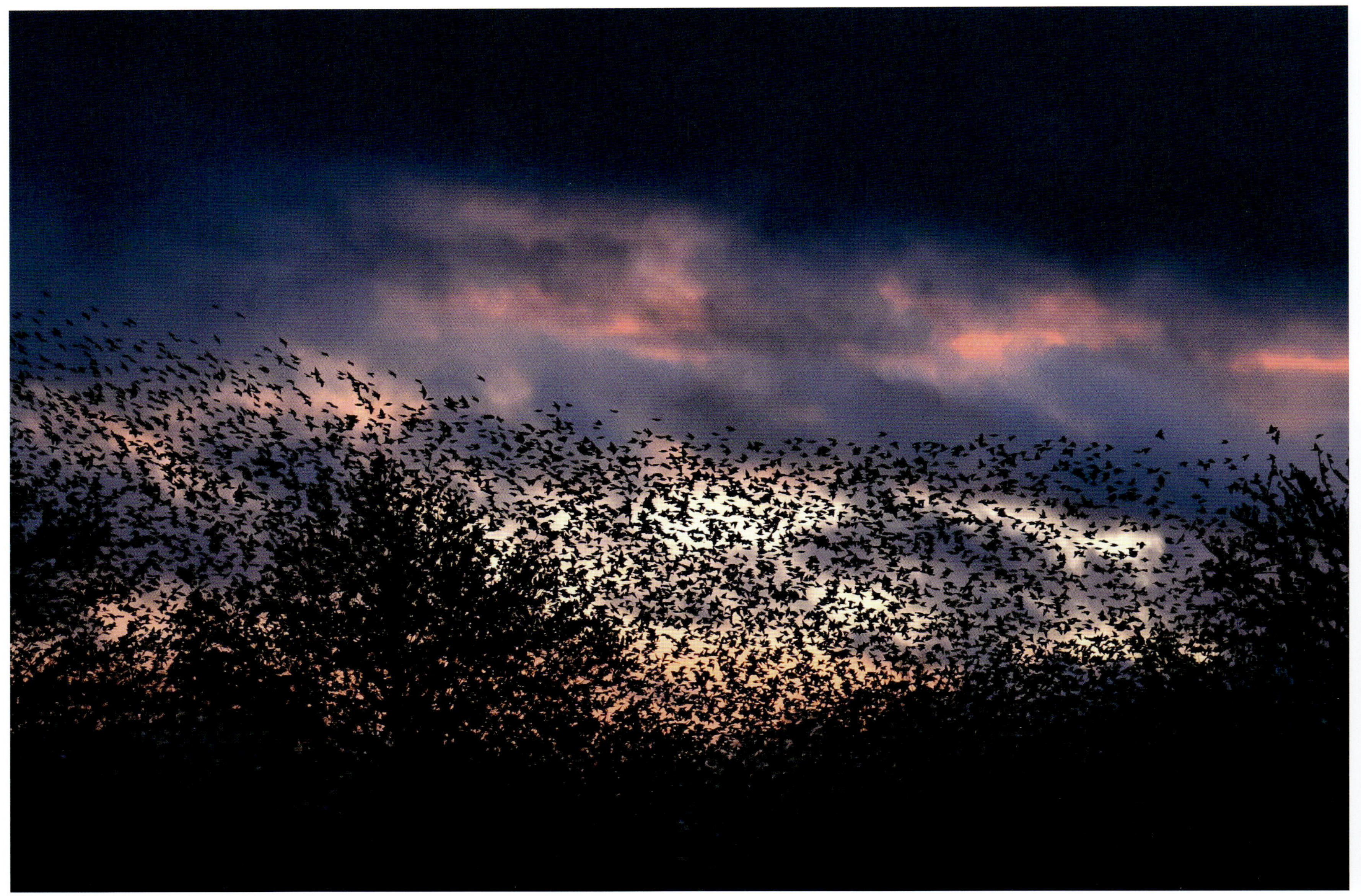

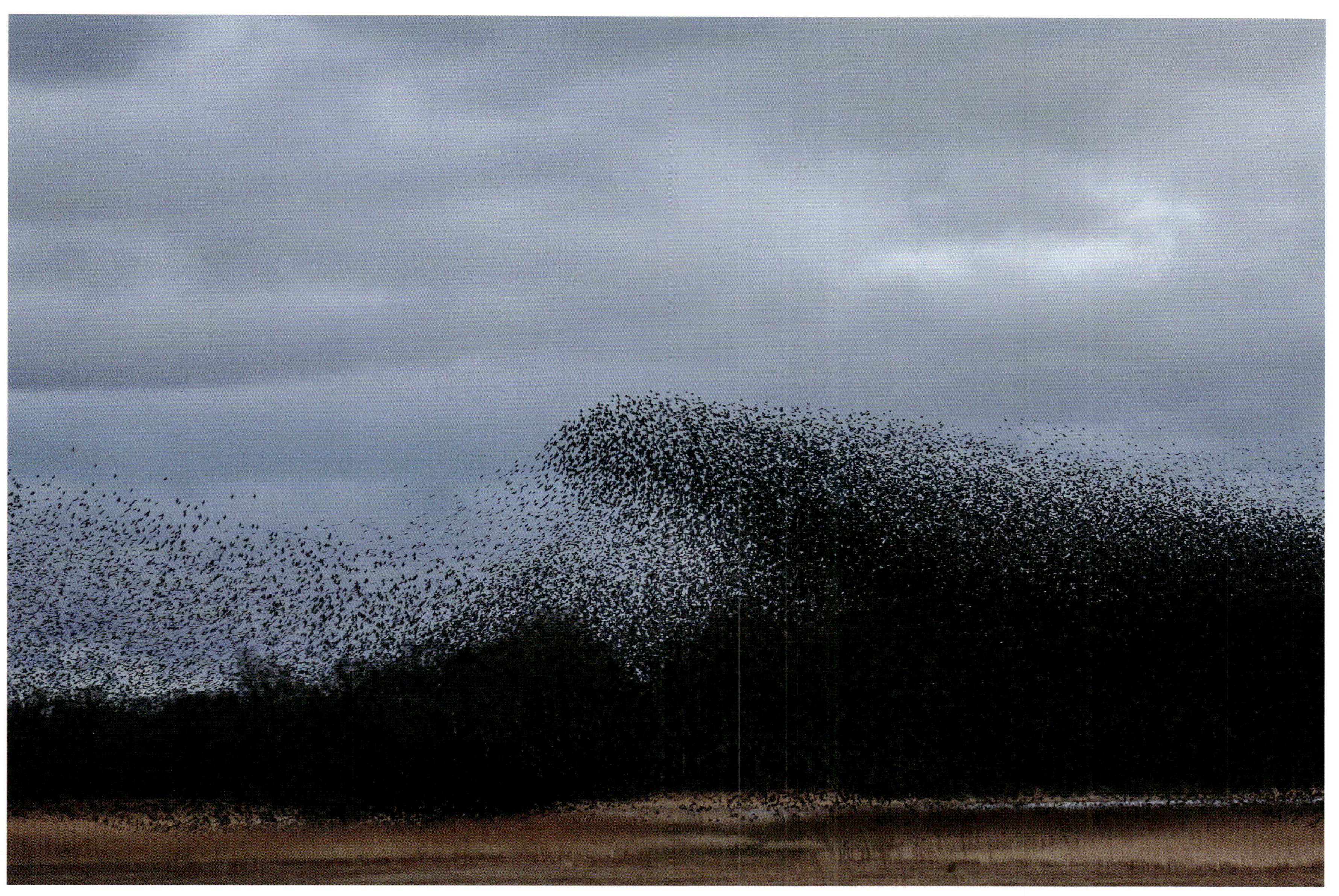

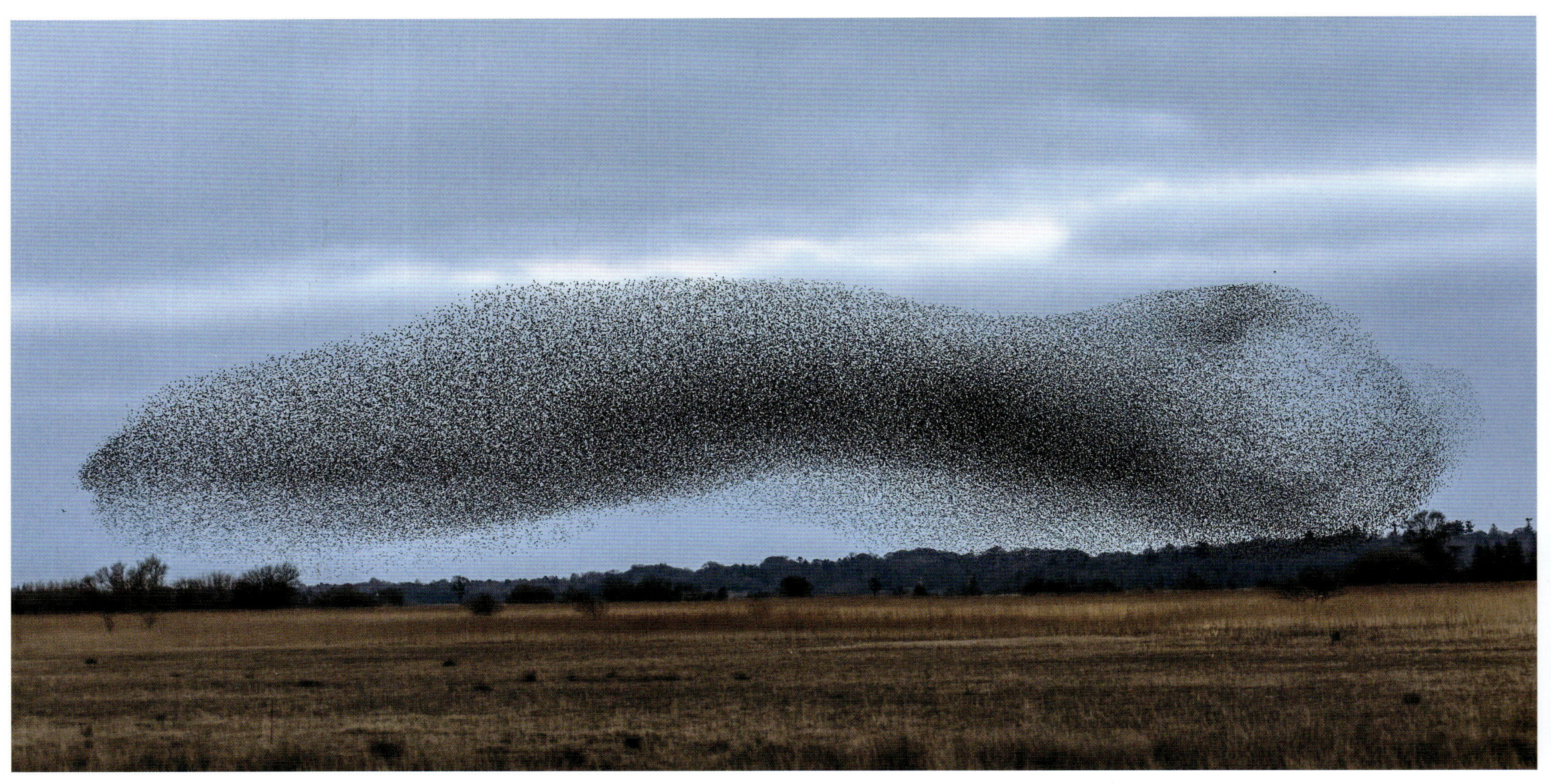

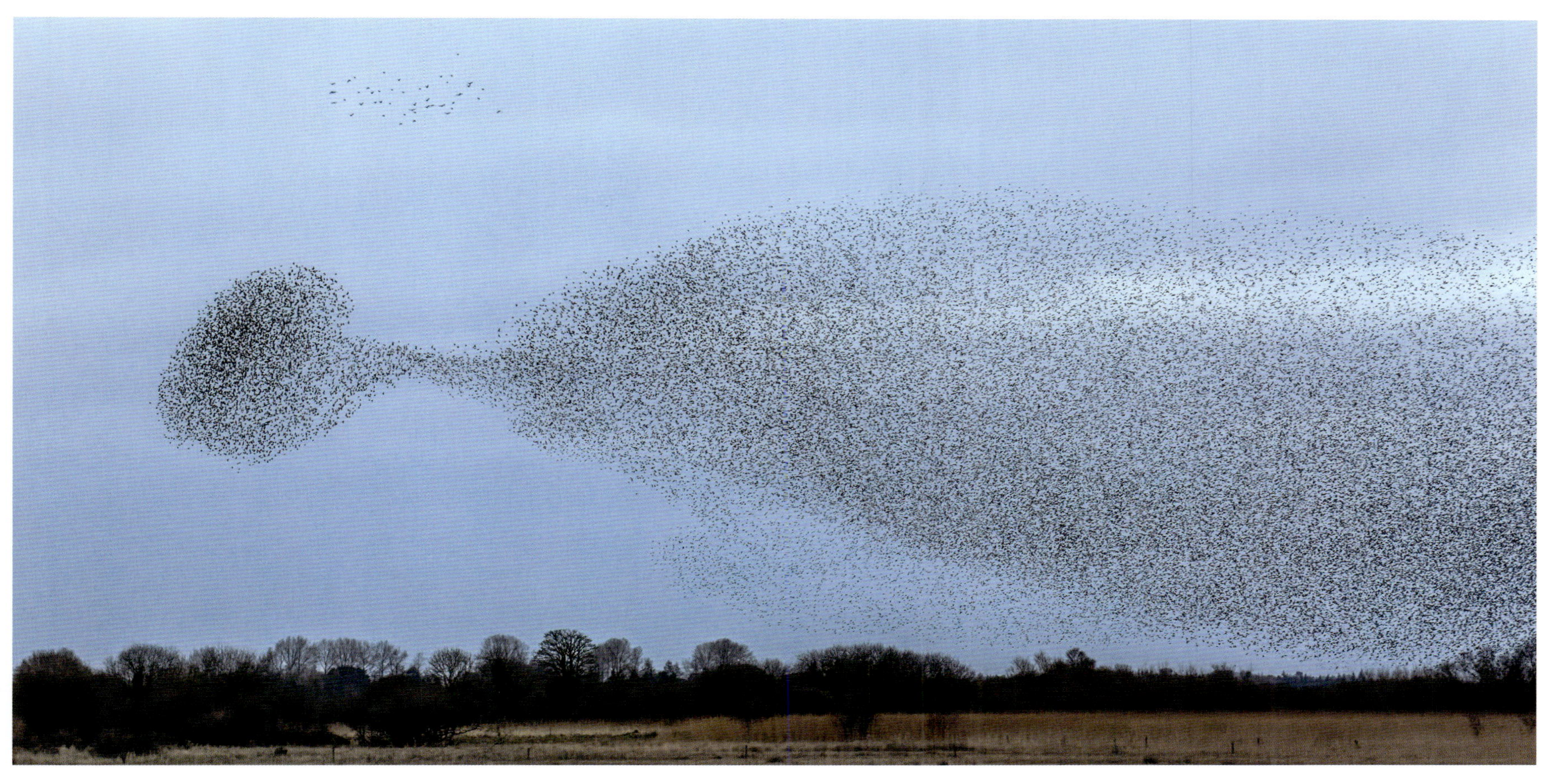

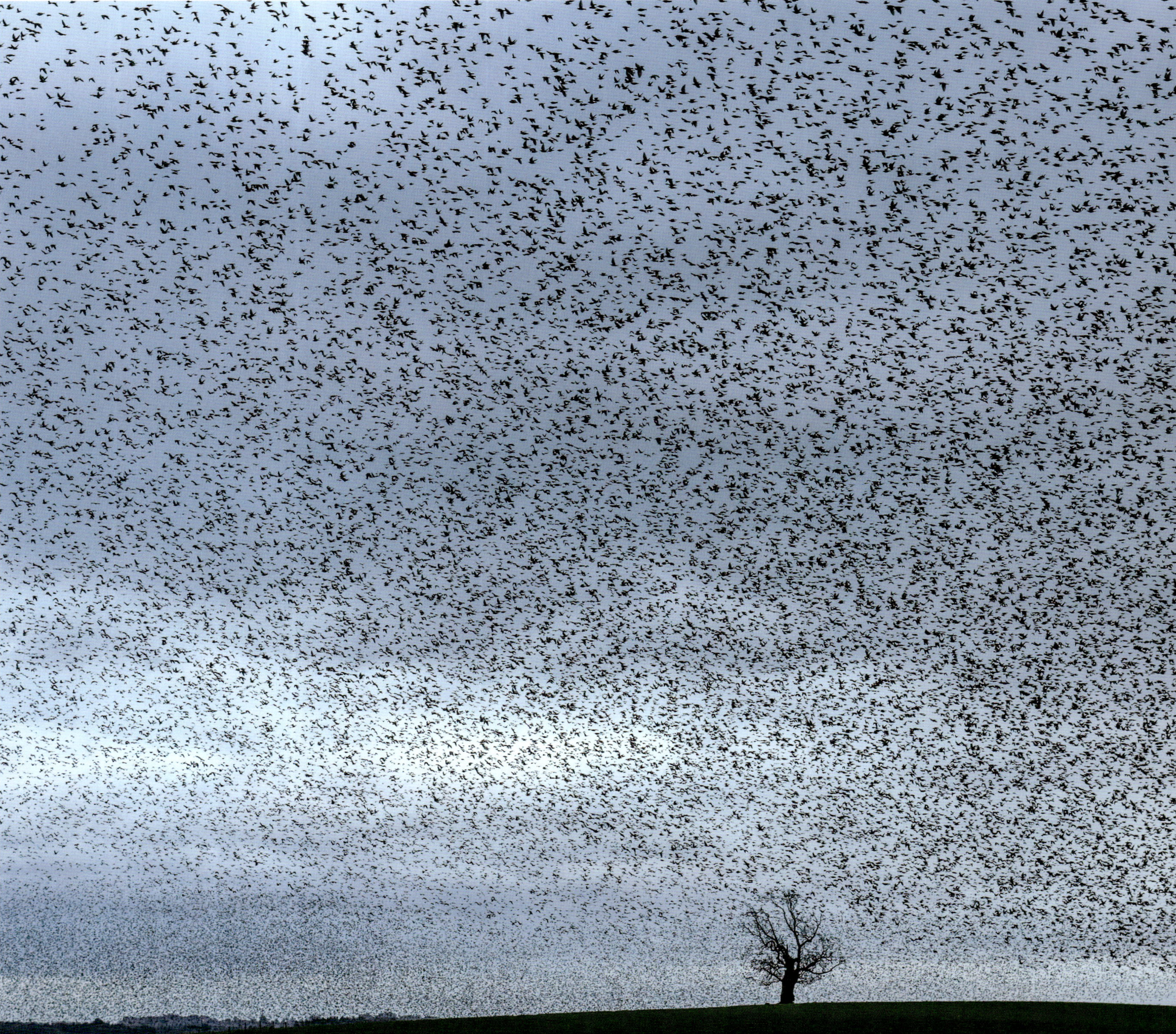

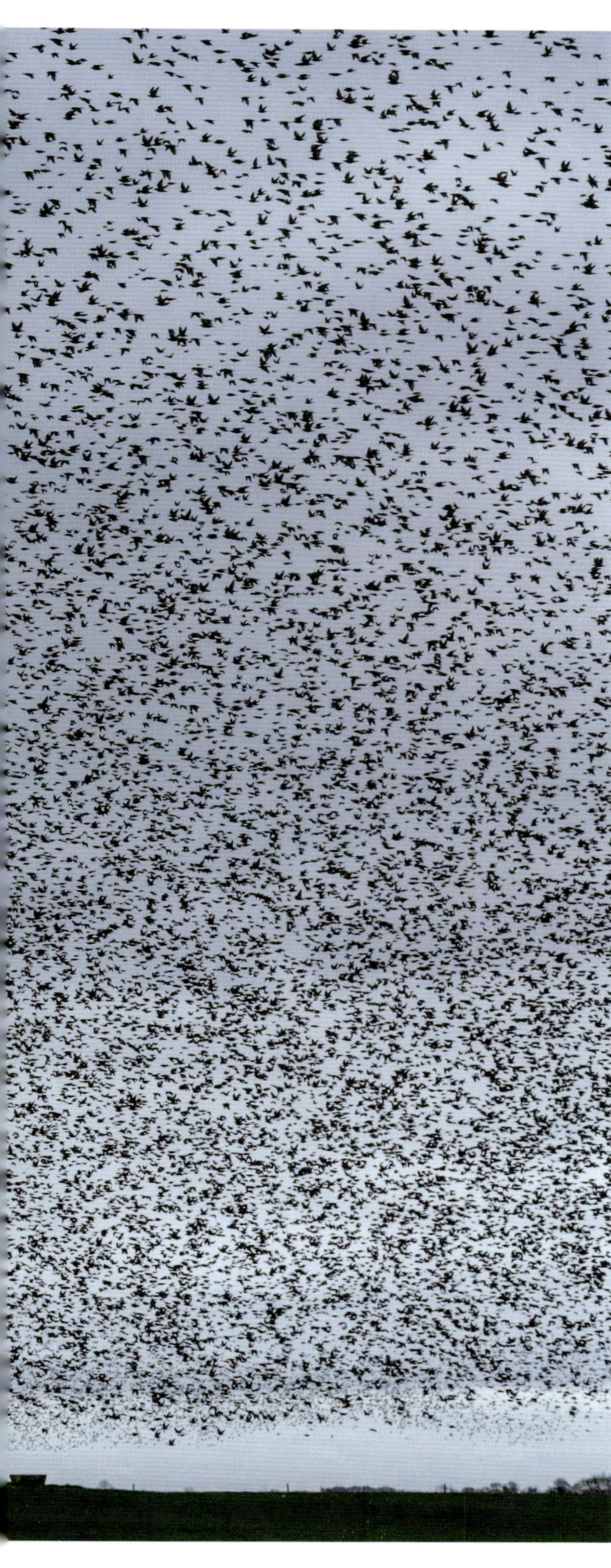

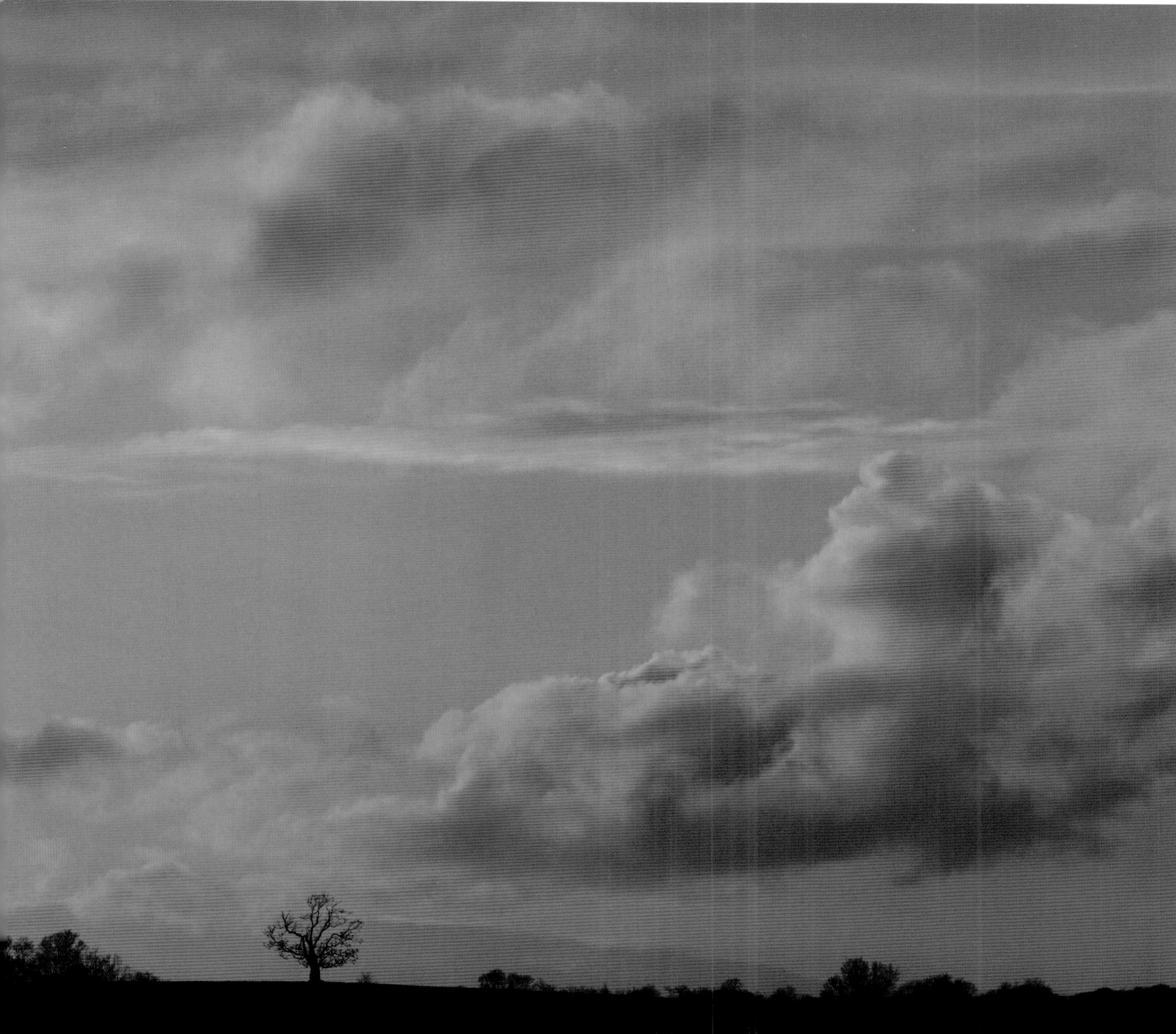

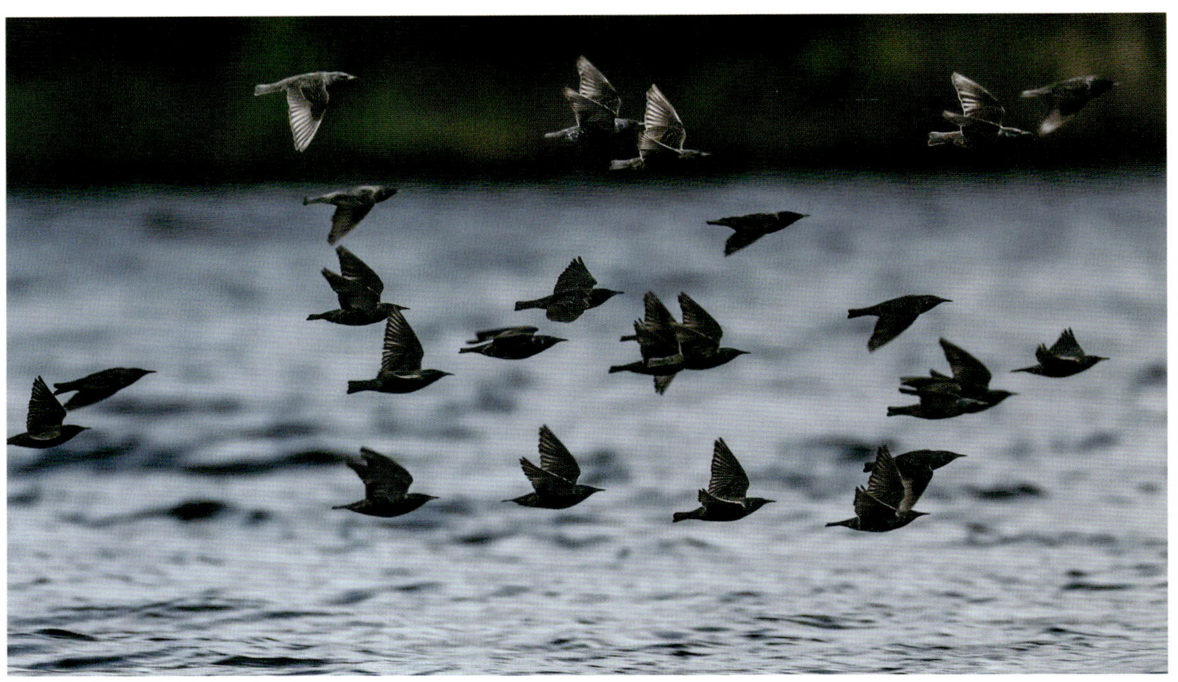

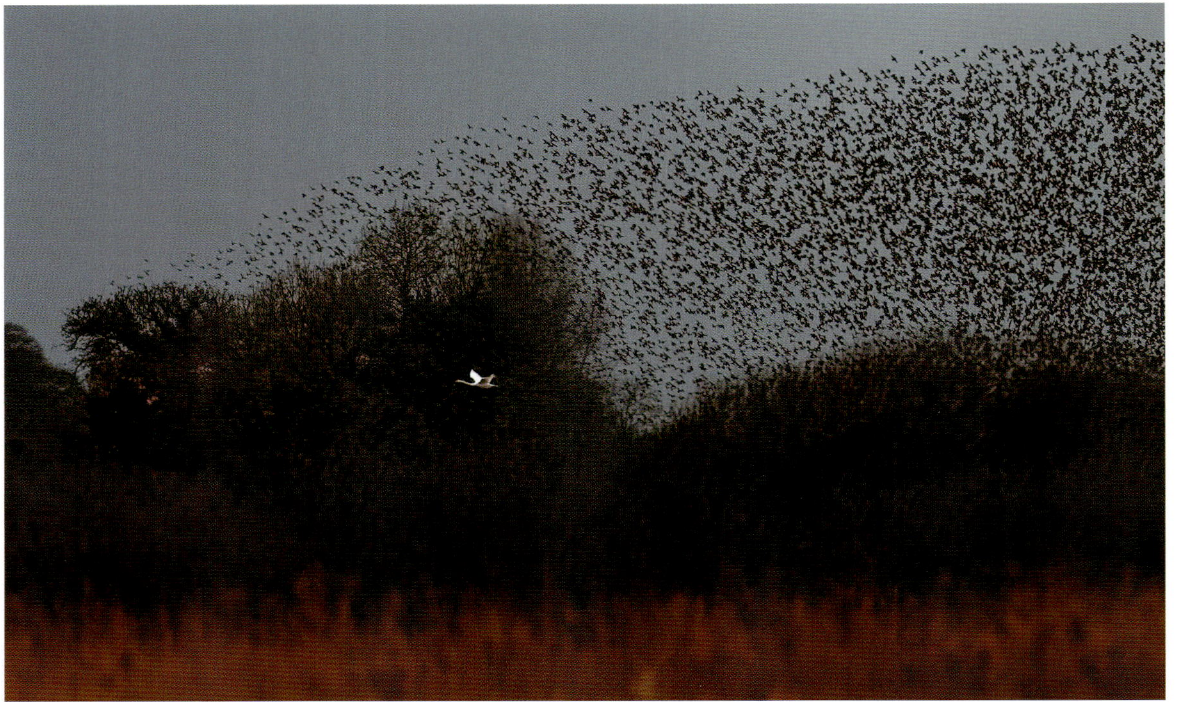

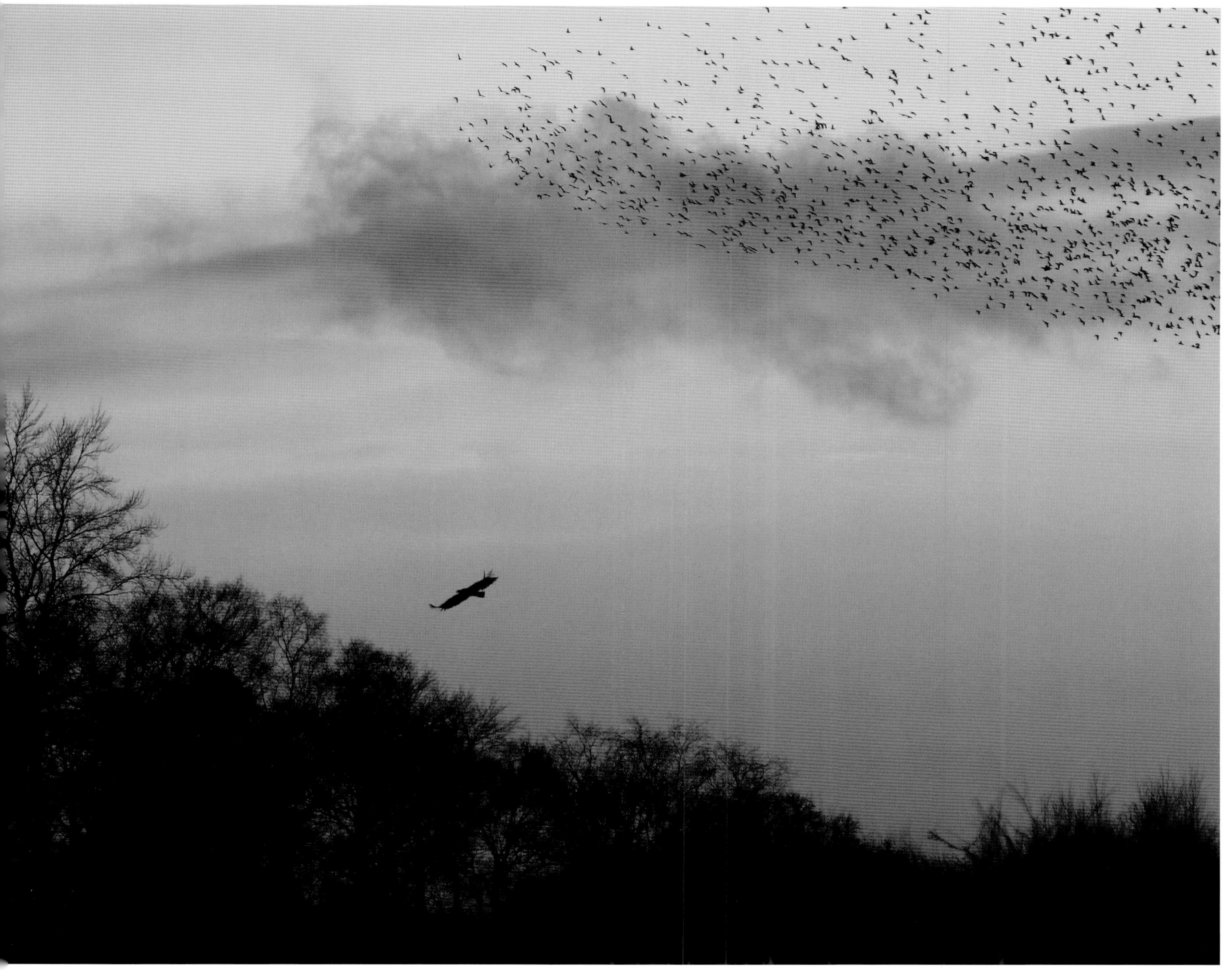

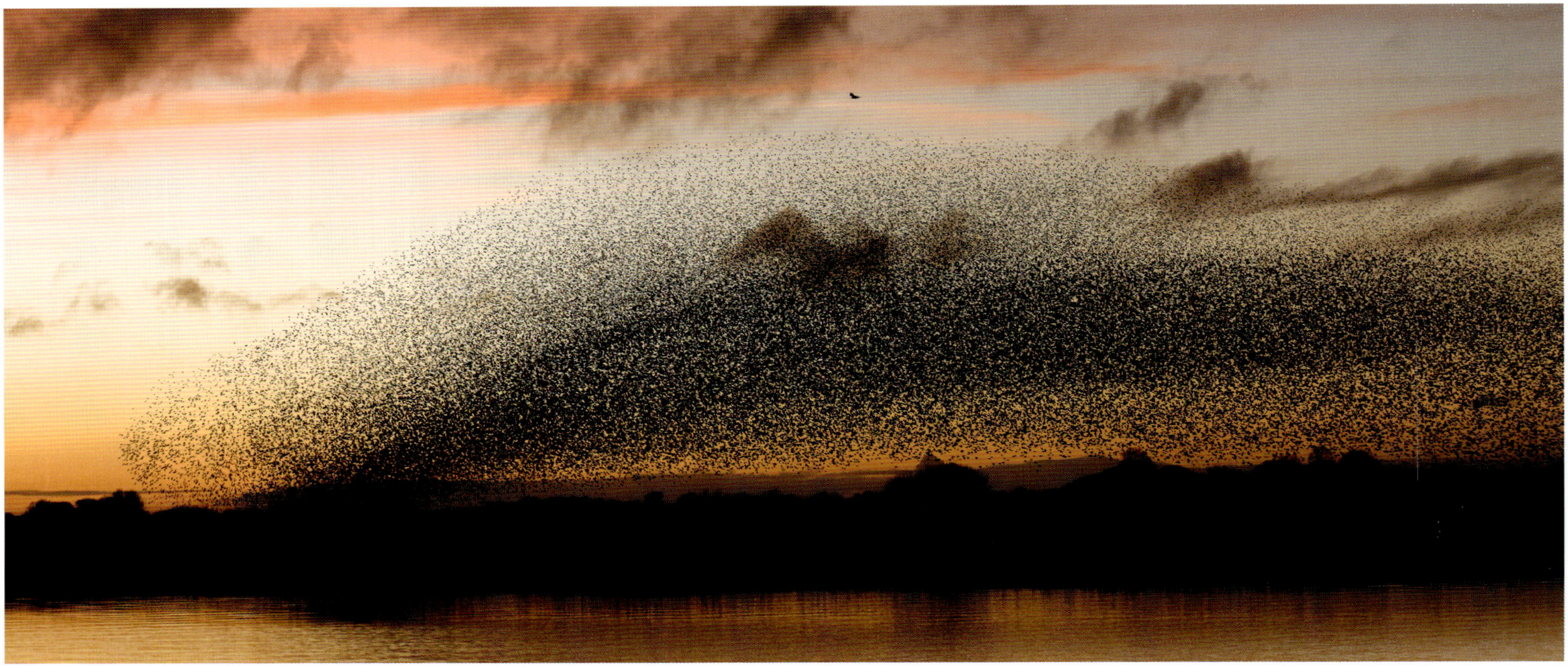

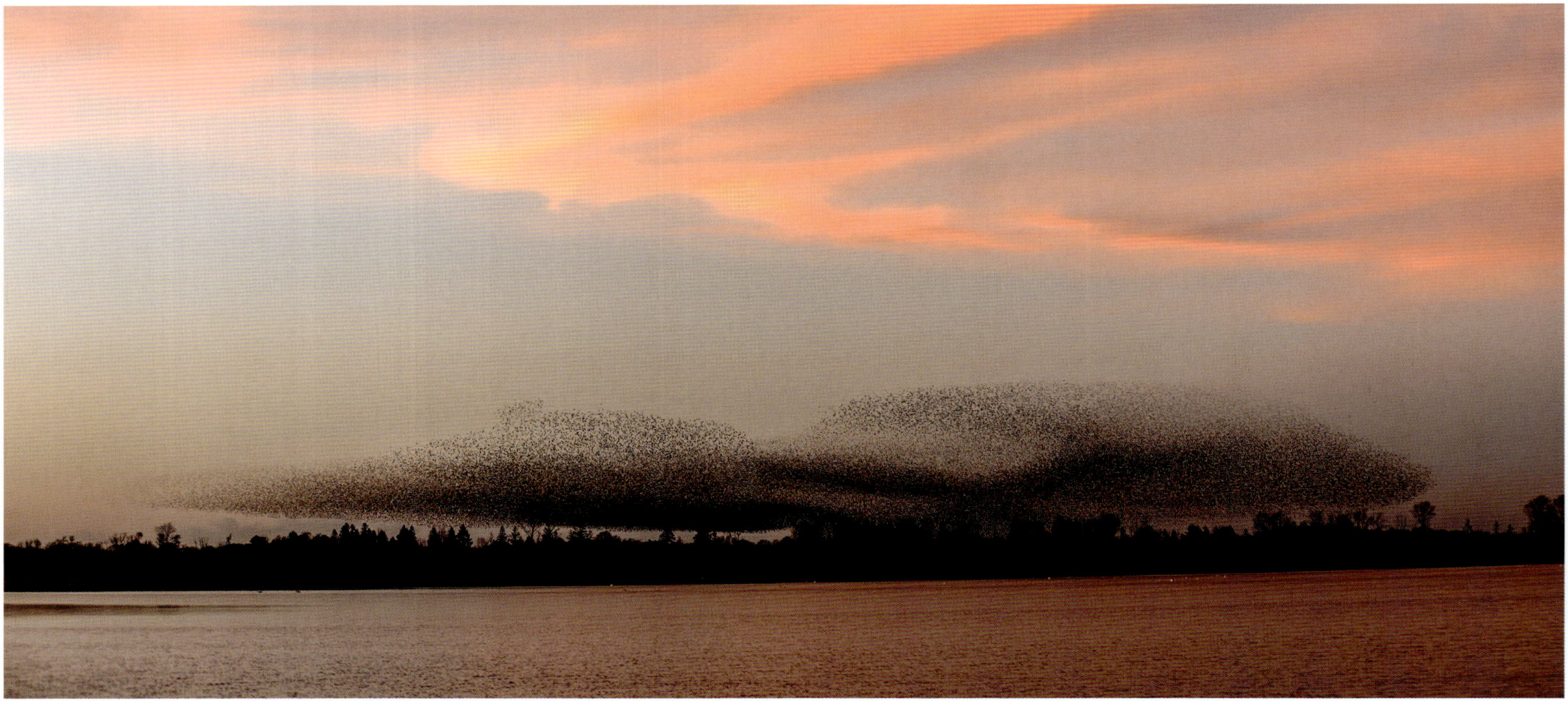

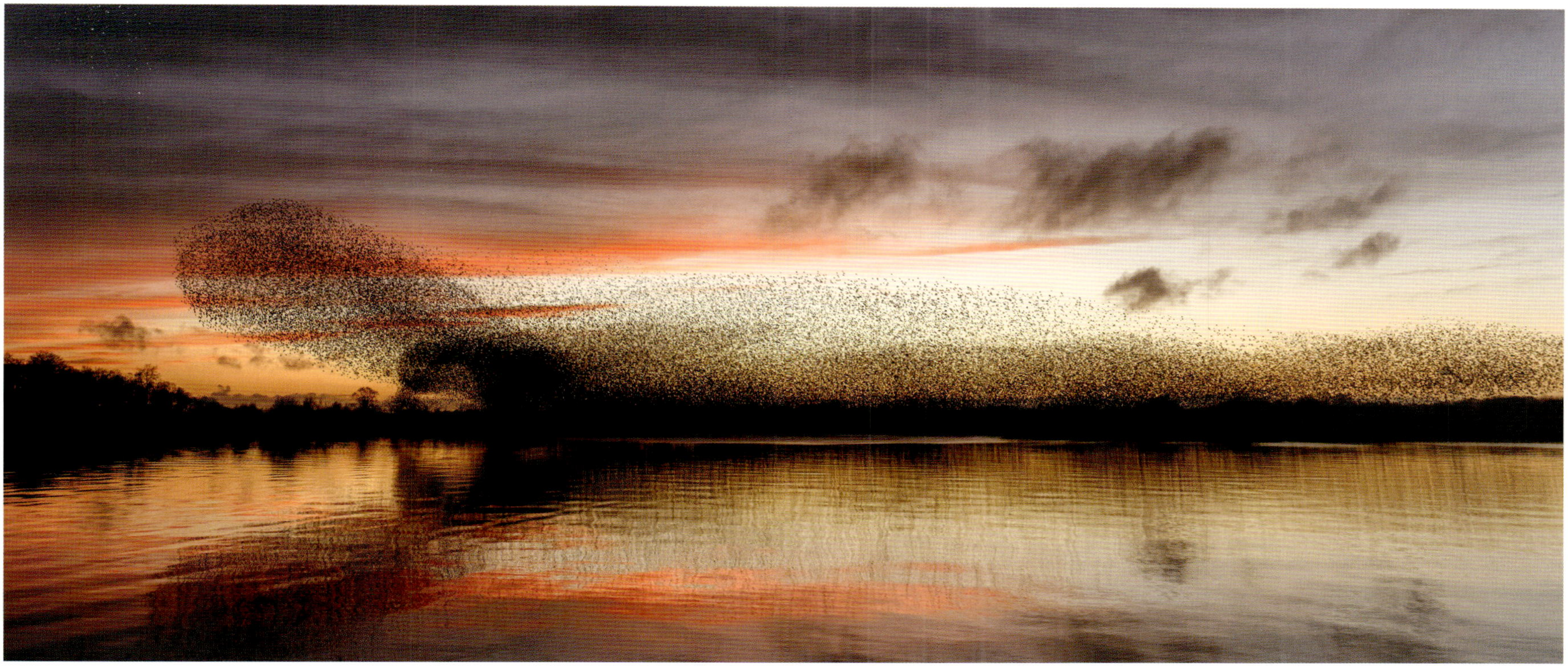

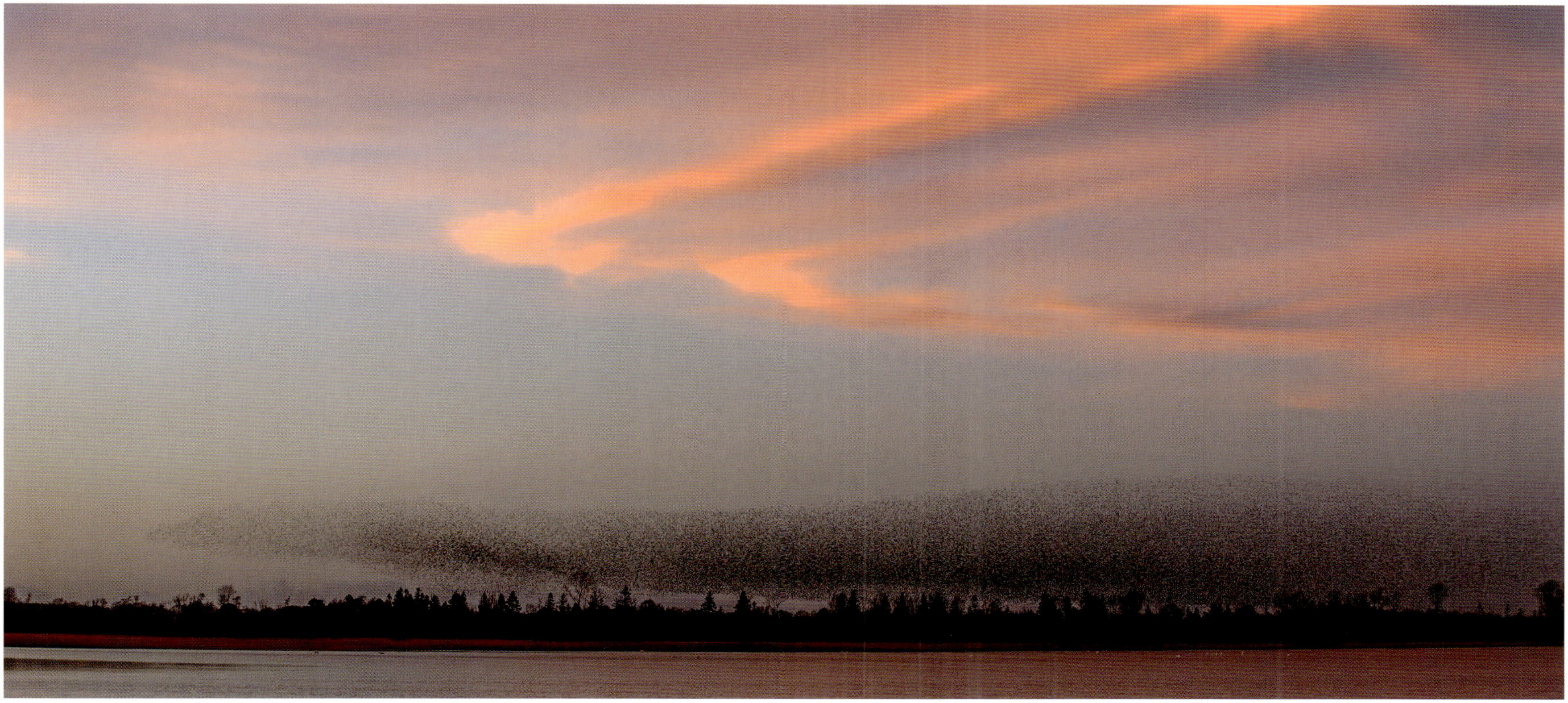

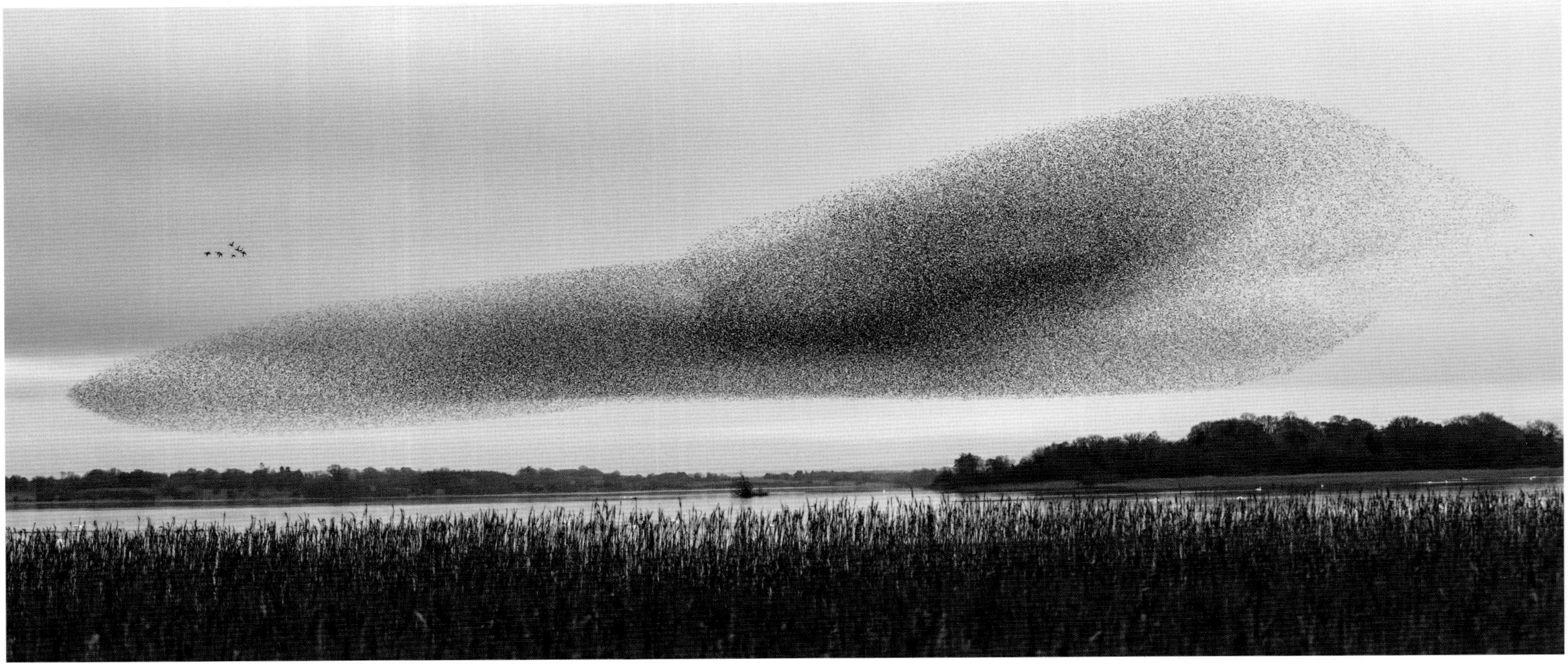

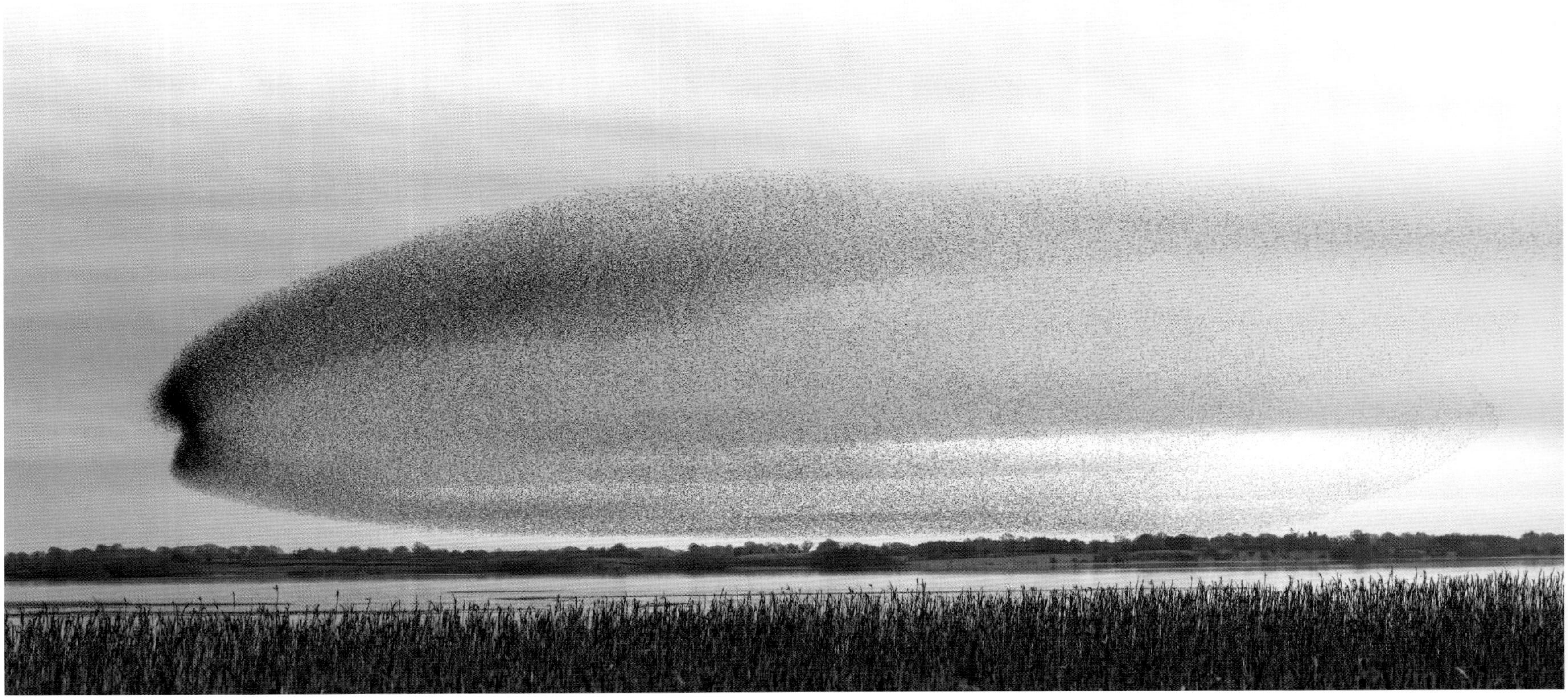

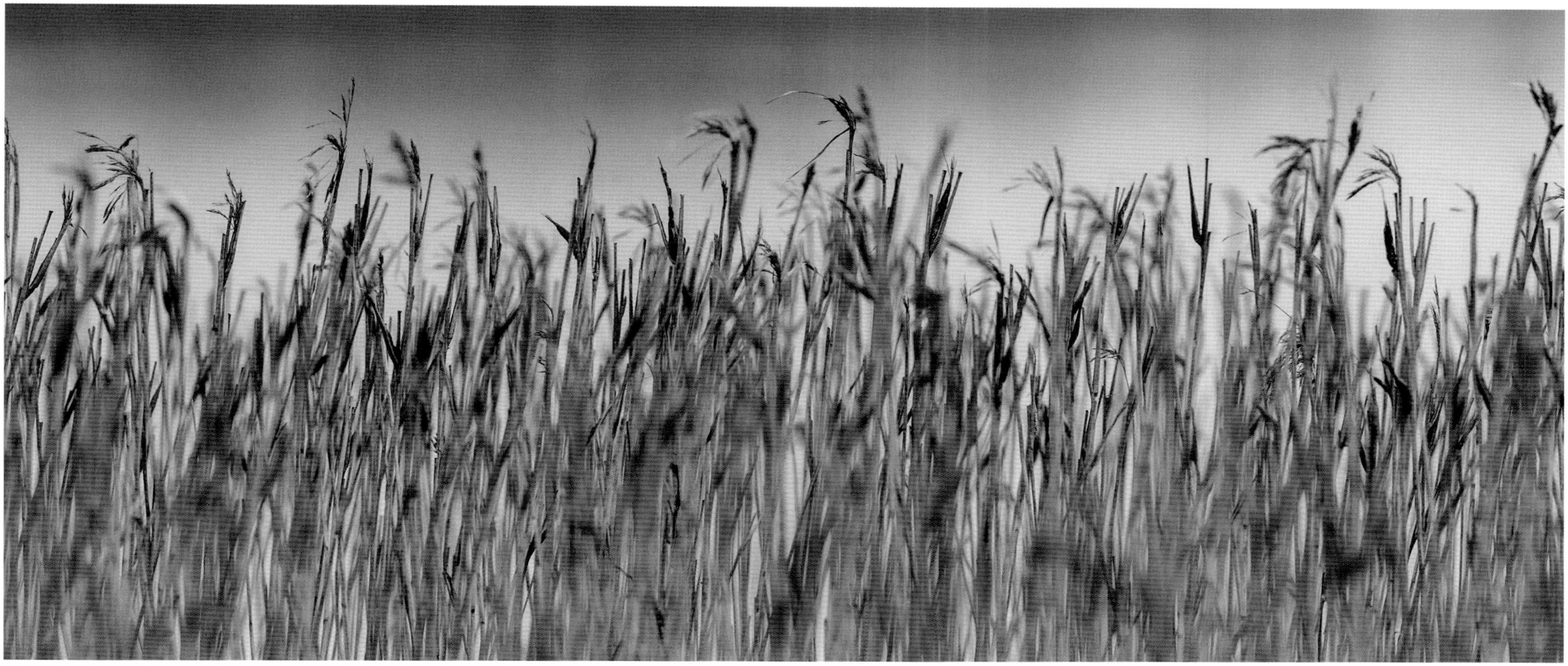

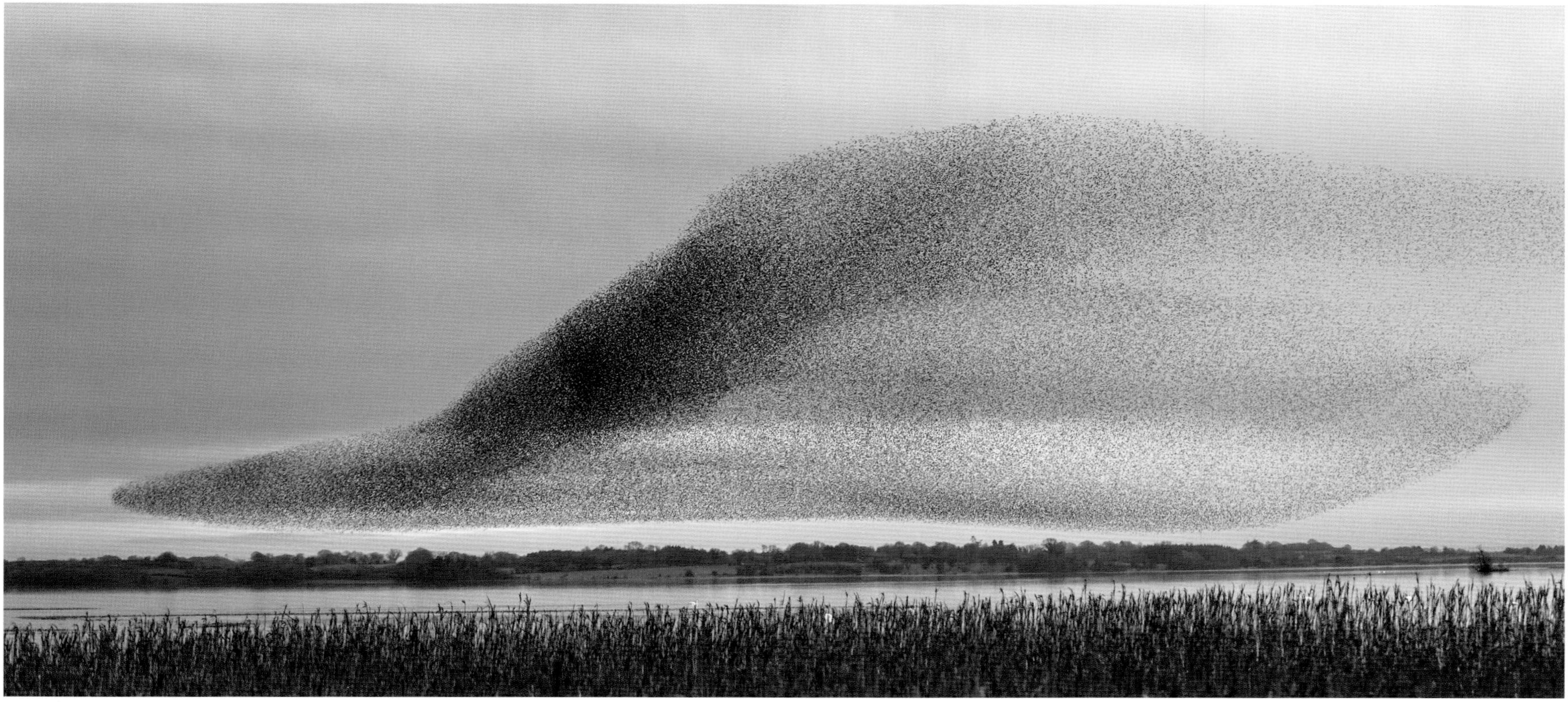

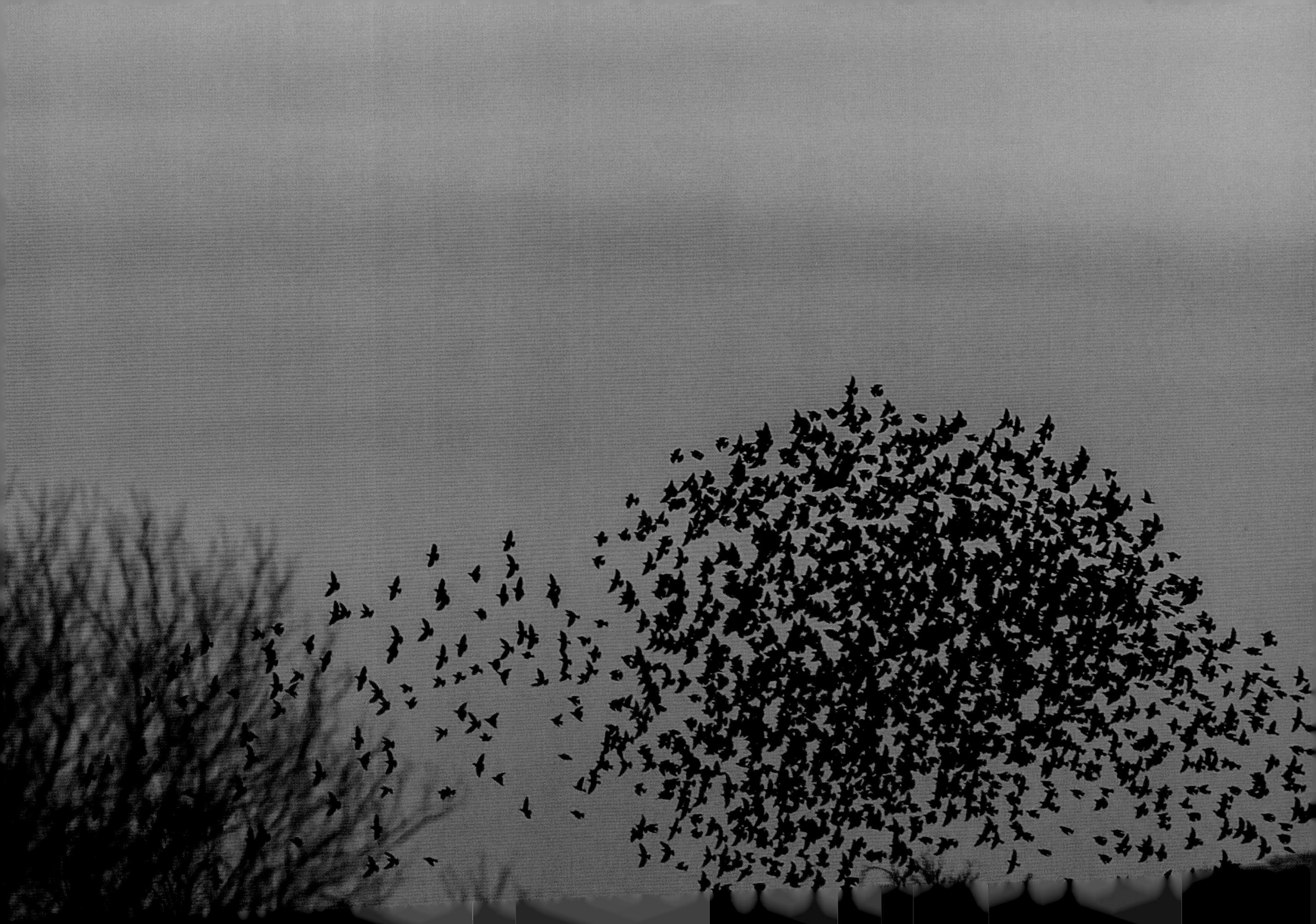

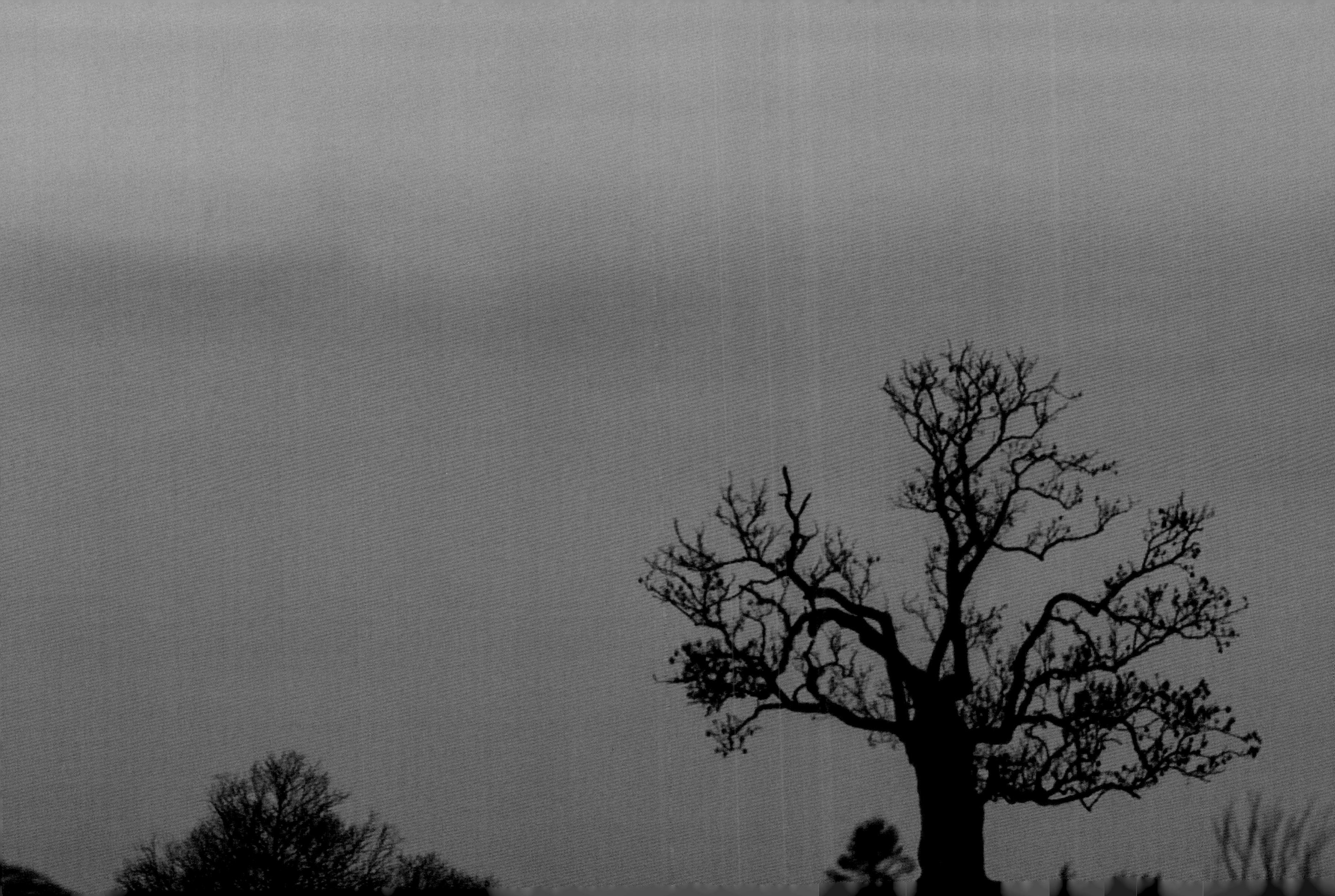

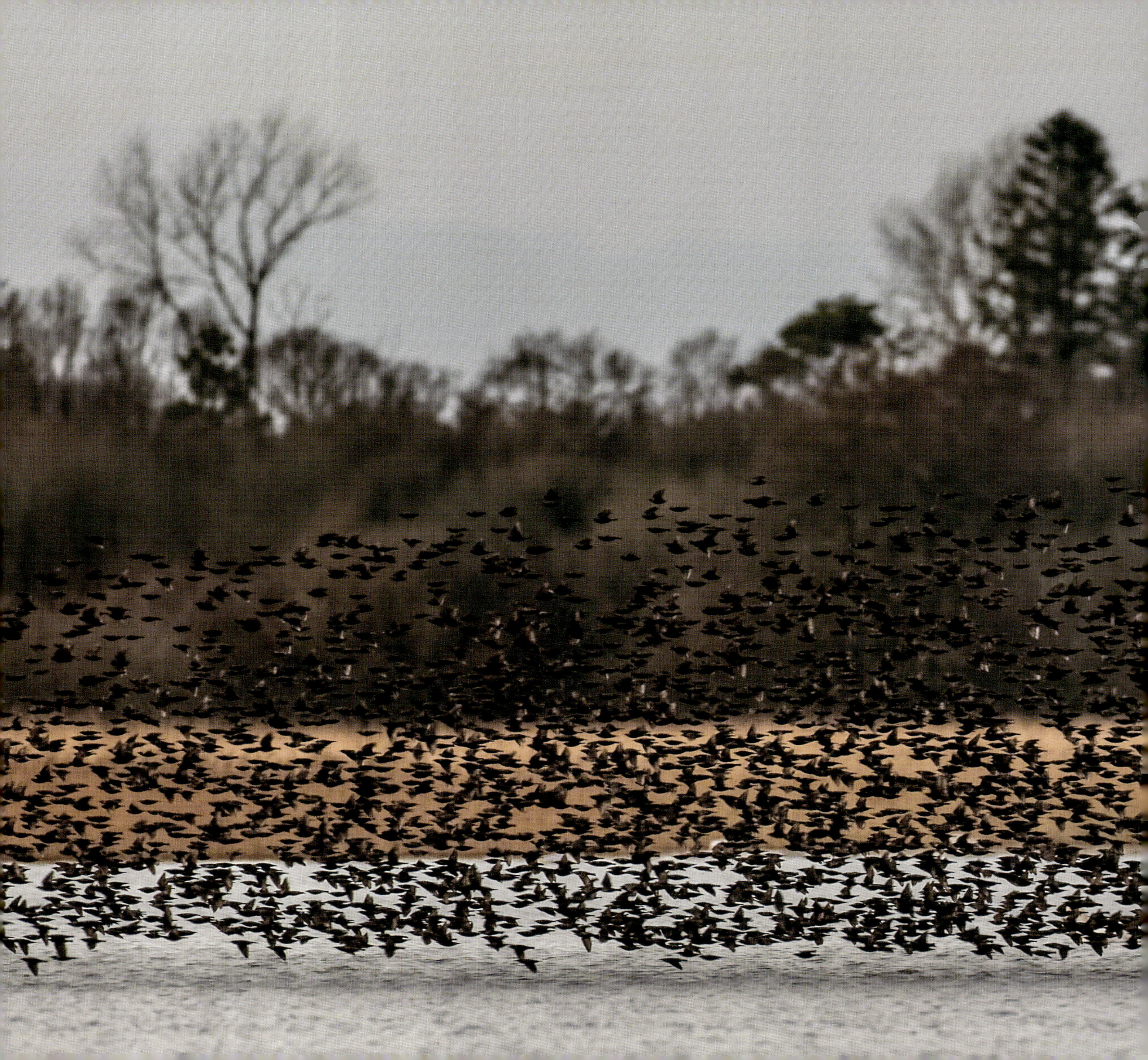

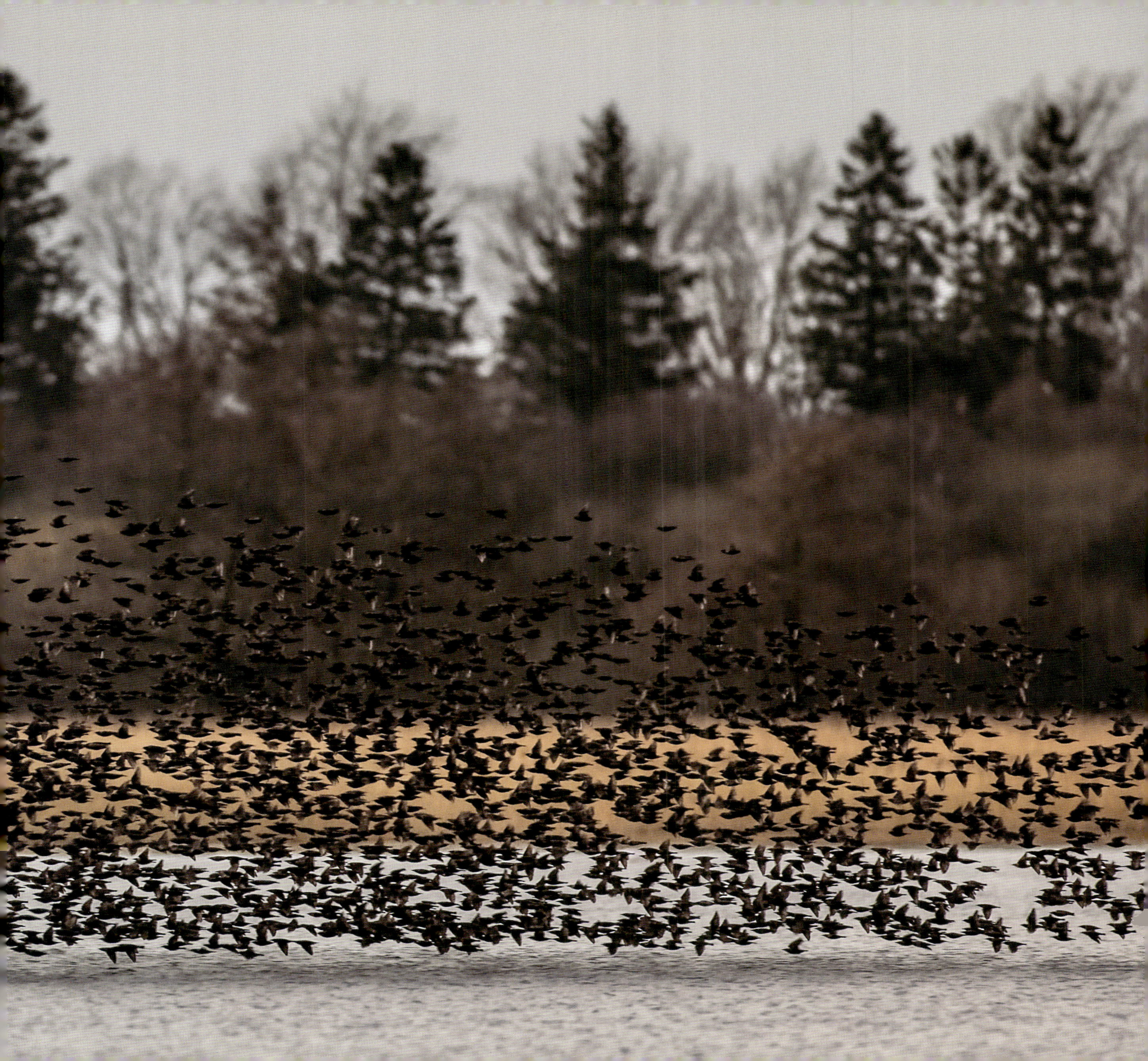

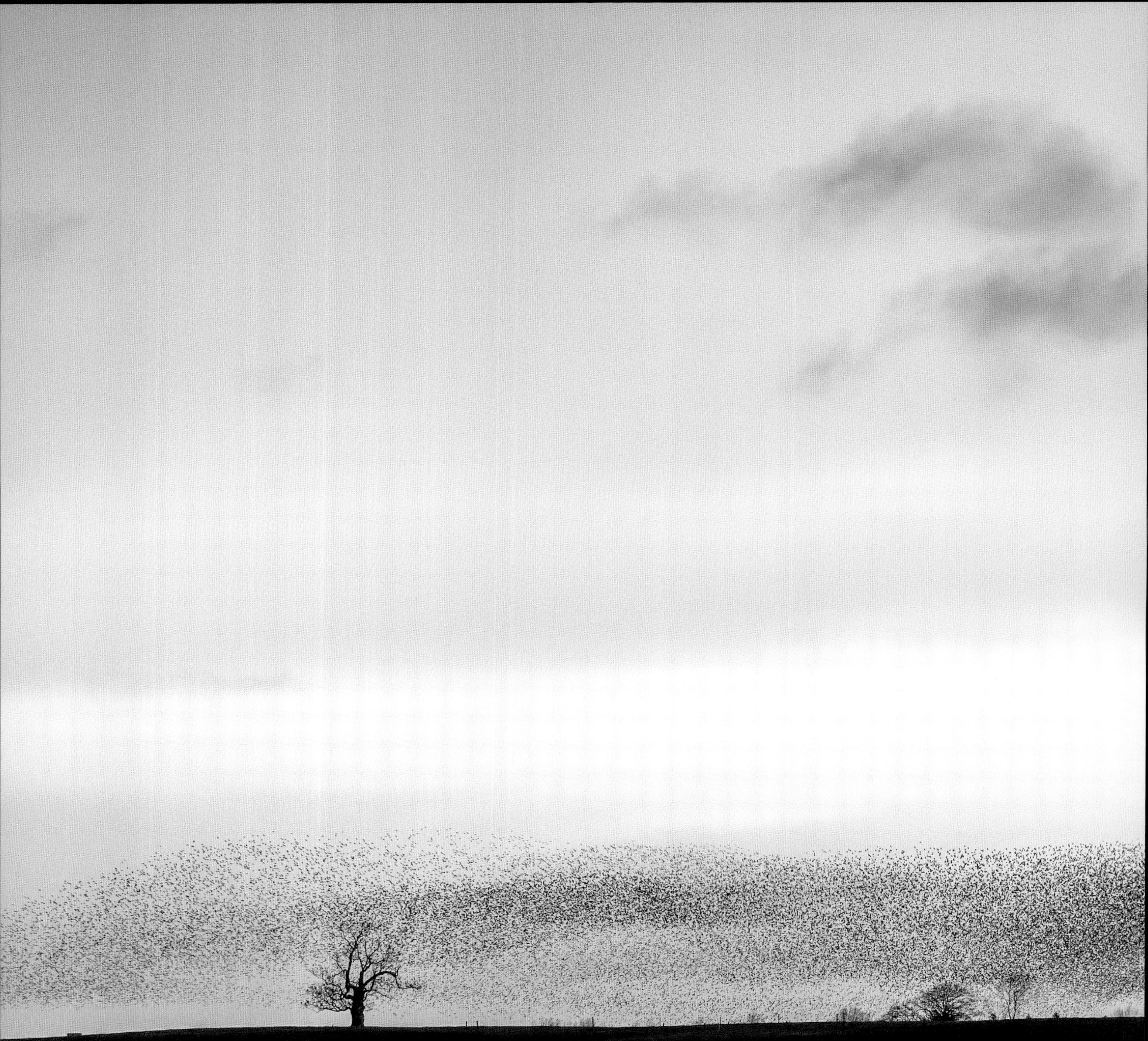

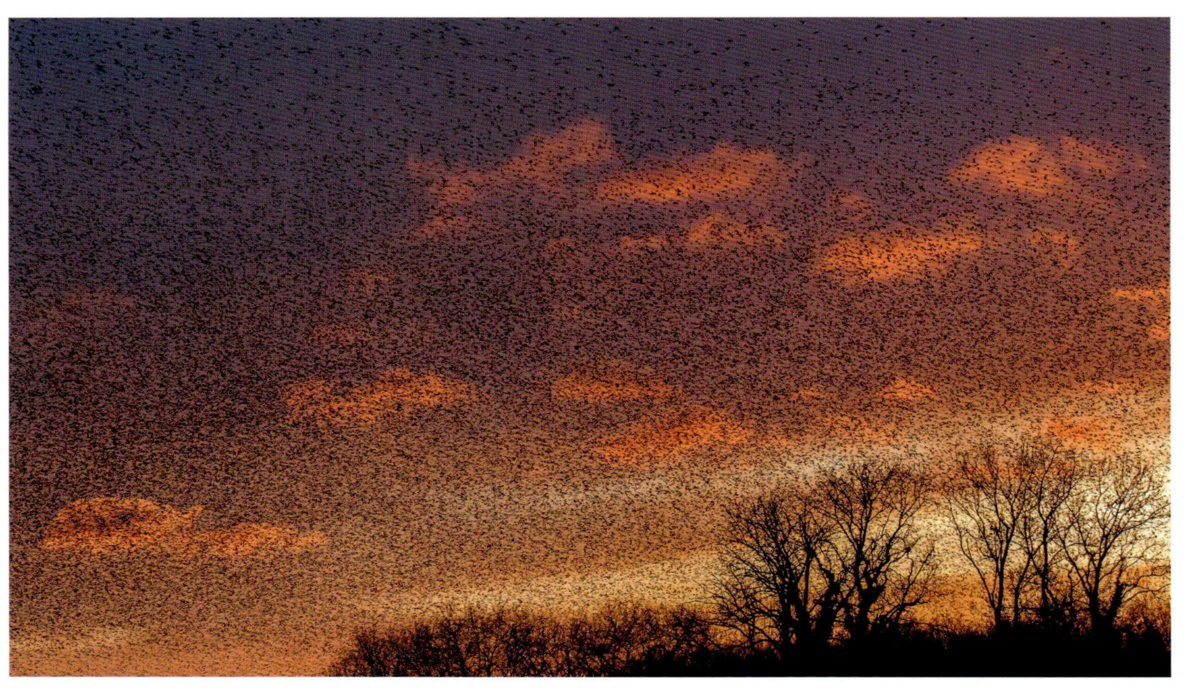

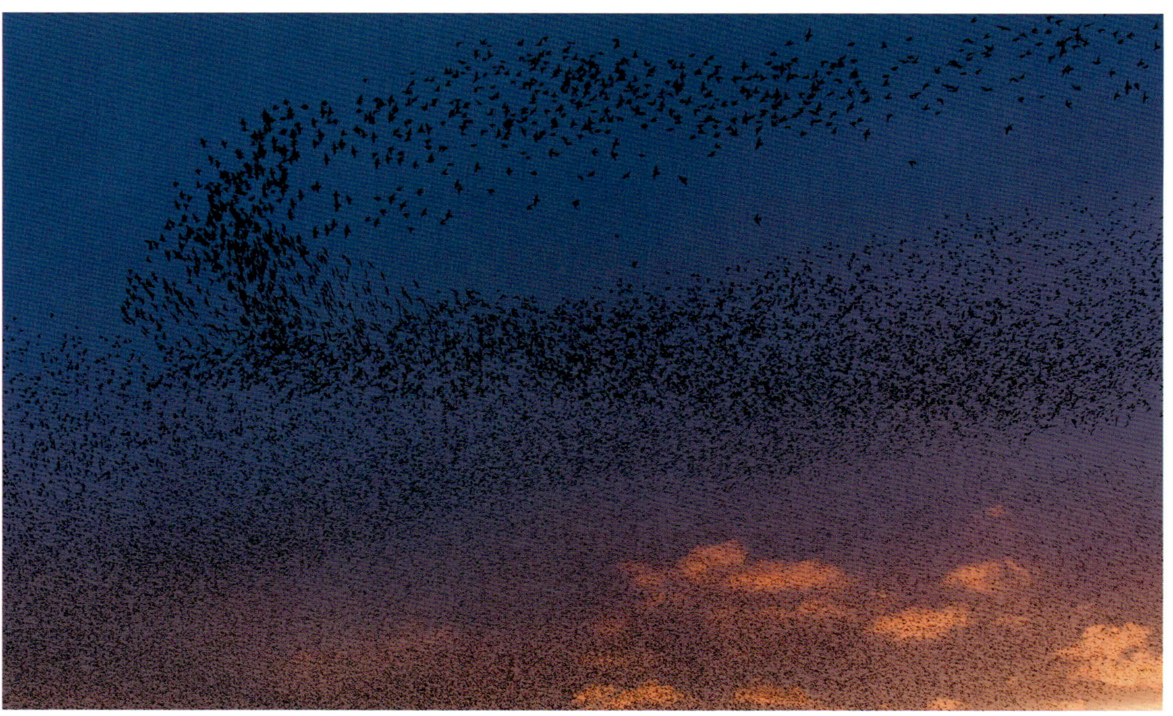

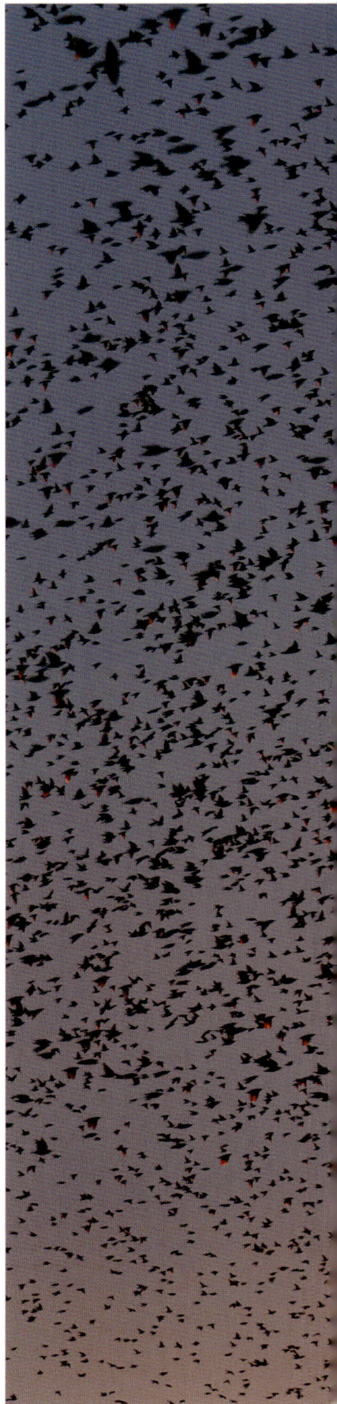

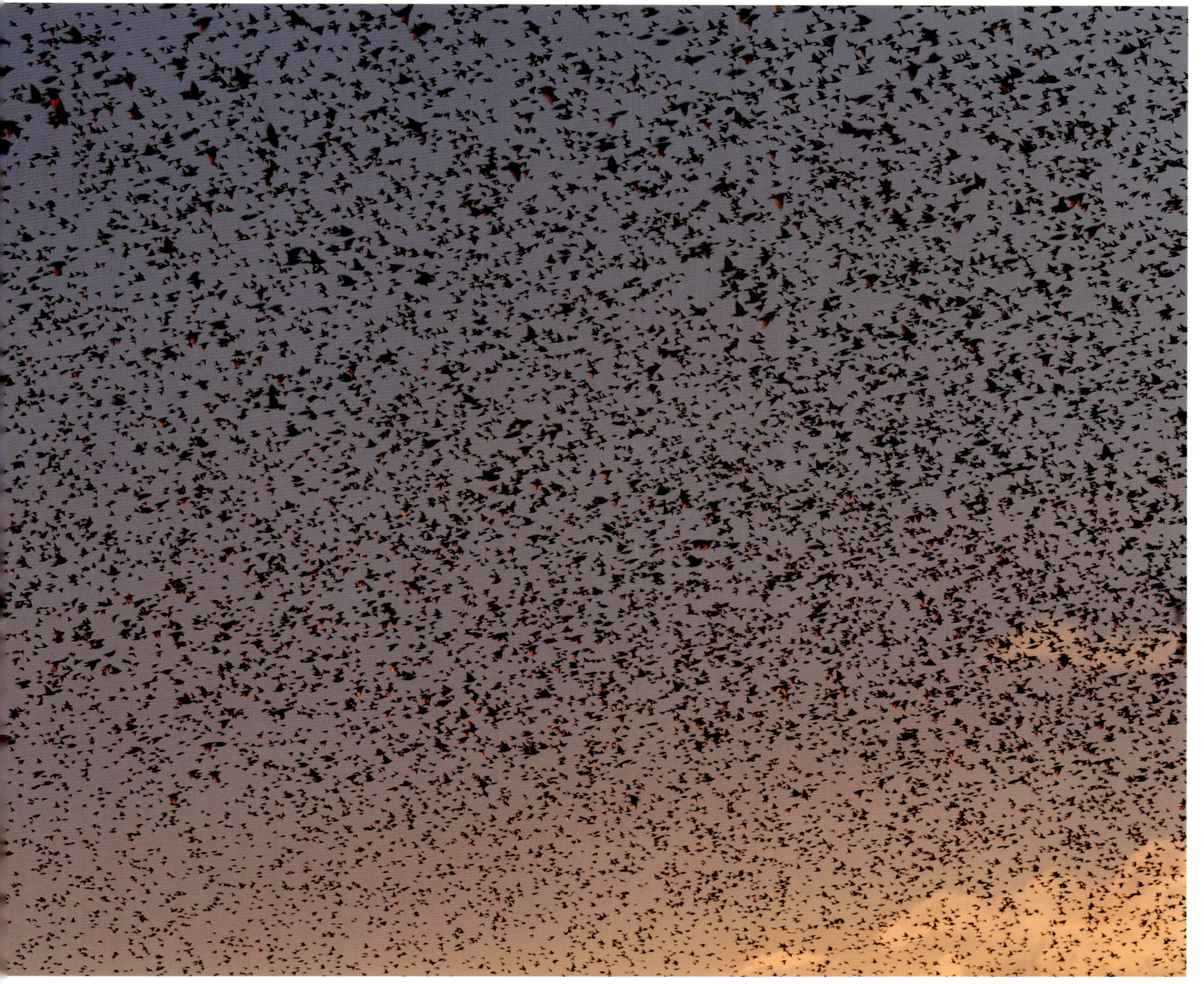

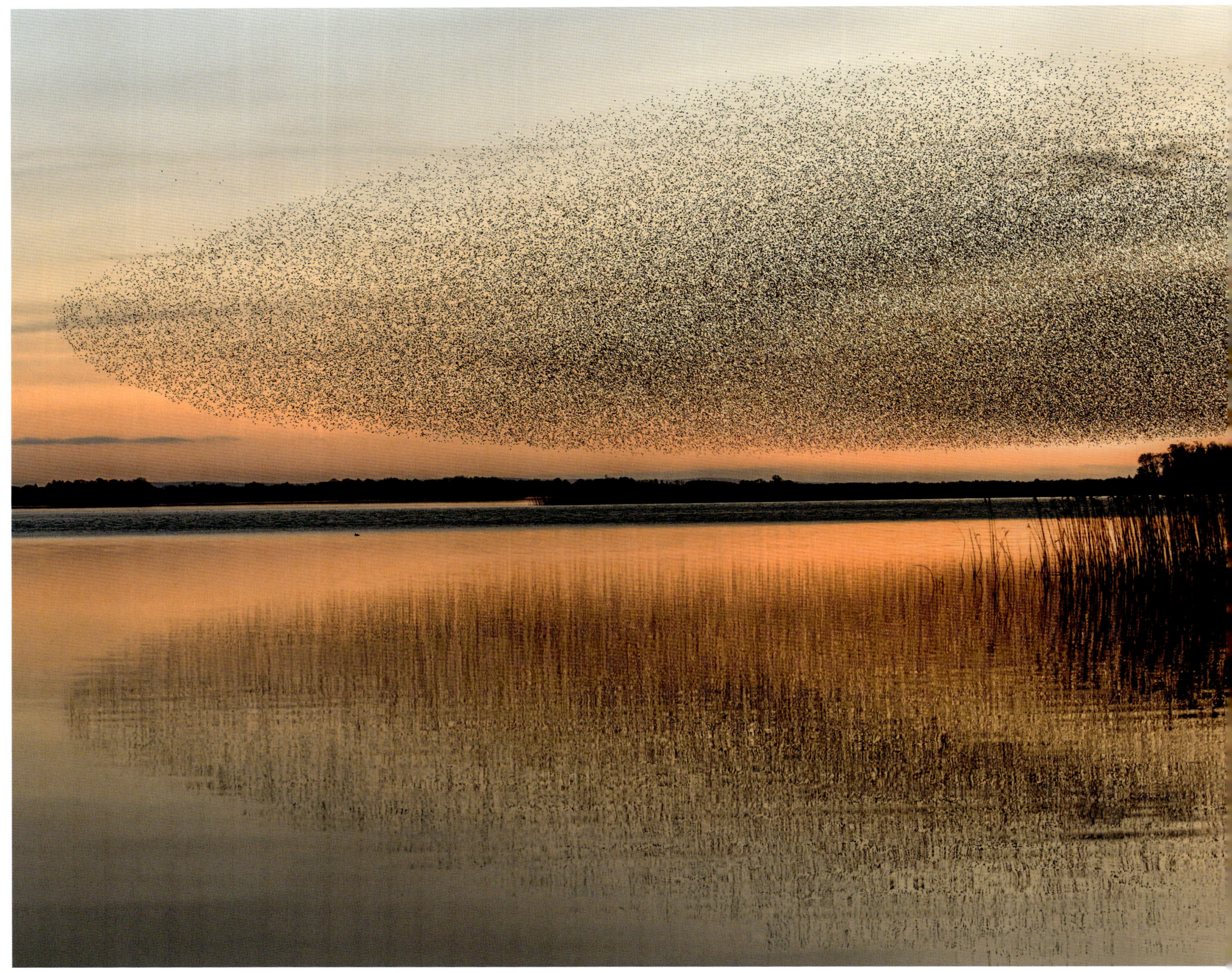

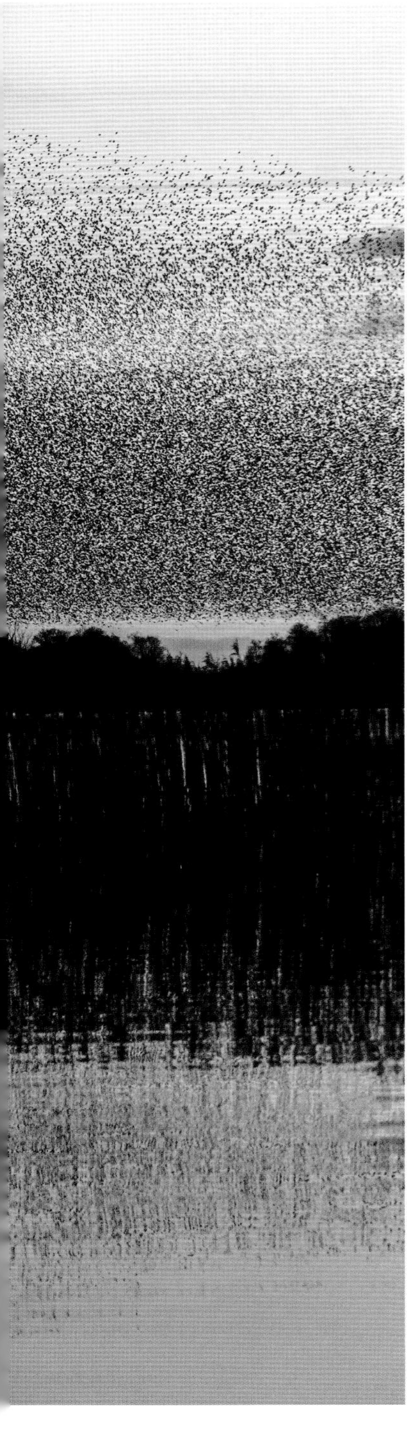

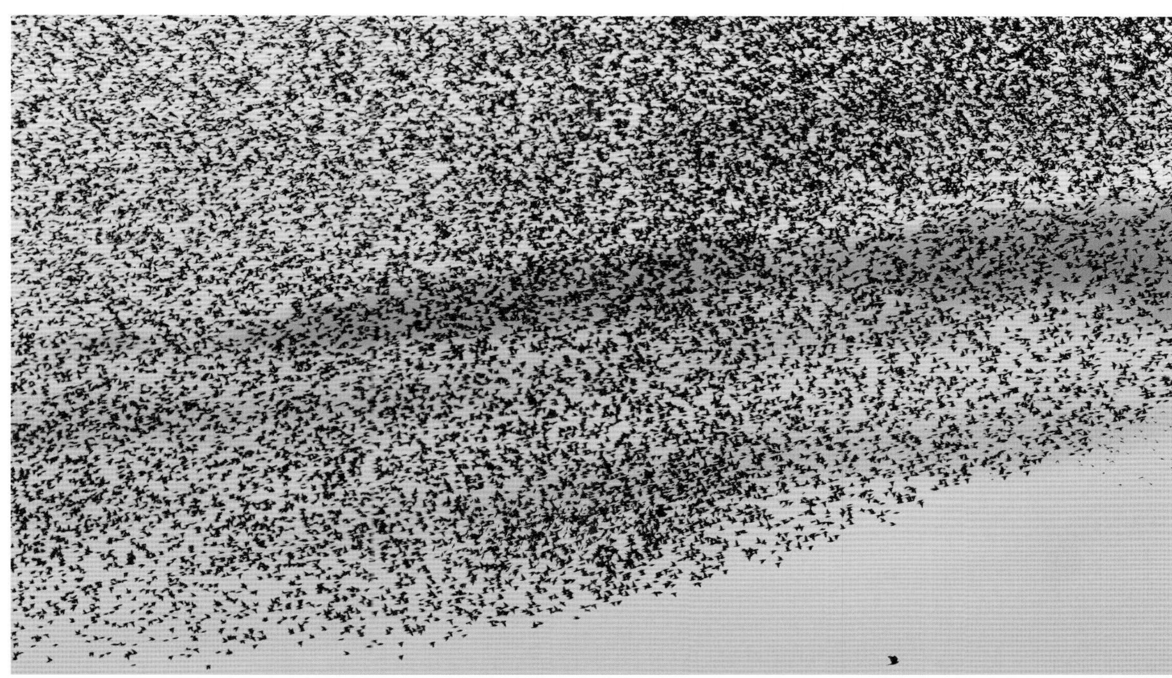

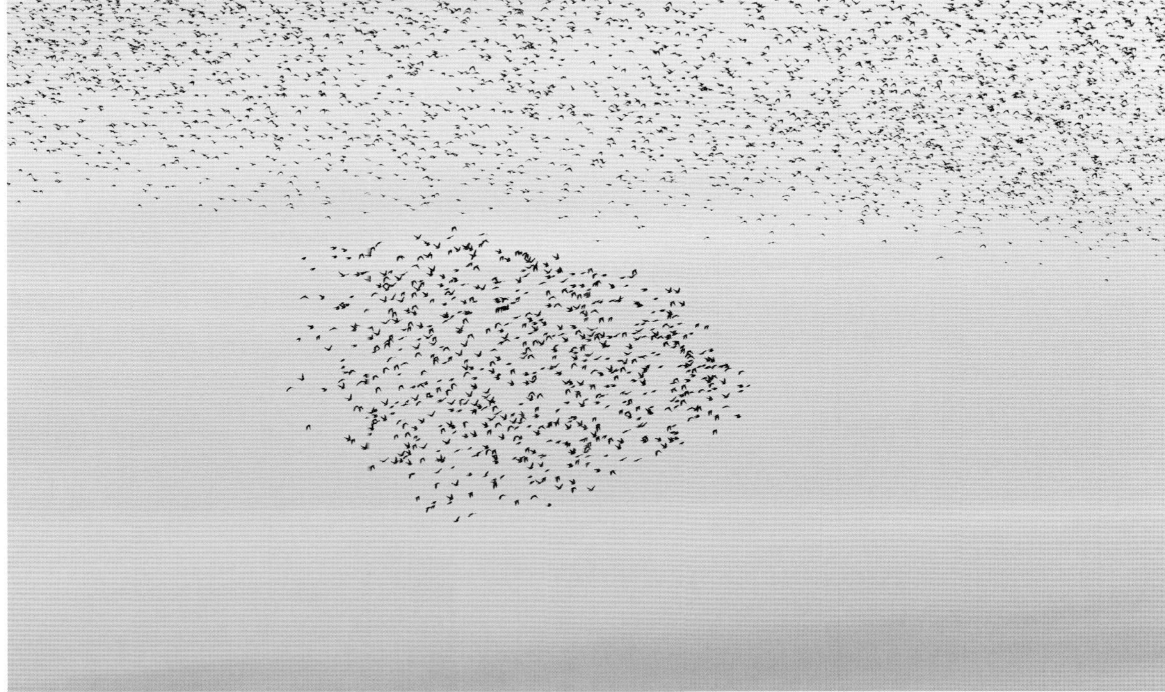

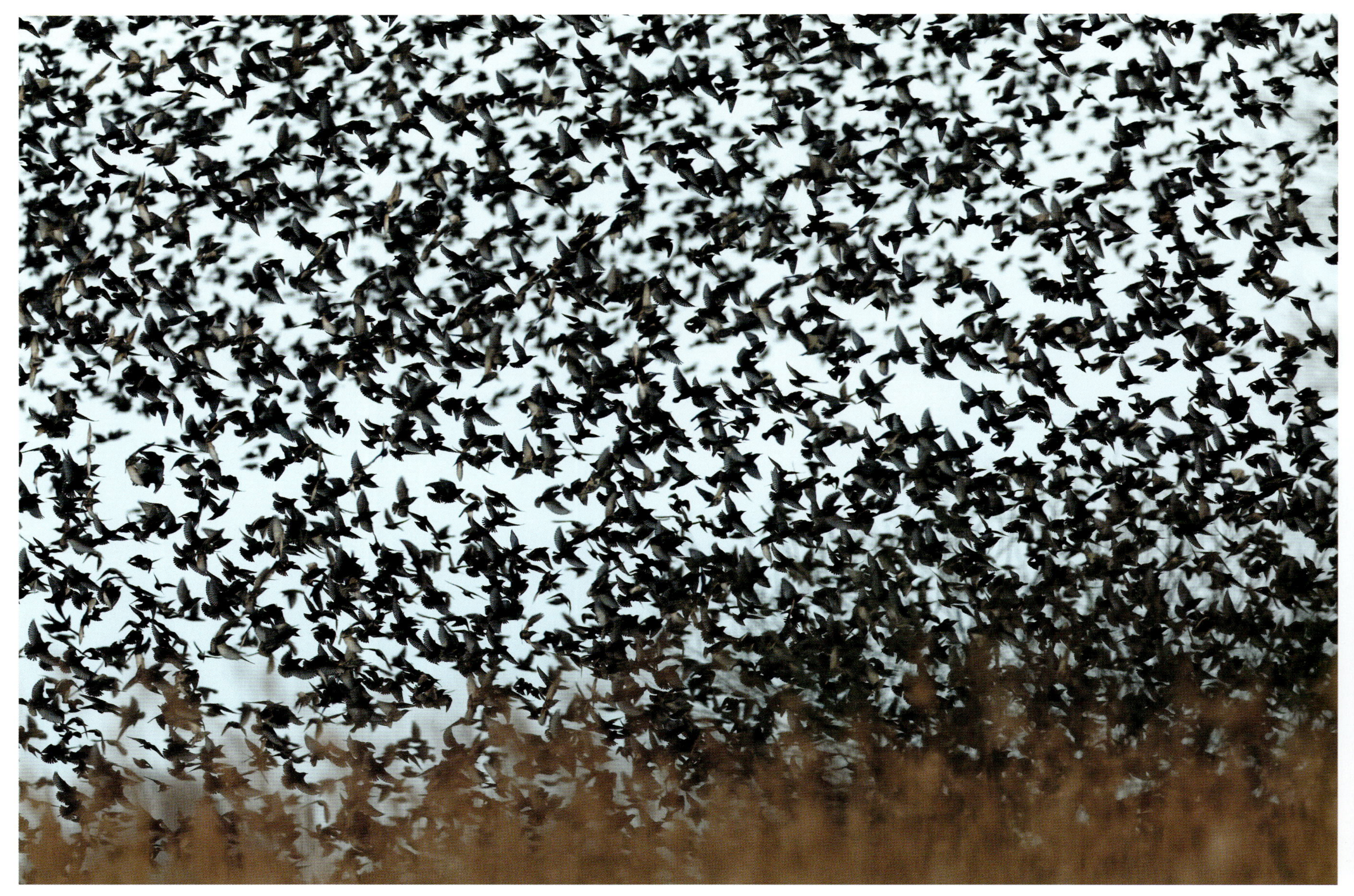

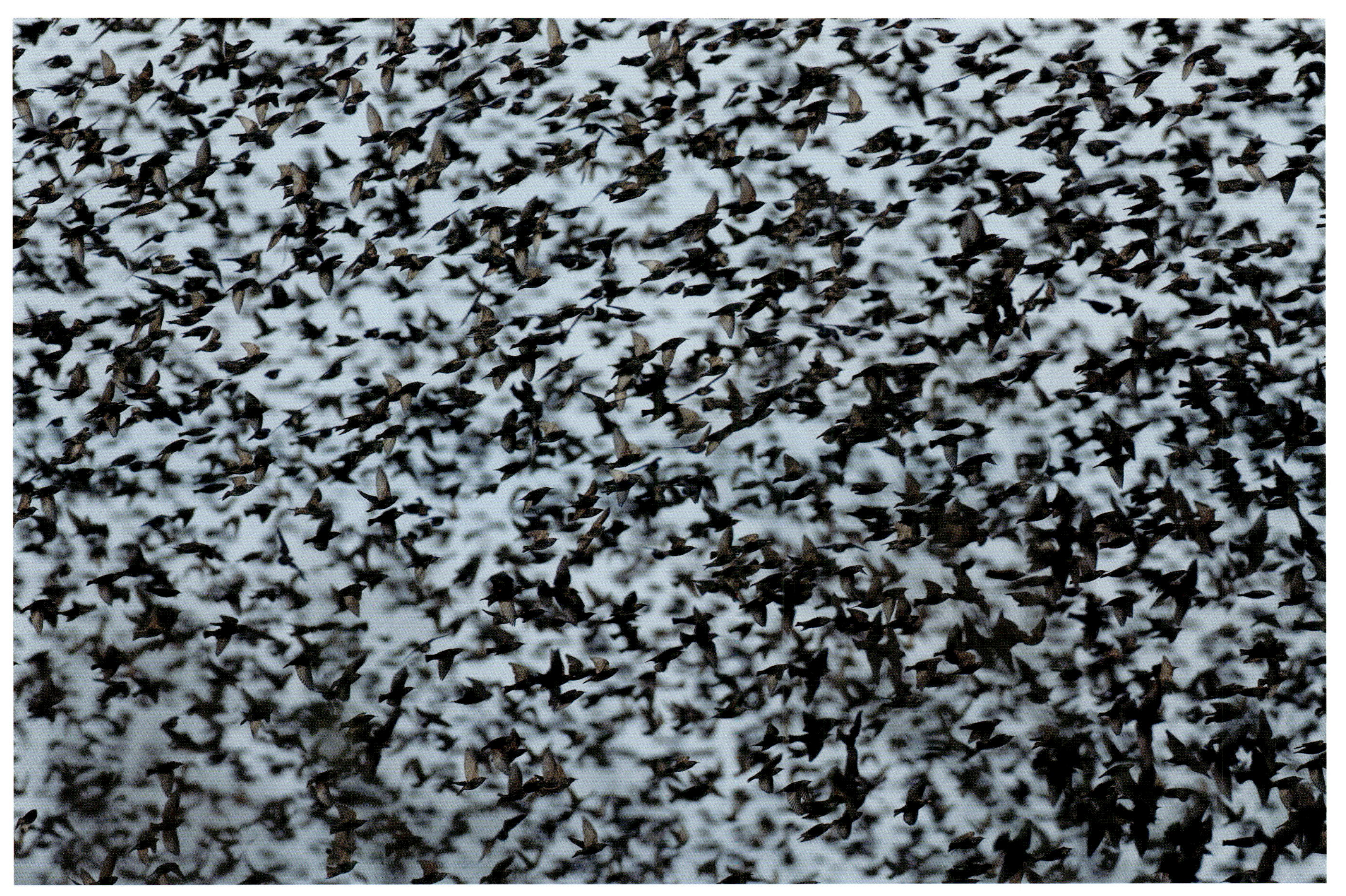

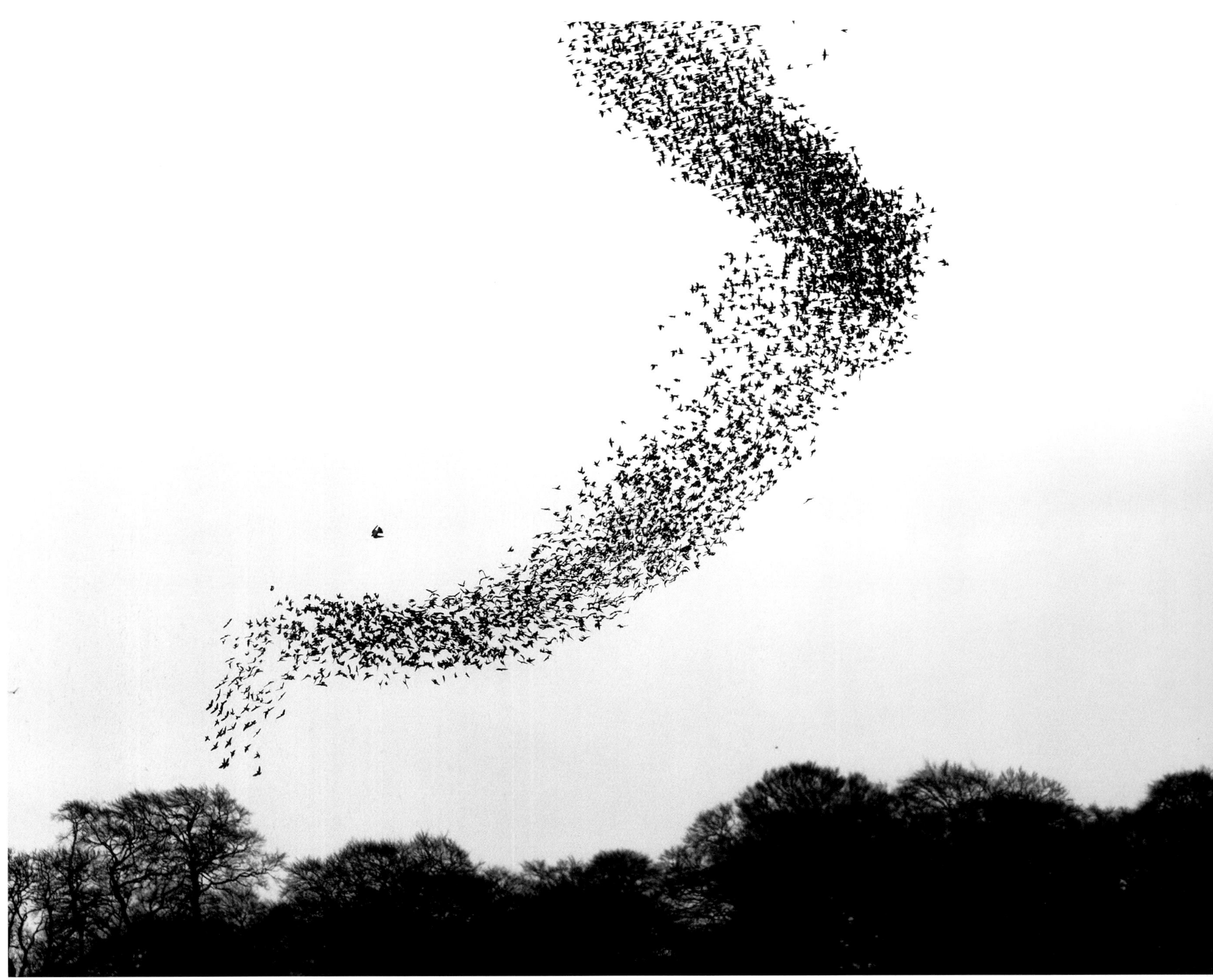

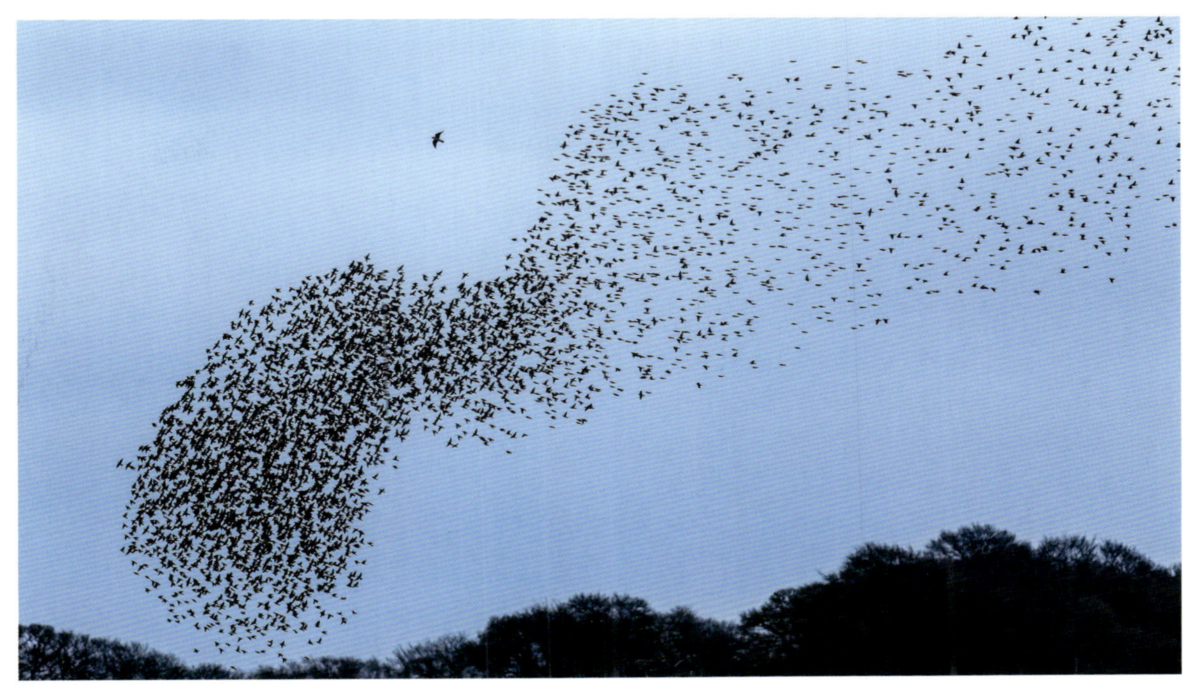

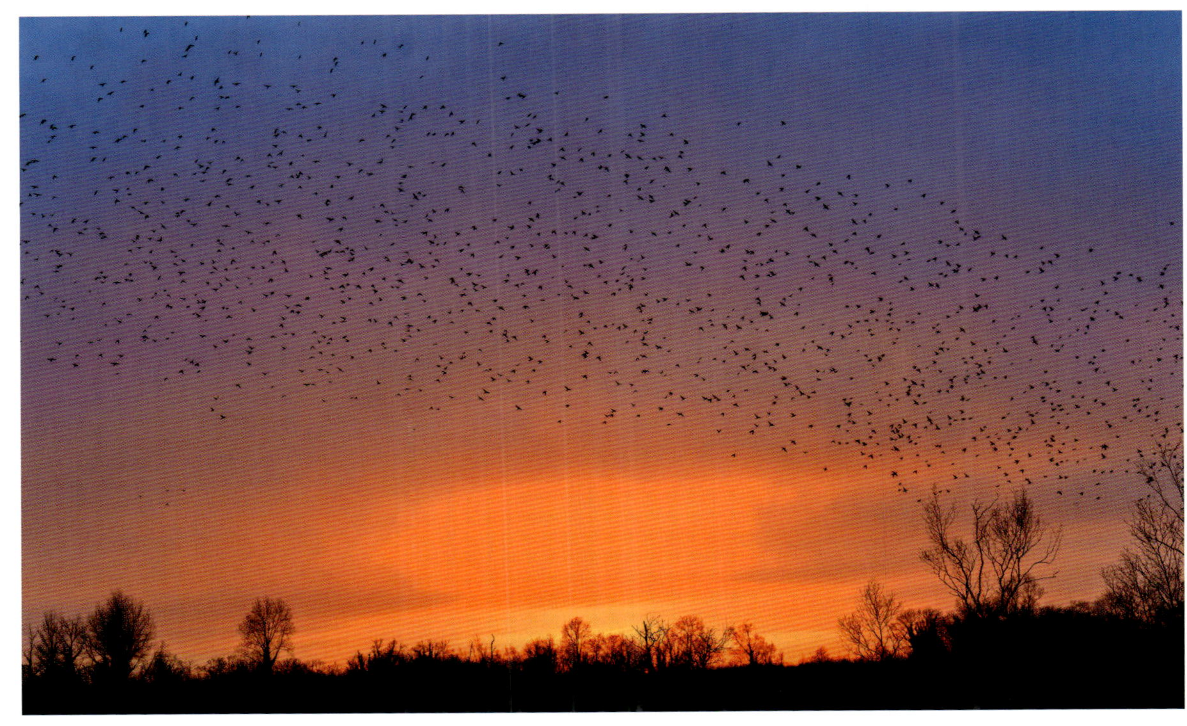

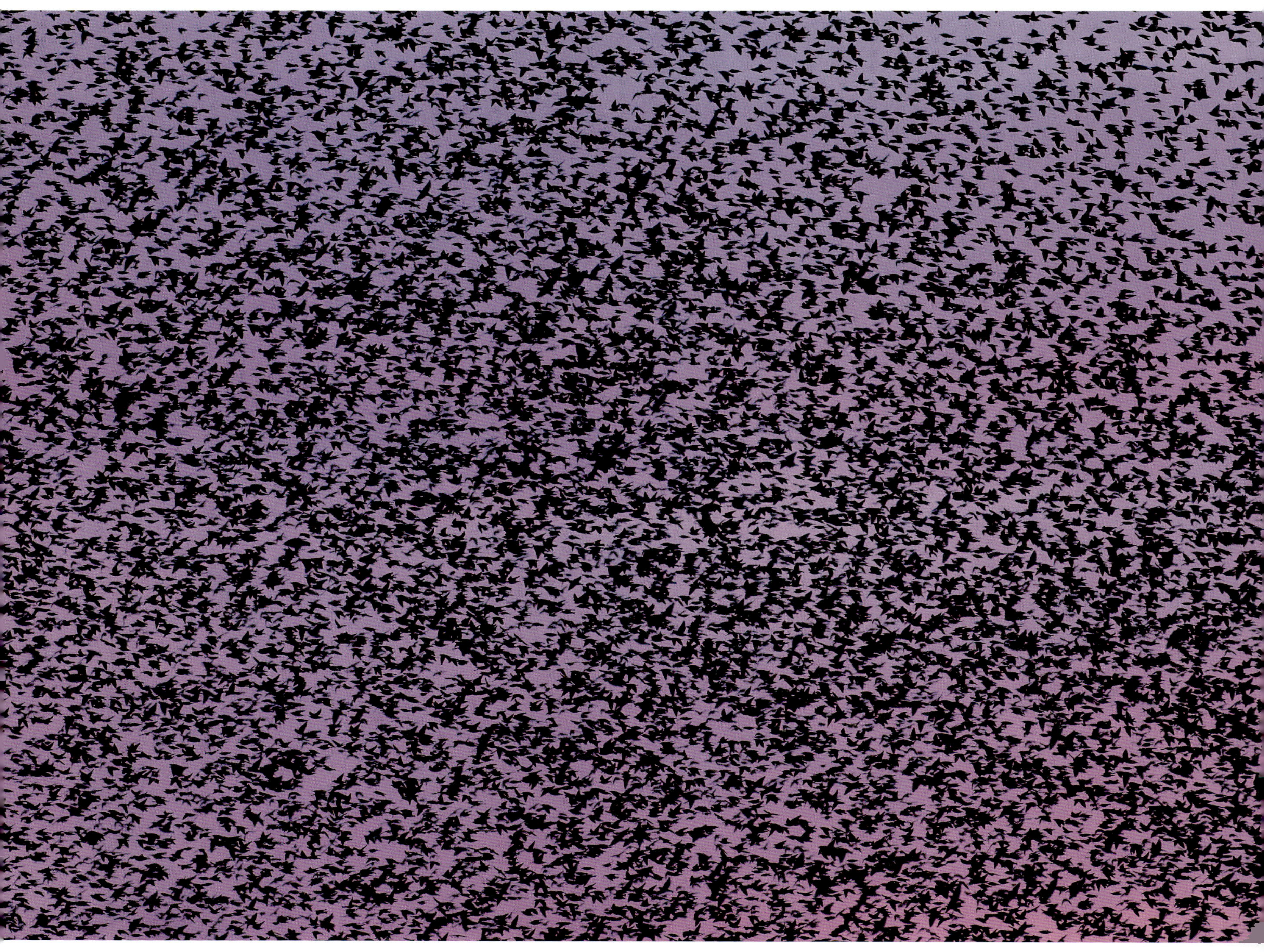

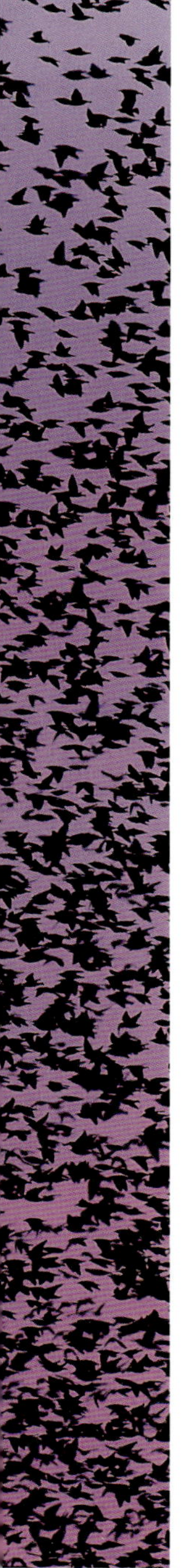
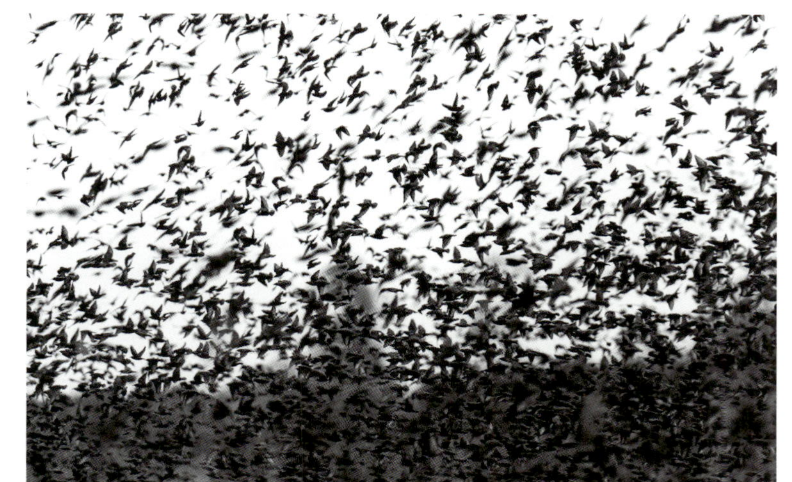
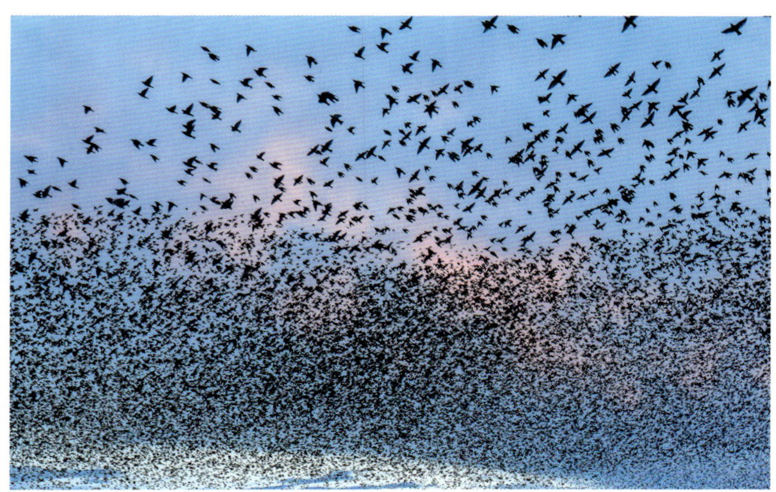

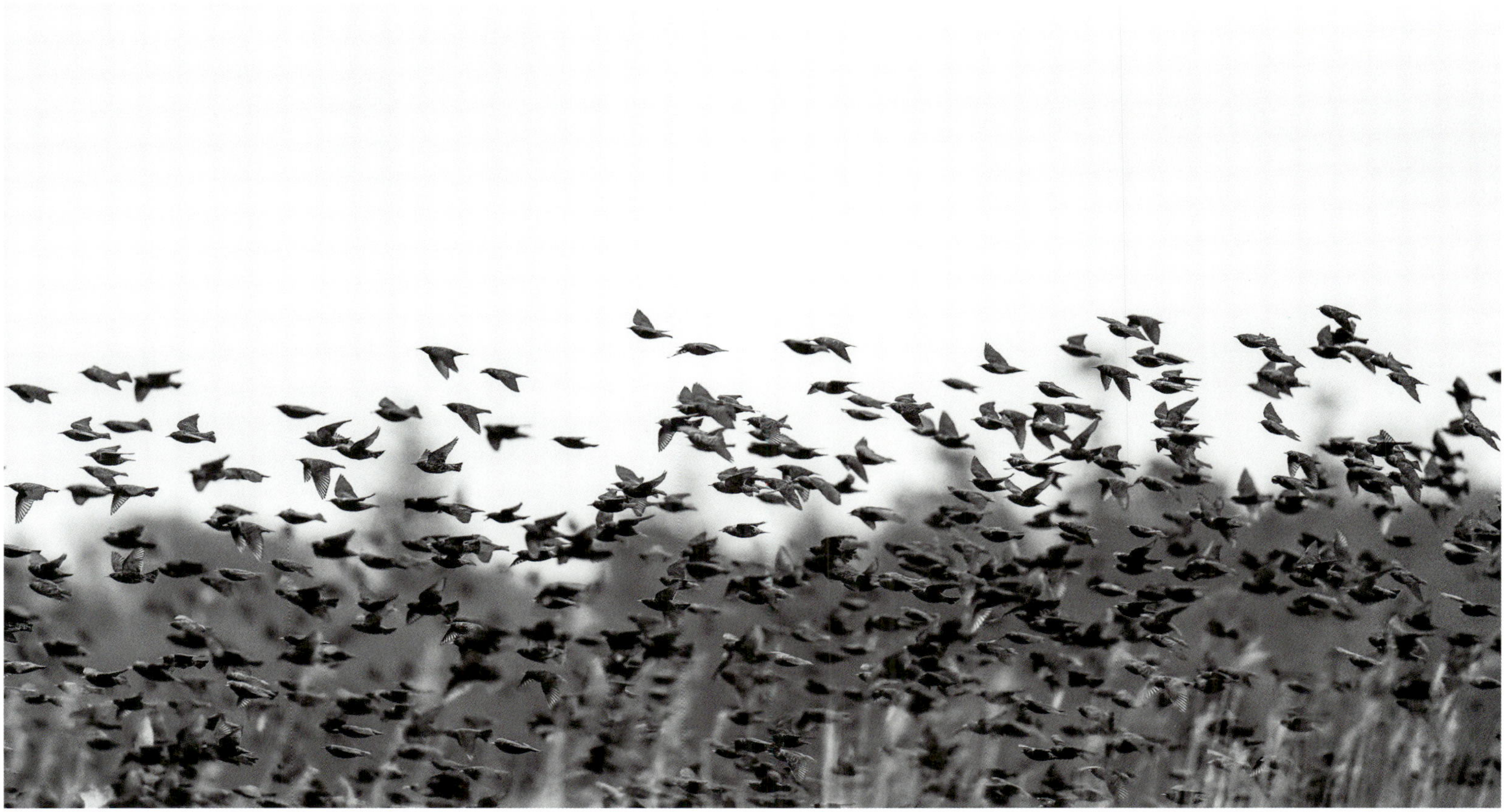

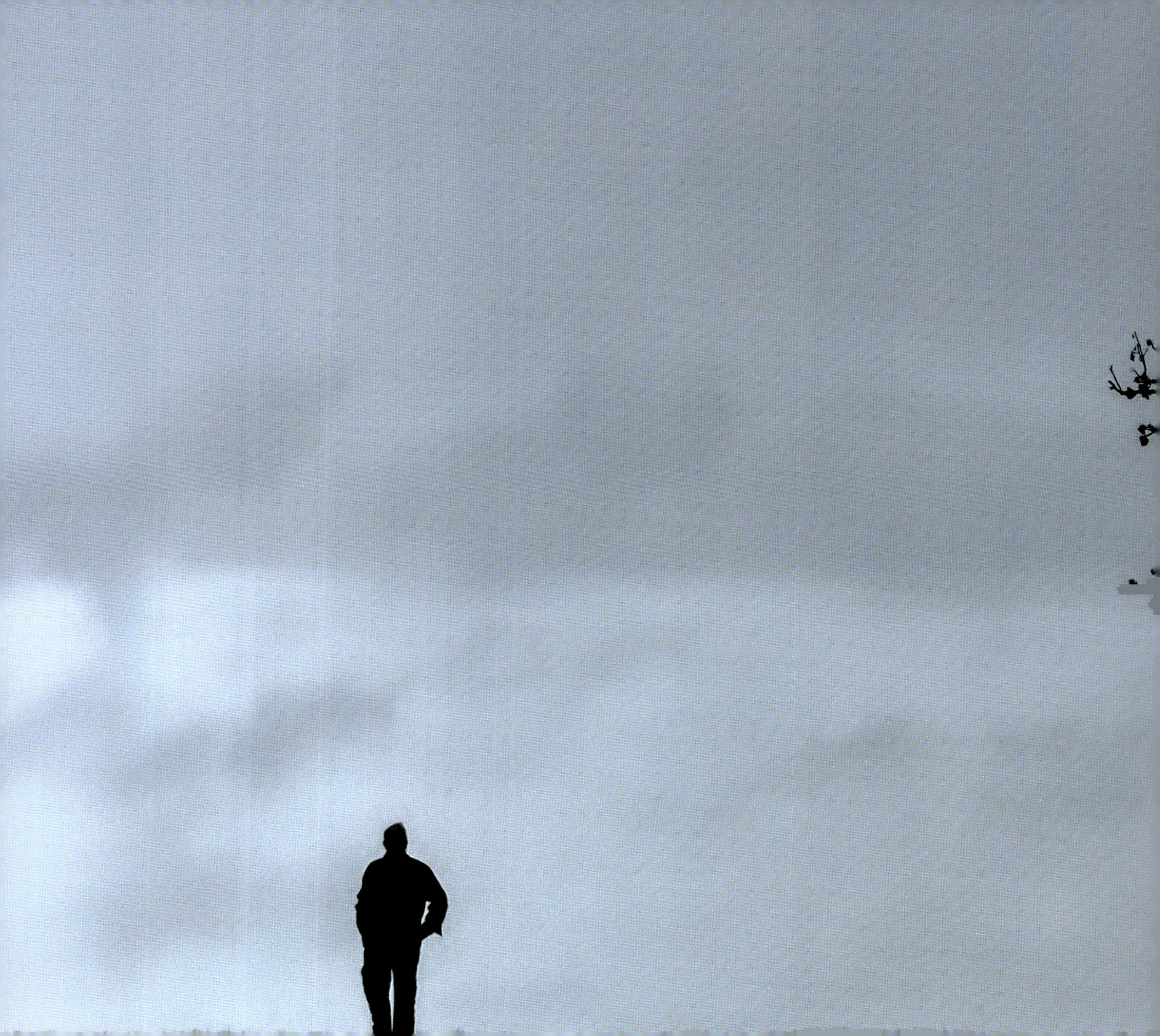

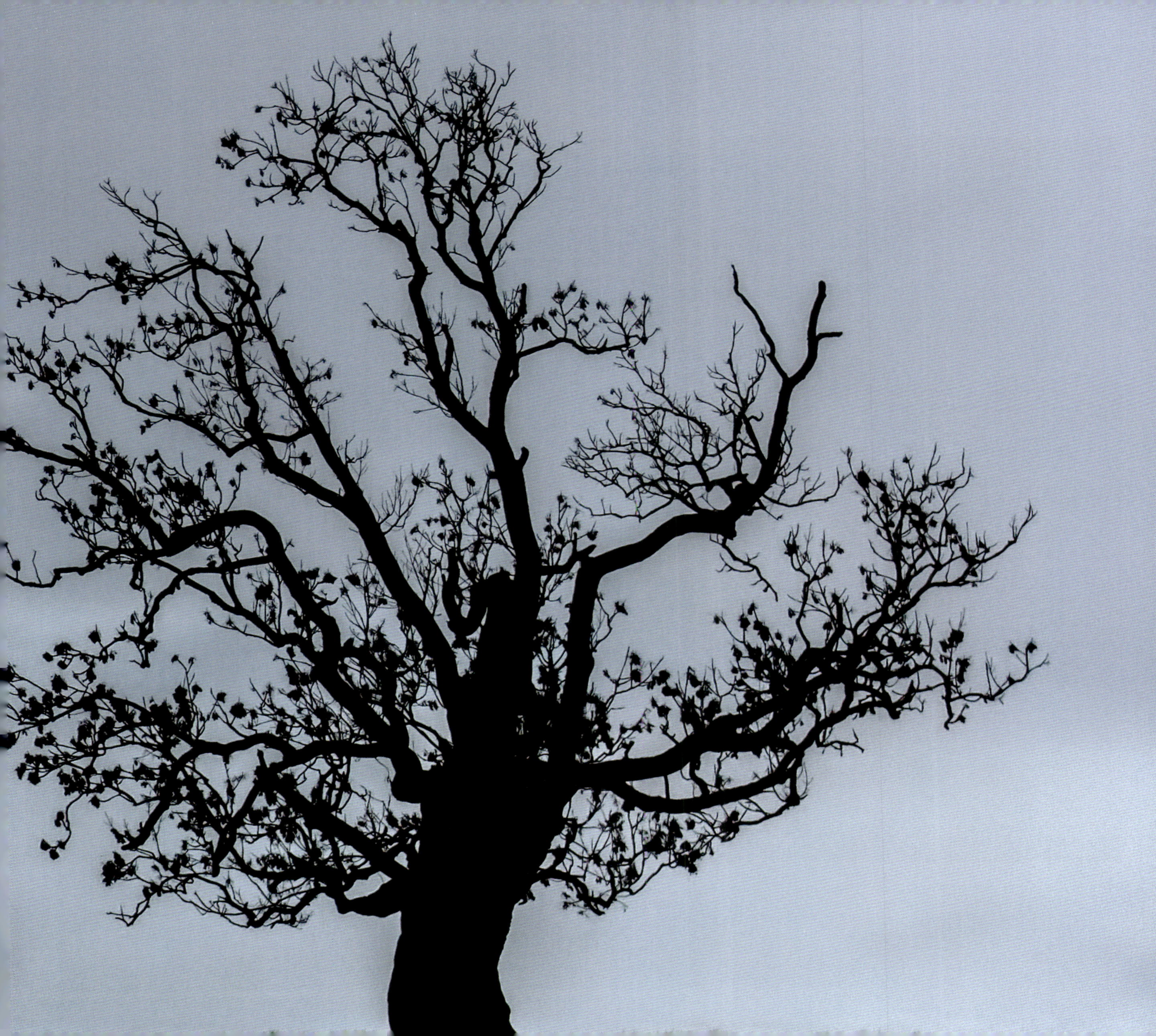

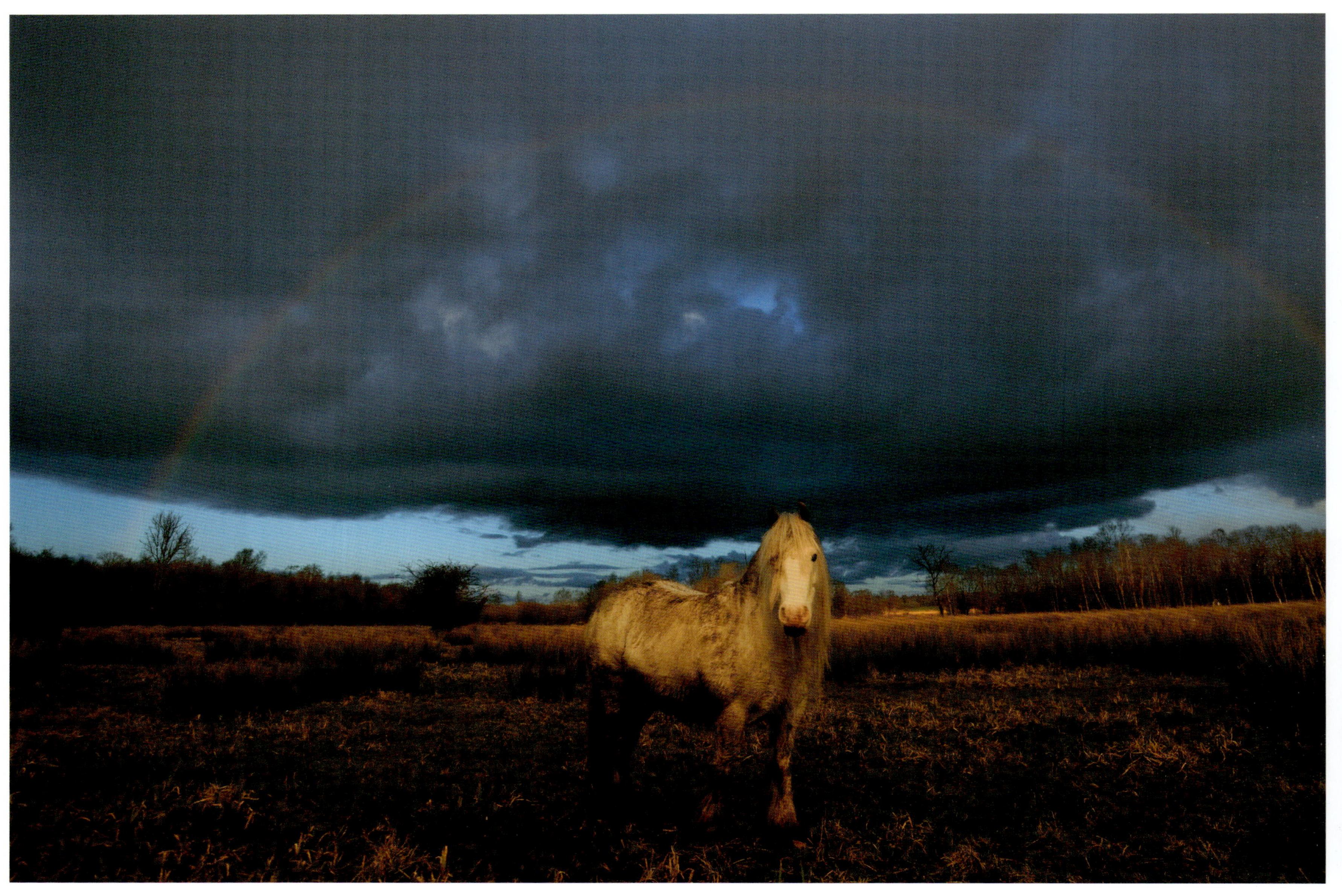

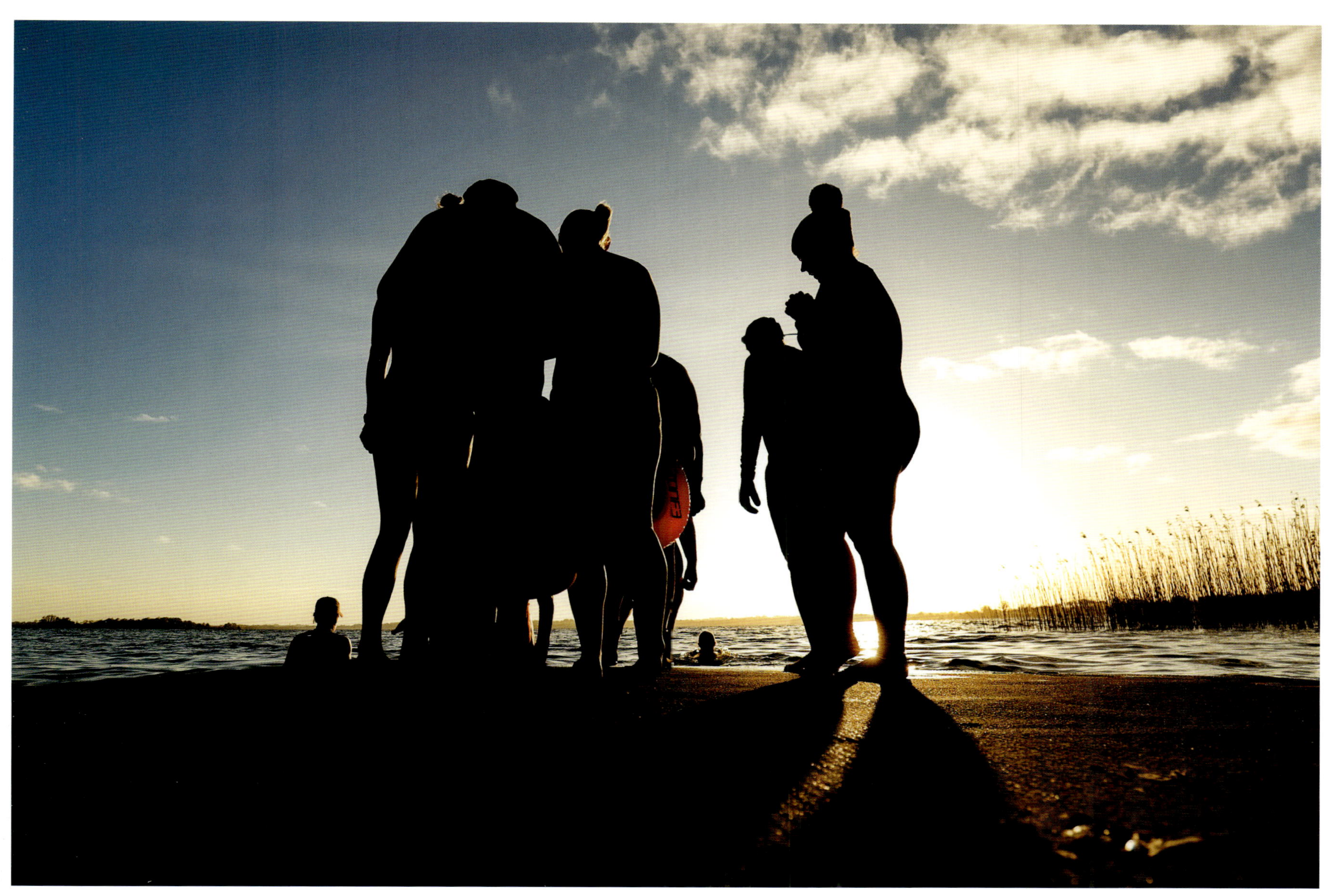

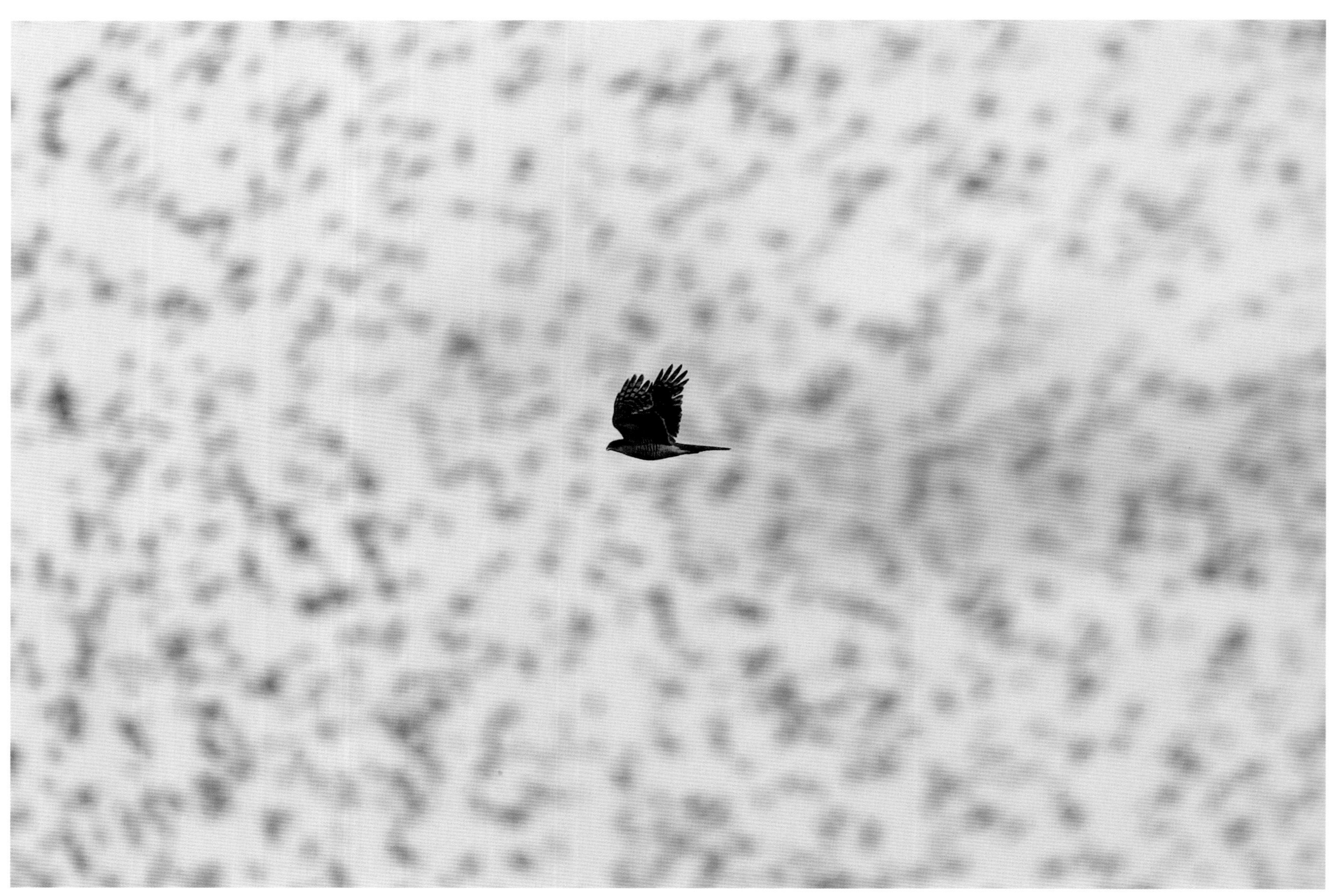

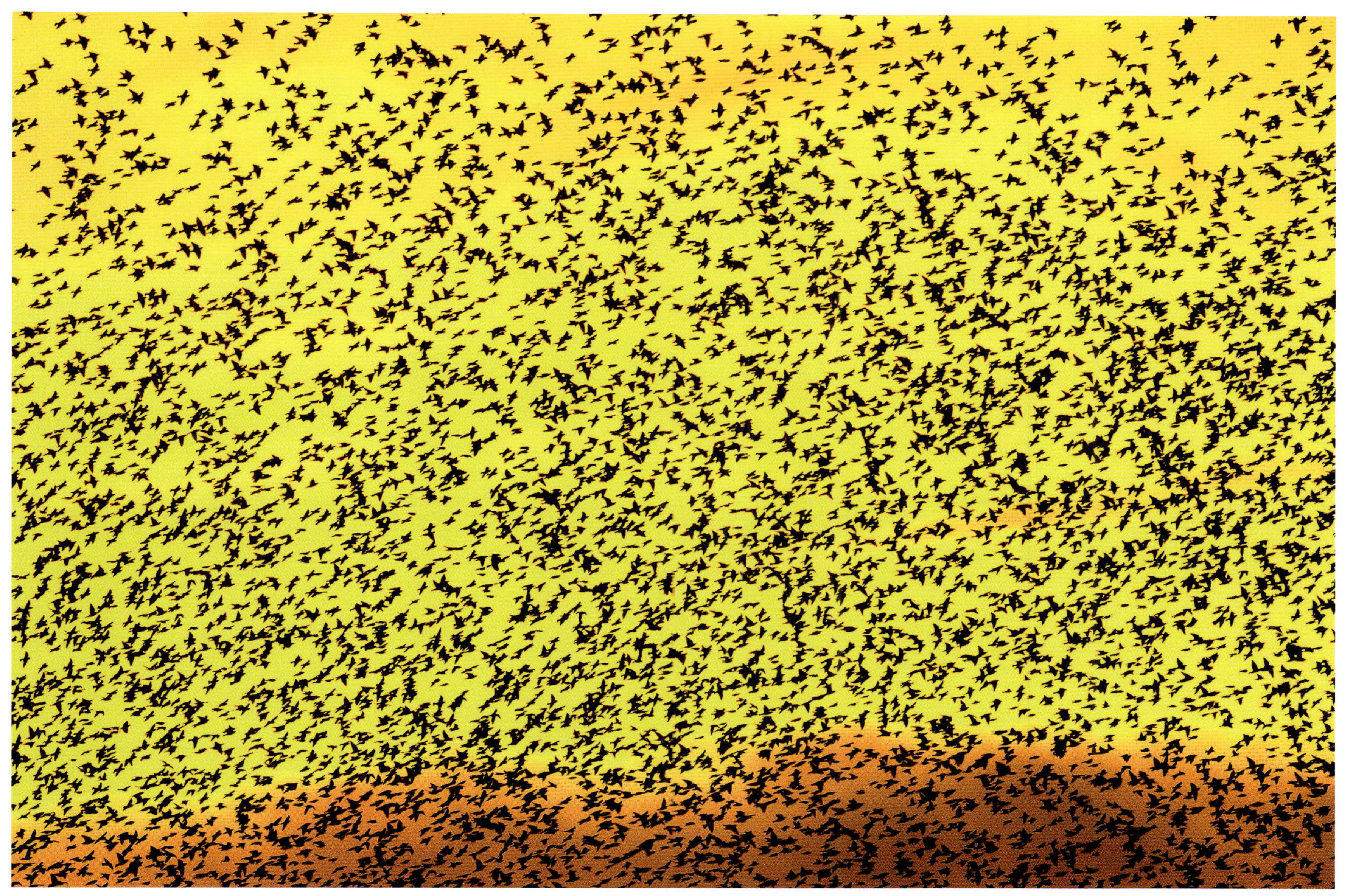

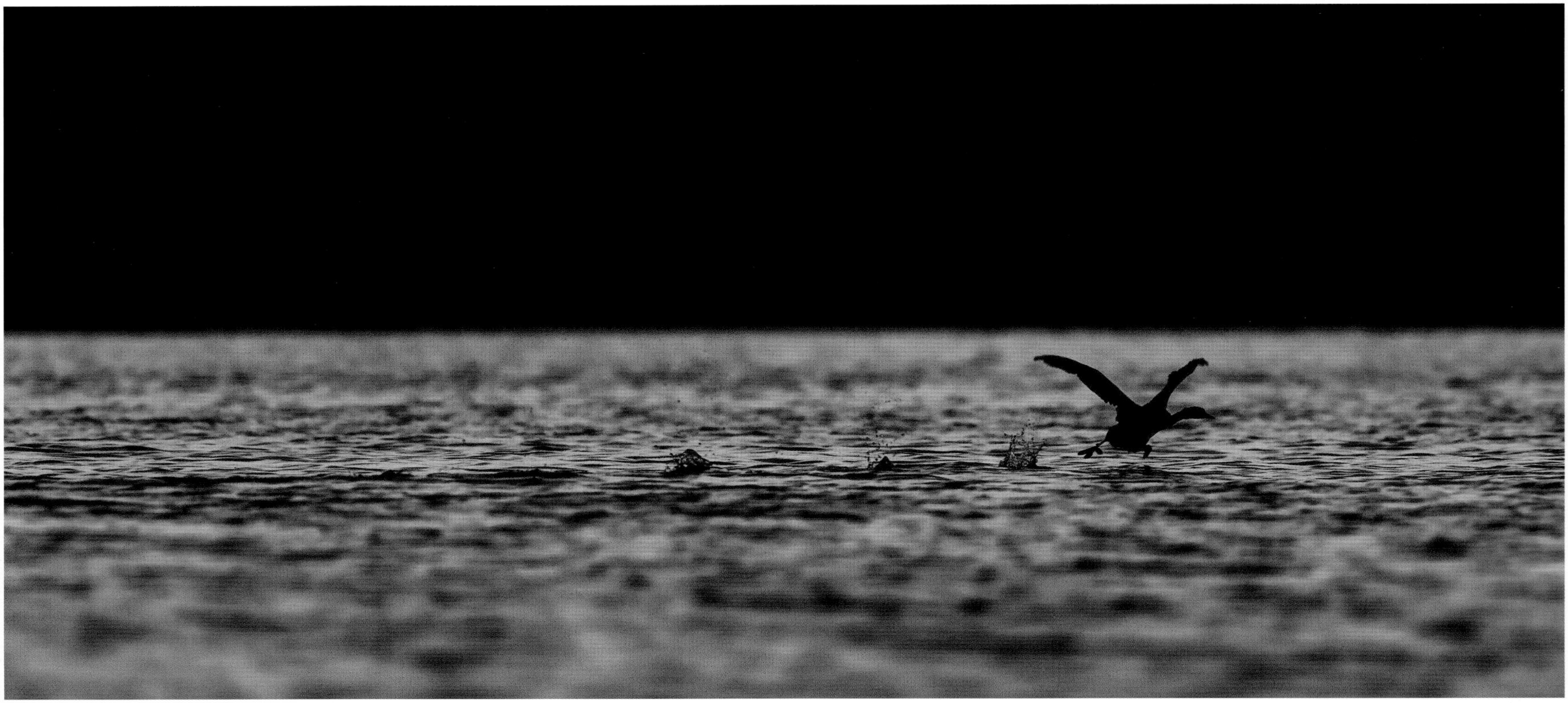

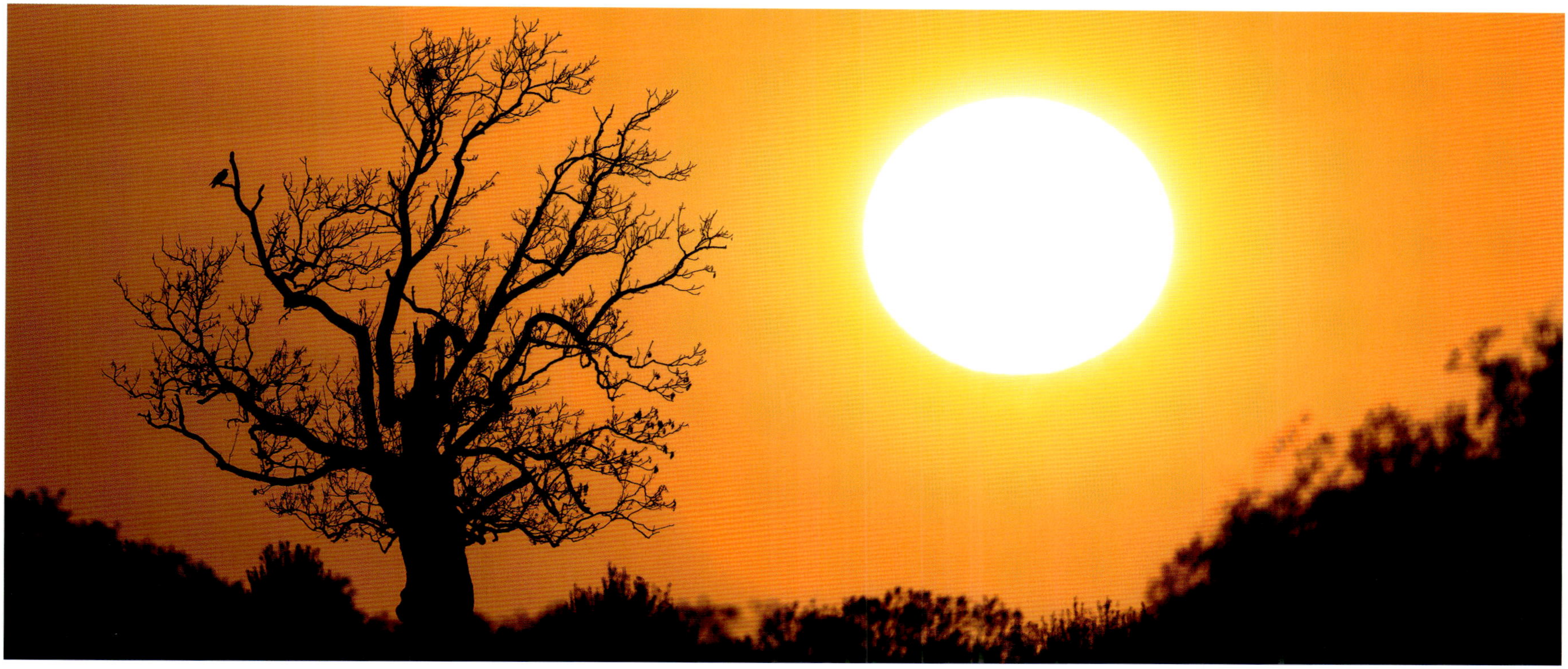

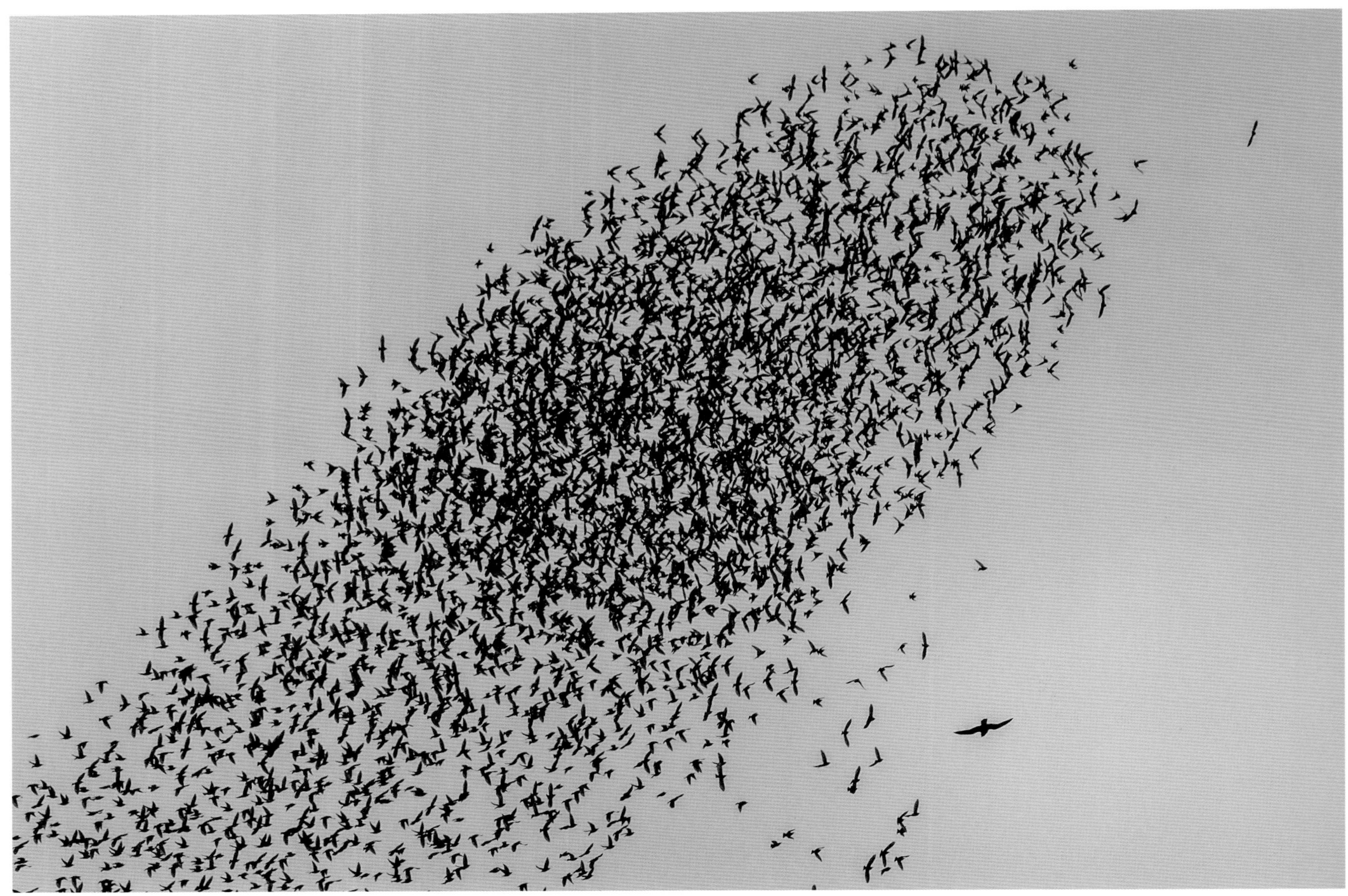

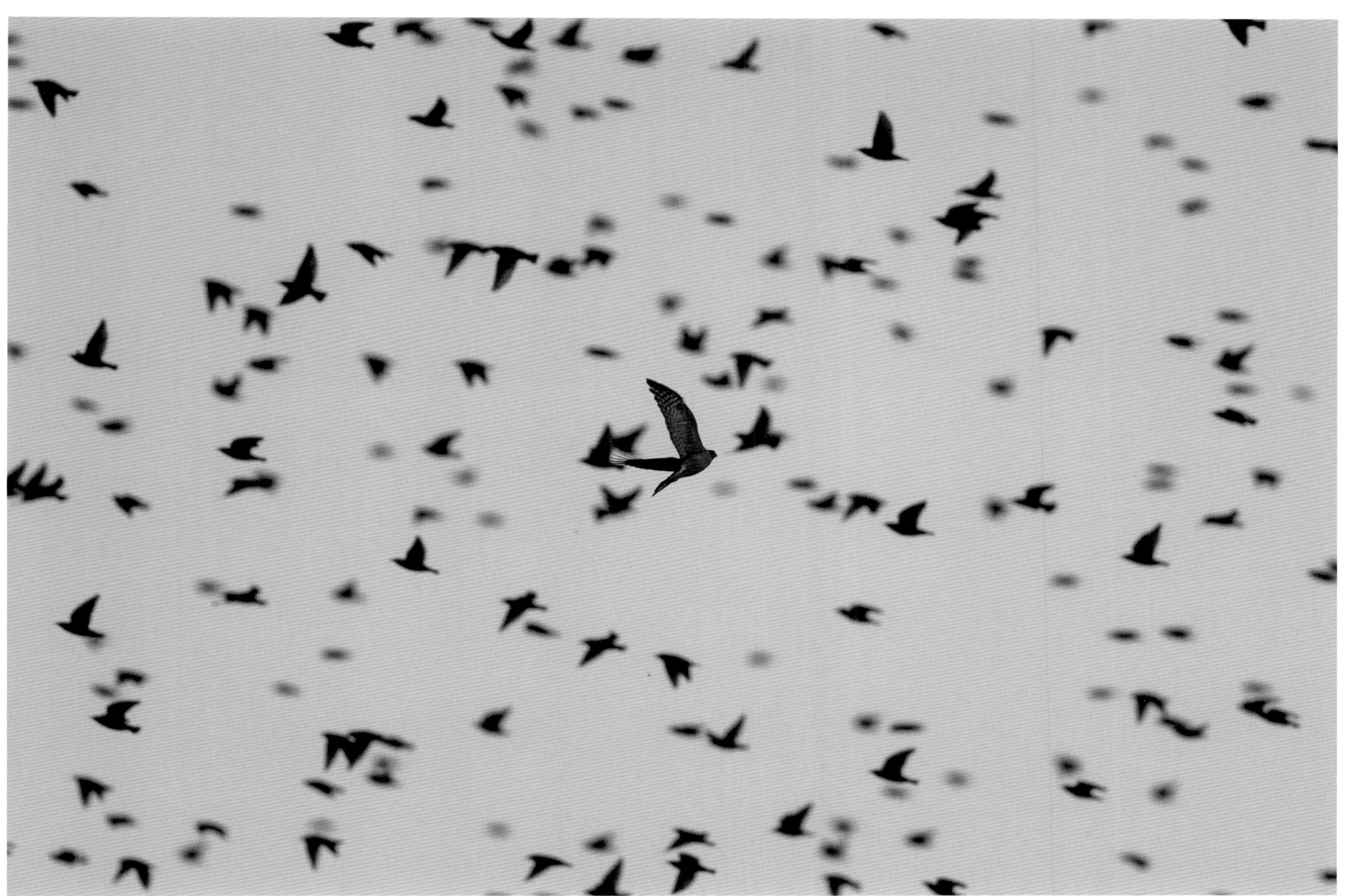

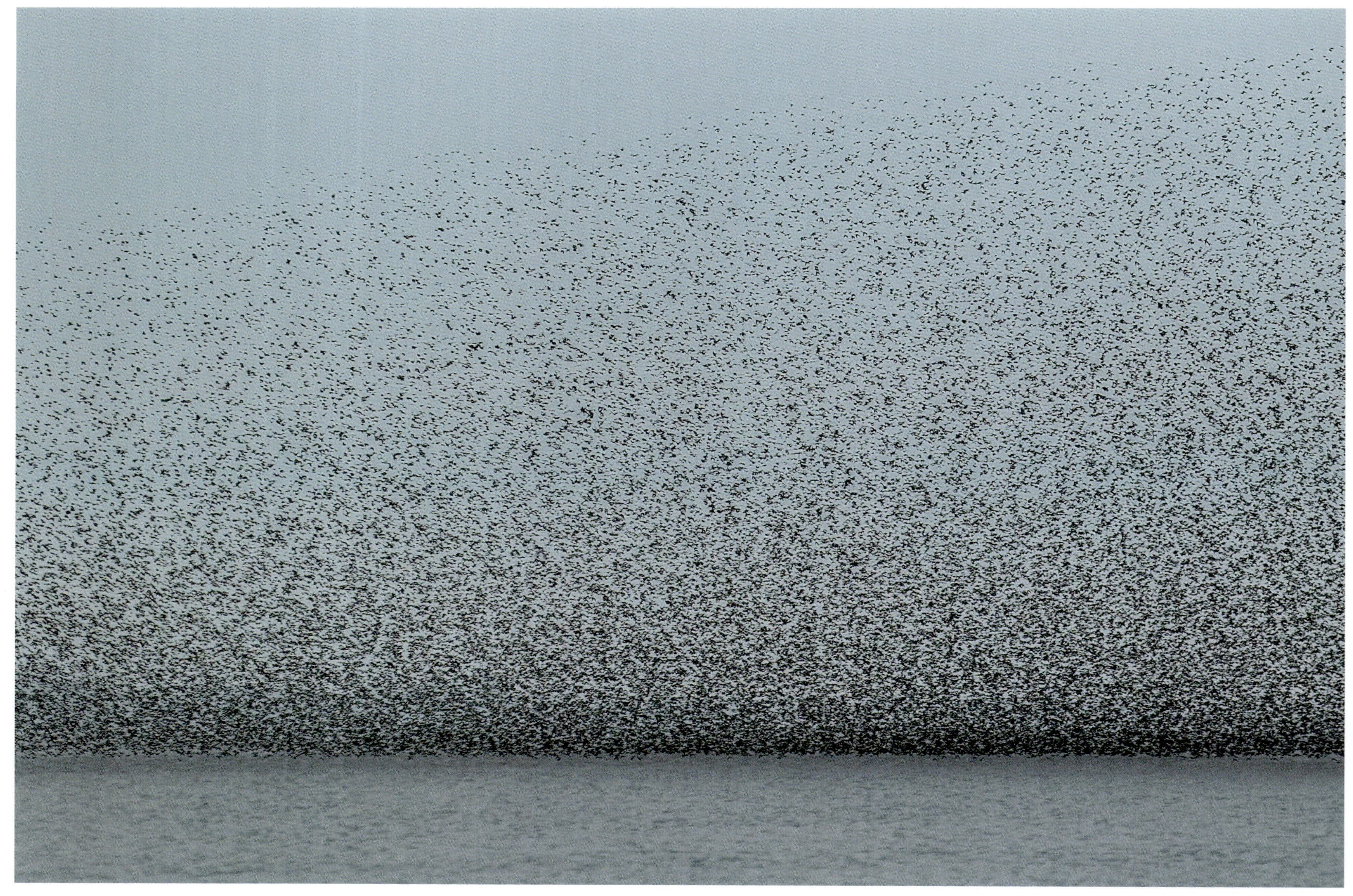

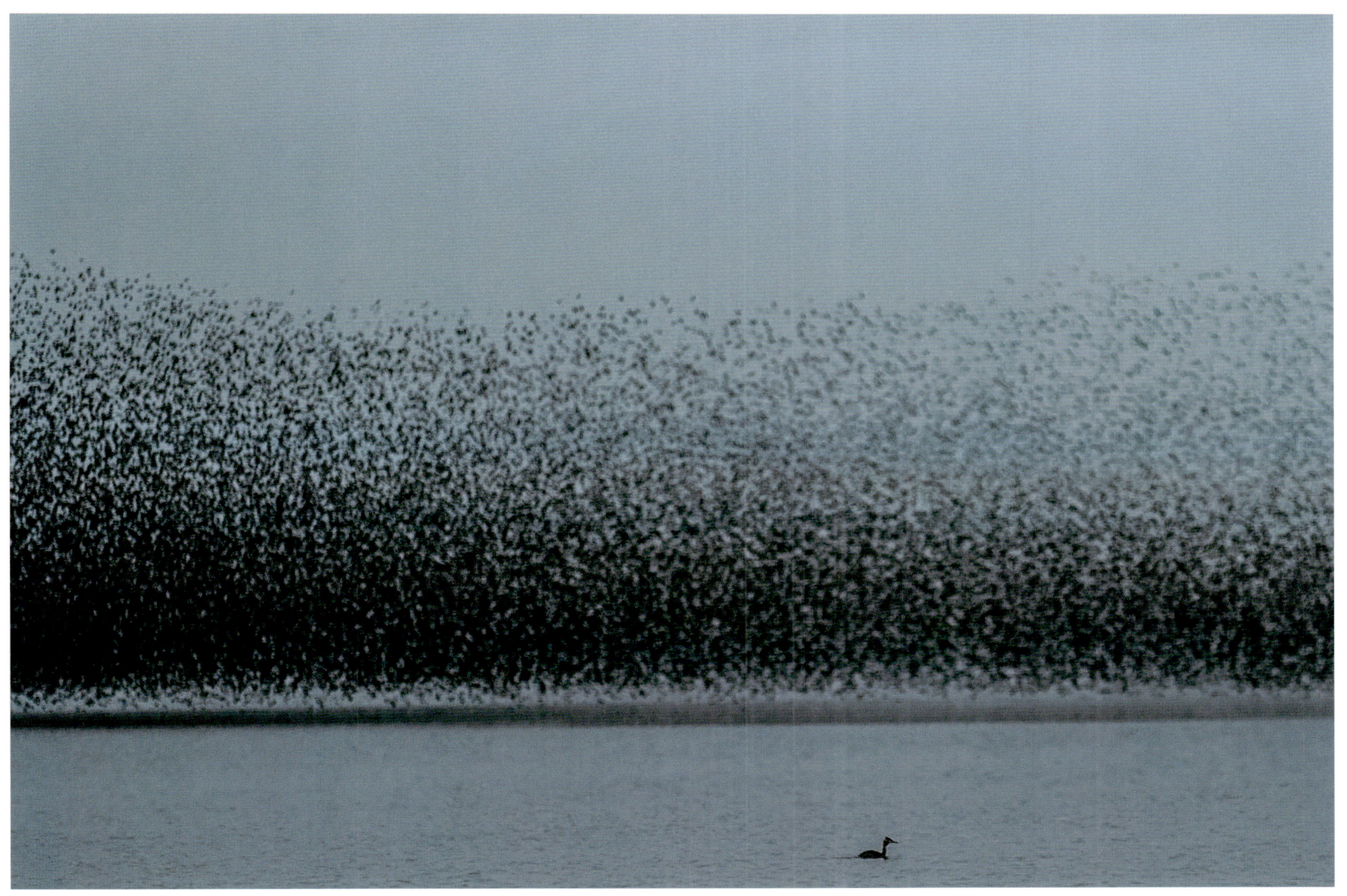

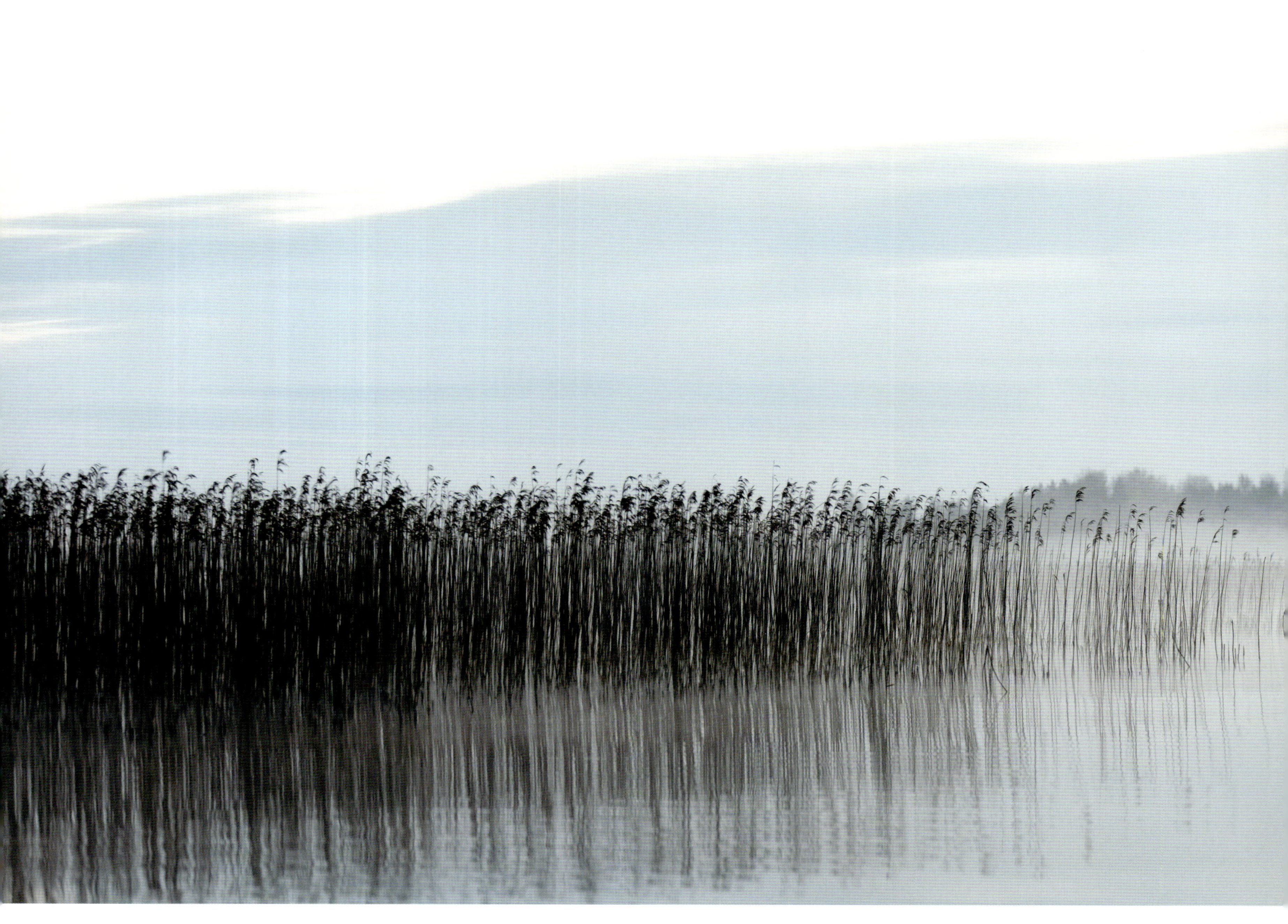

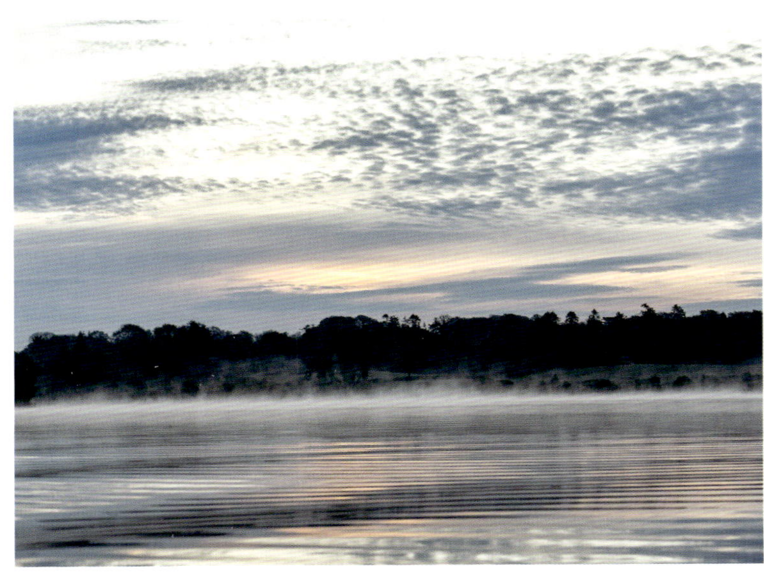
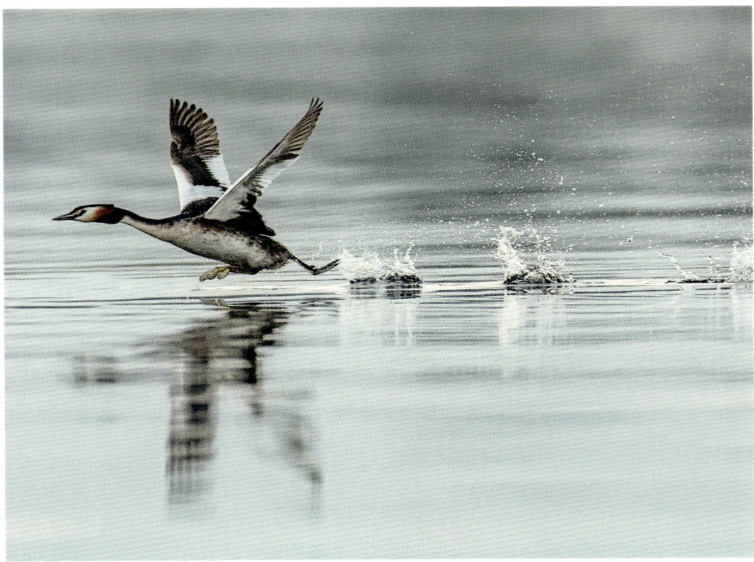
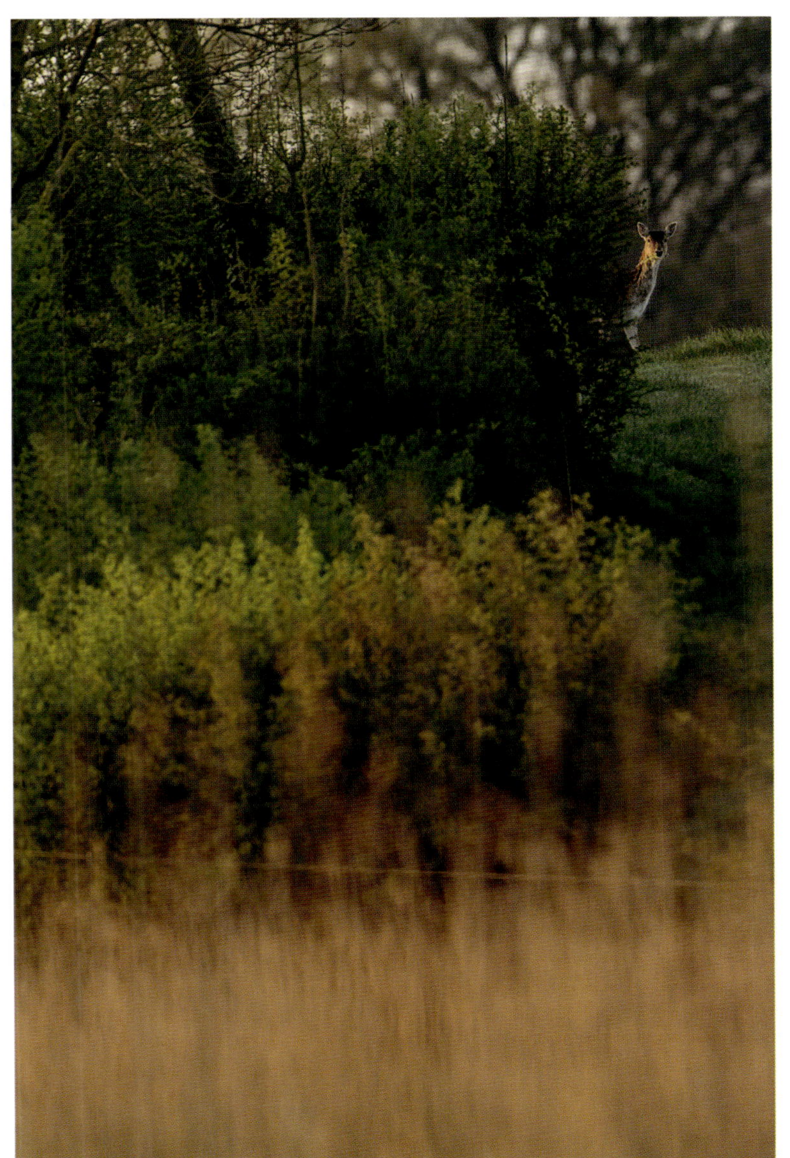

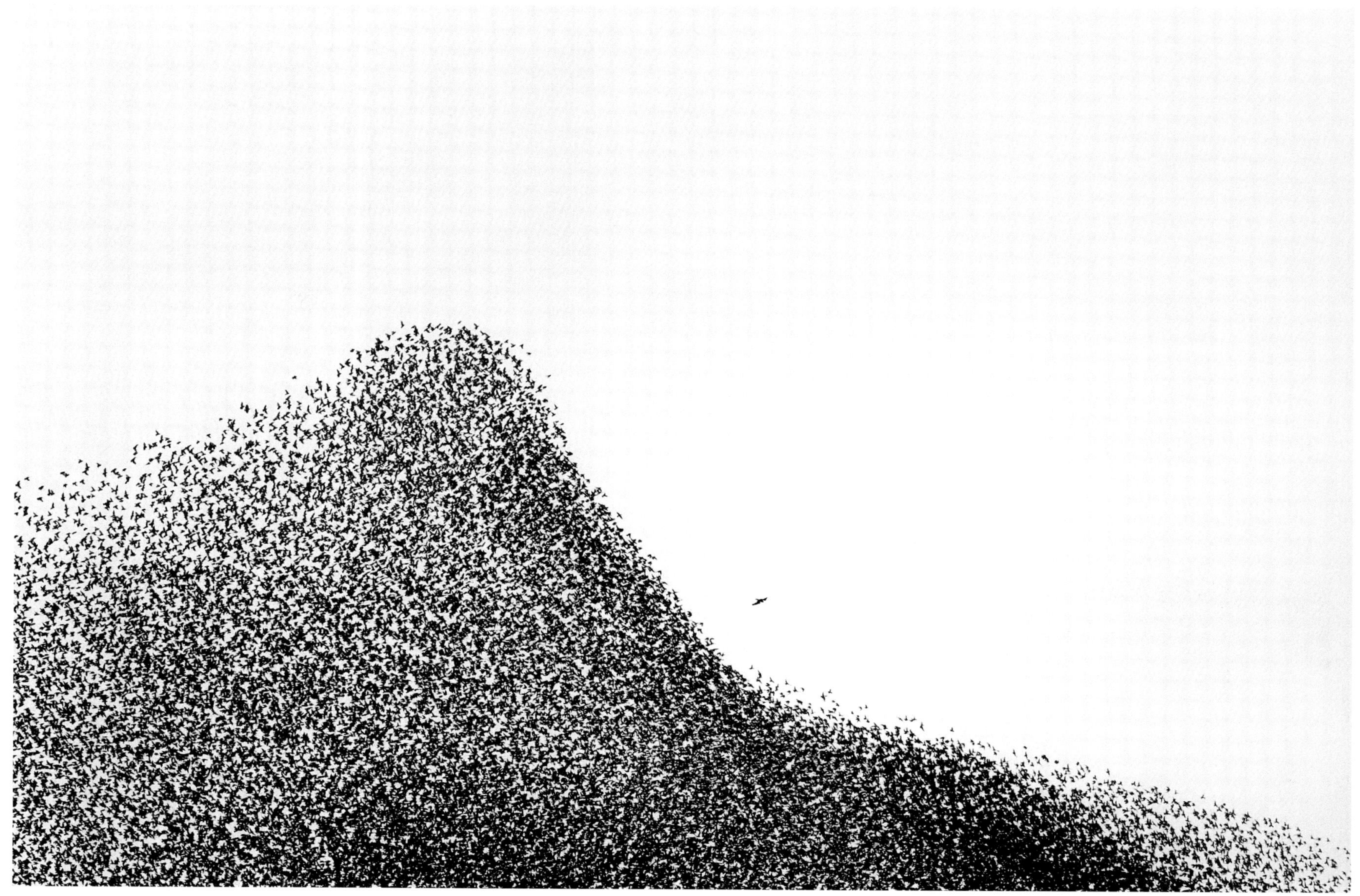

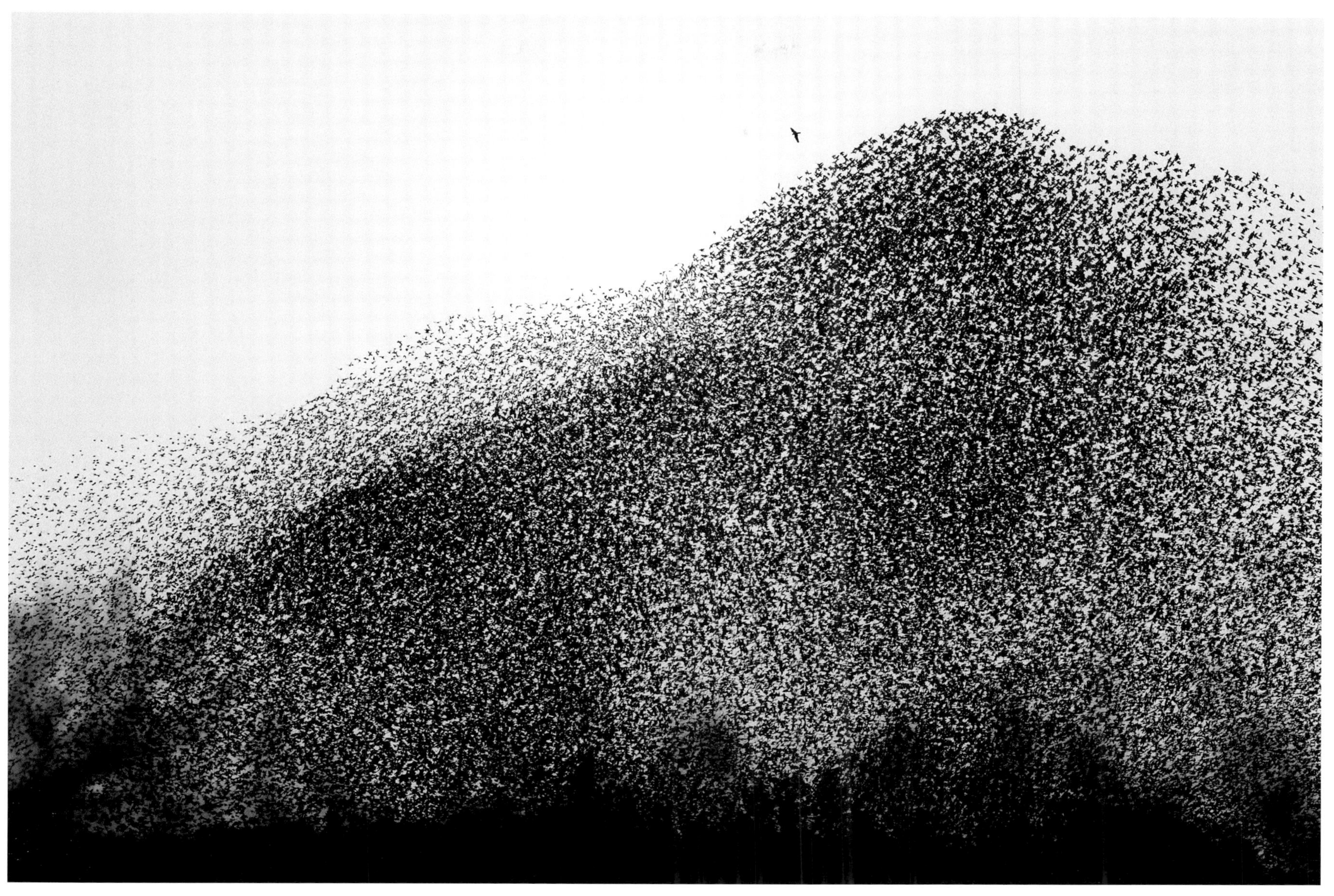

NOTE ON LOUGH ENNELL

LOUGH ENNELL, situated in the heart of the Irish midlands, is one of the lakes that has given rise to Westmeath's nickname 'The Lake County'. Fed by the river Brosna, the lake is a special protection area, known for its bathing spots and vast areas of shallow water. These shallows have produced the reed beds that are so vital to the migrating starlings, and provided the staging grounds for James's photographs. Opposite, James and his daughter Hayleigh have etched the locations of these viewing points, and some of the local landmarks.

NOTE ON EQUIPMENT USED

FOR THE PAST 18 years, I have been a Canon user, relying exclusively on their equipment to capture all my images. The continuous advancements in camera technology have been instrumental in bringing my long-term project to life. Murmurations, which occur at sunset when natural light is at its lowest, present a serious challenge for photographers. However, advances in camera technologies now offer photographers the ability to maintain high shutter speeds in these low-light conditions making it possible to photograph the fast, fluid movements of these mesmerizing phenomena with remarkable clarity and precision. Most recently I have moved to a mixture of the Canon R3 and R5. The lenses I turned to most often for this project were the Canon RF 70-200 2.8, Canon RF 15-35 2.8, or the Canon RF 400mm 2.8.

Butlers Bridge
Ladestown
Bloomfield House
Dysart Island
LOUGH
ENNELL
Lilliput
Brosna River

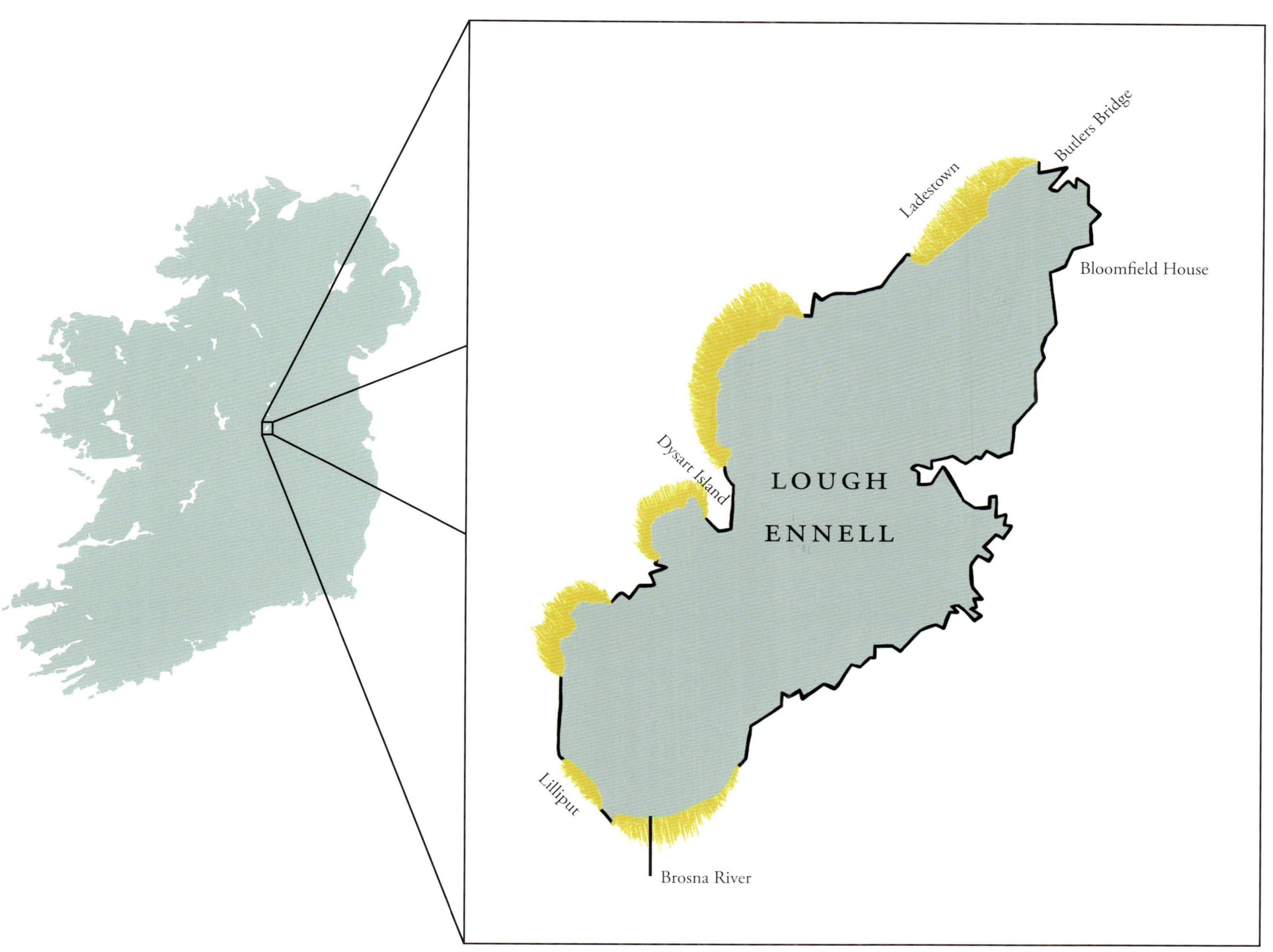

207

ACKNOWLEDGMENTS

To my amazing wife Ann, who never doubted or questioned my commitment to the project. She is the backbone to the success I have had and I would not have been able to achieve any of this without her by my side

To my four beautiful daughters, Hayleigh, Anna, Chloe and Sarah.

My parents, Sean and Ann, my brothers, Shane and David, and sister-in-law Gemma.

My employer Billy Stickland – for his mentorship and encouragement that drives me to be a better photographer – and all the staff at Inpho Photography.

To Canon Ireland, Westmeath County Council, Mullingar Chamber of Commerce and John Geoghegan for supporting the project.

To Colin Hogg, his wife Nora, their daughter and late son, Hannah and Daniel. Colin was by my side for some of the most incredible images captured, including the big bird.

To Enda Maguire, his wife Trudy, and his daughter Anna. Enda was instrumental in providing his local knowledge of the lake and the murmurations.

To all the landowners around Lough Ennell who granted access to the lake.

To Damien Hand and his staff at Hand Imaging, Mullingar, for their expertise in handling the reproduction of the images.

To Brenda Fitzsimons, formerly of *The Irish Times*, for her early support of the project.

To Lilliput Press, with special mention of Stephen Reid, the publishing manager for this book. To designer Niall McCormack and proofreader Kevin Gannon.

Seán Ronayne and John Fallon for their excellent words.